Fundamentals of Electric Aircraft

Fundamentals of Electric Aircraft

SAE INTERNATIONAL

Warrendale, Pennsylvania, USA

400 Commonwealth Drive
Warrendale, PA 15096-0001 USA
E-mail: CustomerService@sae.org
Phone: 877-606-7323 (inside USA and Canada)
724-776-4970 (outside USA)
Fax: 724-776-0790

Library of Congress Catalog Number 2018950020
SAE Order Number R-462
http://dx.doi.org/10.4271/R-462

ISBN-Print 978-0-7680-9322-3
ISBN-PDF 978-0-7680-9547-0
ISBN-ePUB 978-0-7680-9324-7
ISBN-PRC 978-0-7680-9325-4
ISBN-HTML 978-0-7680-9323-0

To purchase bulk quantities, please contact: SAE Customer Service

E-mail: CustomerService@sae.org
Phone: 877-606-7323 (inside USA and Canada)
724-776-4970 (outside USA)
Fax: 724-776-0790

Visit the SAE International Bookstore at books.sae.org

contents

preface

When the opportunity of editing this book was offered to me by SAE International, to put into simple words what an electric aircraft is, I just went for it without hesitation. Having worked for almost two decades in aircraft systems, with specific focus in electrification, I take pleasure in sharing my experience. I wanted to bring together up-to-date information on the state of the art to easily explain what the electric aircraft stands for. Comparing it to traditional aircraft helps envision its future through the road map it is following, rather the runway or "vertiport" I must say: you will find out in the following pages!

Such an opportunity for sharing knowledge offered by this work, at the core of SAE International's mission, helps capturing the lingering challenges behind the electric aircraft, thanks to a better understanding of its metrics, success factors, risks, and impacts. At the same time, the giant strides in innovative architectures and technologies, enabling the push towards total aircraft electrification, are also put into perspective.

An objective and unbiased view of what can be expected from an electric aircraft and what stakes are at play are presented. As one of its primary goals, this title addresses design constraints and timelines foreseen to reach acceptable performance and maturity levels and also what is stymieing progress. Moreover, also put forward is a general view of the progress made to date on the electric aircraft and what to expect in the years to come.

Through authors, contributors, and reviewers, I wanted *Fundamentals of Electric Aircraft* to be a collective effort of aircraft manufacturers, their suppliers (engines and systems), academia, research organizations, and also airlines and certification authorities. This mission can be considered as accomplished and has enabled the portrayal of various viewpoints from across the industry as regards the electric aircraft. This in a way allows for futuristic approaches not to get "grounded" by departing too far from operational down-to-earth realities. Believe me, with the profusion of projects, especially in the urban mobility sector, make-or-break situations are unavoidable. This is just because aviation purely and simply shuns risks and technology gets "authorized" for flight only after risk abatement is demonstrated through stringent means of compliance, with no allowance for miscalculations.

Without the authors, envisioning the electric aircraft from different angles would not have been possible. Contributors and reviewers have spared their precious time for this book, and this has to be hailed because expert information in simple technical language is now readily accessible to all readers.

Getting the best bang for the buck is also a major concern for aircraft manufacturers, engine/system manufacturers and operators while passengers are looking for the best quality of service and cut-price tickets. As an editor, on top of the technical aspects, I wanted to infuse cost-related analyses to get to grips on how performance enhancements and fuel burn savings may bring more value for money as long as new electric architectures and technologies deliver on their promises.

My first thanks go to the authors, Ravi Rajamani, Jean-Charles Maré, and Sven Taubert, who volunteered to collaborate and provided their insights and vision for the electric aircraft. I would also like to extend my thanks to the authors of the references appearing at the end of the chapters, whose in-depth research and findings contribute to the quality of this book's contents. My deep thanks go to the reviewers who threw their expert weight behind comments and suggestions, bringing great added value to this project. I cannot help but name them here:

Monica Nogueira (SAE International)
Amy L. Jankovsky (National Aeronautics and Space Administration (NASA), USA)
James L. Felder (NASA, USA)
Ian A. Halley (Boeing, USA)
Michel Todeschi (Airbus, France)
Marty K. Bradley (Boeing & University of Southern California, USA)
Eddie Orr (Rolls-Royce, UK)
Wayne Pearson (Honeywell, USA)
Norman Pereira (Federal Aviation Administration, USA)
Todd Spierling (United Technologies Aerospace Systems, USA)
Isidoro Martinez (Universidad Politécnica de Madrid, Spain)
D. Scott Barninger (American Airlines, USA)
Bruno Stoufflet (Dassault Aviation, France)

Also, key information and expert knowledge shared by Serge Roques (Safran, France) has allowed for relevant chapters to gain in clarity.

I would like to very warmly thank Monica Nogueira (SAE International) who initiated this project and was always there to support it. Without her help, this book would not have made it to the readers' bookshelves or tablets.

Company names are mentioned in this title as long as their product or technology squares up with the objective to explain what an electric aircraft is and how it may evolve. Demonstrators and technologies reviewed must not be considered exhaustive. Moreover, the credit they take for their achievements can in no way cast any shadow on projects and research not discussed here. In fact, difficult decisions were made when choosing contents, so that balanced and concise descriptions are presented to the reader.

The aim is to quench the technical curiosity of the reader, whether a beginner or a seasoned engineer. I am sure the reader will learn a lot and figure out the electric aircraft as they sputter into motion and see how they plan to occupy and breeze through our future skies, more or less closer to our eyes but far away from our ears. While *Fundamentals of Electric Aircraft* will undoubtedly bring clarifications to college and university students, engineers, experts, and academia in aviation and other transportation sectors can get a close view of forms and functions of technologies at stake. Strategic business planners can also derive good input from the information presented here.

This book provides not only theoretical knowledge of what is boiling down from developments and research around the electric aircraft but also facts and figures on key parameters of electrified systems allowing comparison with conventional solutions. These are not unfathomable black boxes but hardware and software tailored to fit cramped spaces and swinging into action probably just about somewhere close to where the reader may be seated during a "more-electric" or future "electric" flight!

Readers, thirsty for more knowledge, can count on future book installments that SAE International and myself are planning to offer on the electric aircraft. Therefore, your feedback will be of great help.

So dear reader, just now flip this page and delve into the realm of the electric aircraft and see what they are all about for now, and how making them fly in a variety of sizes, shapes, and routes is not as easy as it may look like. But the overwhelming trend is crystal clear: they will stick. The inexorable future reality of the electric aircraft is on the move.

Pascal Thalin

CHAPTER 1

Introduction and a Brief History of Electric Aircraft

Ravi Rajamani
drR² Consulting

Pascal Thalin
Chair and Member - SAE Electric Aircraft Steering Group

1.1 Background

A totally electric aircraft is in the making and will become reality in the foreseeable future. This is true thanks to a whole range of innovative solutions developed within the aerospace industry, and also enabled by the progressive replacement of hydraulic and pneumatic systems by those powered by electrical energy in today's aircraft.

The focus for wide-body aircraft manufacturers is not so much on building "electric" aircraft, but on developing the so-called more electric (ME) aircraft. This means that manufacturers try to move as many supporting aircraft systems as possible to electric power. Although many systems—computers, navigation instruments, actuator controls, lighting, ventilation and in-flight entertainment systems, for example—are already electrically powered, new applications like engine starting and environmental control systems are now going electric on modern aircraft.

In the Boeing 787 Dreamliner, for example, the greatly expanded electrical system generates twice as much electricity as previous Boeing airplane models, with a number of hydraulic and pneumatic flight systems being replaced by electrical systems [1.1]. This relatively recent trend is set to grow given the advantages of more flexible electrical systems. They offer fuel savings thanks to greater efficiency, improved reliability, and easier maintenance, thereby allowing overall performance enhancement.

On average, electrical systems use up one-quarter of an aircraft's energy supply today, chiefly to power functions with low or medium consumption. Tomorrow, we will need to significantly scale up the capacity of electrical generation and conversion systems to power all the aircraft's electrified functions, on the ground and in flight. At the same time, we will have to continue making systems lighter, more compact, and easier to maintain.

For airlines, this paradigm shift is driving fuel savings, reducing the carbon footprint and NOx emissions and cutting maintenance costs while also improving dependability. Nevertheless, regarding incremental electrification on existing aircraft, or on totally new aircraft with still conventional architectures, the ME trend will at some point stall due to shortage of optimization possibilities. Therefore, radically different architectures are necessary.

1.2 Electrification Trend

The aerospace industry is rethinking new concepts, departing from conventional architectures, for the design of radically new aircraft that is either hybrid or fully electric. The key to meeting these challenges is a reduction in aircraft weight and the design of lighter, innovative, more efficient, and reliable propulsive and non-propulsive systems. Thanks to the development of hybrid-electric (HE) and electric engines, partially or totally relying on electricity for power and thrust, and increasing system electrification, such new aircraft will be more energy efficient, have a smaller environmental footprint, and cost less to maintain.

The first thing that will be evident about these aircraft is that they will look dramatically different from what we know today. Put very simply, aircraft currently have engines attached to the wings or the fuselage. Without relying on electrification, these engines can be replaced with newer, more efficient engines, as Boeing and Airbus have done recently on their narrow-body 737 MAX and A320neo programs. With these new engines, efficiency gets improved by more than 15%, but the fundamental architectures remain unchanged. So, in terms of the overall systems engineering, this change is really just a retrofit, and not something radically new. In comparison, electric propulsion will usher in an entirely different paradigm. Instead of one or two engines attached to the wings, all-electric (AE) and HE propulsion may require a row of engines arranged across the wings of the plane in a "distributed propulsion" architecture, or the motors might be buried within the body itself, which in turn may look like a giant flying wing. All these radical design concepts are being considered to gain ever greater aerodynamic efficiencies.

These dramatic changes will require a complete redesign of the plane. It's not a matter of simply packing in a lot of batteries somewhere in the plane and running cables to the engines. AE and HE propulsion will have to get completely integrated into the airframe, with novel new ways of supporting the requirements of energy production, storage, distribution, and conversion on board.

In small aircraft, there's already considerable research and development (R&D) being funneled into these concepts since about 2010. We have already witnessed test flights of one- to four-seat aircraft—like the Airbus E-Fan (2014), Ehang 184 (2018), Volocopter VC200 (2016), Lilium (2017), Joby Aviation S4 (2017), Airbus Vahana (2018), and Kitty Hawk Cora (2018). Many other companies are working on an entire range of concept aircraft from single-seat motor gliders from Pipistrel to regional jets from Zunum and Eviation.

Regarding HE aircraft for up to 100 passengers, the CEO of Airbus, Tom Enders, has stated that he reckons the industry will be capable of building them somewhere by 2030—at least for short, local trips.

1.3 Early Electric Flights

Using electricity to power aircraft is not a new phenomenon. The earliest documented use of electric motors in aviation seems to be in France at the end of the nineteenth century where electric motors were used to drive propellers powering airships. Gaston Tissandier and his brother, Albert, were avid hot air balloonists. Gaston is reputed to be the first to fit an electric motor—from Siemens—to a dirigible and achieve electrically powered flight [1.2]. This took place on October 8, 1883, in Auteuil, France (**Figure 1.1**). In his historical work on airships, published in 1886 [1.3], Tissandier describes the balloon as:

> *...28 meters long and 10 meters wide and 9.2 meters in diameter in the middle. It is equipped, at its lower part, with an appendage cone terminated by an automatic valve. The fabric is of percaline, made waterproof by a new varnish of excellent quality. The volume of the balloon is 1,060 cubic meters.*

FIGURE 1.1 Tissandier electric airship experimental flight in Auteuil, France, on October 8, 1883.

Courtesy of thisdayinaviation.com

And, describing the electrical components, Tissandier says,

> *The engine consists of a Siemens dynamo machine specially constructed, and having a 100 kilogram force with a weight of 45 kg. The propeller is two-bladed; it is coupled to the machine by means of a gear transmission. It is 2.80 m in diameter with a speed of 180 revolutions per minute. The dichromate battery of my construction is made up of 24 large-area zinc elements with high throughput.*

He then goes on to describe in detail the first flight. The power density in early electric components, including accumulators (batteries) and motors, was not high enough to power heavier-than-air aircraft. In fact, it was partly for this reason that it was not until the 1970s that true electric flight was possible.

The first person to achieve this milestone was a German, Fred Militky, who teamed up with the aircraft maker Brditschka in Austria to convert an existing HB-3 motor glider to electric power by using an electric motor. The AE aircraft, designated MB-E1, was flown by Heino Brditschka on October 21, 1973, from Wels Airfield near Linz. This was powered by Varta Ni-Cd batteries and a 10 kW Bosch DC motor [1.14]. Probably because of the state of the art of the electric technology, sustained flight was not a reality; the MB-E1 achieved a 9-minute flight that first day (**Figure 1.2**). Interest turned to solar flight after this.

FIGURE 1.2 Motor glider MB-E1, designed by Fred Militky and flown by Heino Brditschka, October 1973.

Courtesy of aerokurier.de

1.4 The Solar Years

Solar flight is an interesting and continuing chapter in the electric aircraft saga. While the most eco-friendly of all options, because all the power can be generated from solar energy, it is also the least practical for manned flight. But for various unmanned missions, such as high-altitude, long-duration reconnaissance and communication missions, it is an ideal alternative. In fact, the National Aeronautics and Space Administration's (NASA) Helios Prototype solar-powered aircraft, designed by AeroVironment, reached an altitude of nearly 100,000 ft in 2001, before it was destroyed in turbulent air two years later [1.4]. The Helios had 14 motors and was essentially a flying wing with the entire 184 m^2 (1980 ft^2) area covered in photovoltaics. Reserve power was stored in lithium batteries (**Figure 1.3**).

The 1970s and 1980s were an active time for solar flight. The first solar-powered flight happened on November 4, 1974, in California, by Sunrise 1, an unmanned aircraft built by Roland Boucher's company, AstroFlight. Paul MacCready, the founder of AeroVironment, was the first to cross the English Channel with the Solar Challenger in 1979. The Solar Challenger had no batteries, so without the sun, flight would have been very difficult! This is not true of the Solar Impulse II, the first manned solar aircraft to circumnavigate the world. The Solar Impulse, built by the Swiss team of André Borschberg and Bertrand Piccard, who also shared the piloting duties during the flight, completed the 5-month long flight in more than 16 months because they were grounded in Hawaii for nearly a year to make some repairs to the aircraft. The Solar Impulse II took off from Abu Dhabi on March 9, 2015, and landed back there on July 26, 2016, flying for 42,000 km (26,000 mi) [1.5].

As mentioned above, solar flight is completely impractical for a manned flight. Take the Airbus A380-800, for instance. It has a wing area of roughly 845 m^2 (9128 ft^2), and a horizontal tail plane area of about 205 m^2 (2207 ft^2). The total area of the upper fuselage, assuming that every square inch is used for photovoltaic (PV) cells, is roughly 800 m^2 (8611 ft^2). Because solar radiation will not be directly incident on the entire fuselage, we can assume that only 65% of the solar radiation is useful, which makes the total effective area of the planar surfaces and the fuselage about 1570 m^2 (16,891 ft^2). The conversion potential for PV cells is not very good, and even with no losses, they can only generate about 0.2 kW/m^2 of power. With all these assumptions, an Airbus A380, with its entire upper surface covered in the highest efficiency thin-film solar cells will generate

FIGURE 1.3 NASA/AeroVironment Helios Prototype.

Courtesy NASA

at best 314 kW of energy in the most favorable conditions possible. To effectively carry its own weight and that of any passengers, it will probably have to be made of balsa wood. This tells us that solar energy is really not practical for commercial aviation. While it is true that the first manned electric flight used only a 10 kW motor to carry one person, the MB-E1 was tiny compared to the A380 and was designed only for one person!

So, the only practical means of achieving a manned electric flight is through rechargeable batteries or a hybrid propulsion system. But is battery power viable for commercial flight, other than short flights involving a few passengers? We argue below that it may take time to get there.

1.5 All-Electric and Hybrid-Electric

It was not until the turn of the twenty-first century that a renewed interest in electric flight resulted in an infusion of R&D funding in this area, as well as the growth of startups working on different kinds of electric aircraft. These now range from personal mobility devices to four-seaters, with plans for larger aircraft as well. NASA, for example, is working on a revived X-plane research program that is converting a four-seater general aviation aircraft to an AE configuration. The growth of electric aircraft has coincided with the growth of electric automobiles, and it is the economies of scale engendered by the latter that will prove to be a boon for the electric-aviation industry as well.

Electric propulsion comes in two main flavors: all-electric (AE) and hybrid-electric (HE). The latter can be further divided into parallel HE and serial HE. Other classifications have been proposed with many subtle variations, but every one of them will fall into one of these two (or three) categories. The AE model consists of electric energy stored in batteries driving motors that are attached to fans that provide the propulsive force. Serial HE schemes have the same propulsive element, that is, electric motors, but the source of the electric energy is generally chemical transformation of one kind or another, typically from a fuel-powered turbine driving a generator. There may or may not be a battery in the middle of this power train. In case of a parallel HE, the chemically powered turbines provide both the propulsive force and—through generators—the electric energy needed to drive one or more electric motors. There are many instances of all three architectures that are either flying or being developed. The AE design is the most prevalent because it is the easiest way of converting an existing conventionally powered aircraft to an electric design. Hybrid designs take a lot more engineering and most products are slated for the future.

In this exposition, we are only concerned with manned aircraft. There have been many electrically powered drones and unmanned air vehicles (UAV), but those are not under consideration in this title. Gliders were the first aircraft to be converted to electric propulsion by the addition of motors to assist in takeoff. For example, the paper by Urs and Vezzini [1.6] lists a number of gliders that have been fitted with motor-driven propellers. At least one, the Silent Club, was being produced in 1997. A number of European companies, like Antares, have notably been involved in such conversions. According to Wikipedia, the first airworthiness certificate for an electric aircraft was granted to the Lange Antares 20E in 2003 [1.7].

Batteries are not the only source of electric power for aircraft. Fuel cells have been used to generate electricity, and this would be a form of HE propulsion technology. Boeing worked with a company in Spain to develop a demonstrator, based on converting a Diamond HK-36 Super Dimona motor glider as a research test bed. This was powered by a Proton Exchange Membrane (PEM) fuel cell with stored hydrogen as the source of fuel, and the test flights took place in early 2008 [1.8]. Today, the most active R&D program in fuel cell-powered flight is being conducted in Germany by a consortium of companies and DLR, the German government's equivalent of NASA. The four-seater aircraft, called HY4, developed by this consortium, is based on the Pipistrel motor glider, and is planned to be flown commercially by H2FLY [1.9]. The first flight of the HY4 took place in late September 2016 from the Stuttgart Airport.

A more detailed exposition of the history of electric propulsion with discussion about the various technologies involved is available in a recently published book by one of the authors of this chapter [1.10].

Meanwhile, the U.S. and European government agencies have been busy funding research in this area. NASA, working with Google as a sponsor, conducted the first Green Flight Challenge in October 2011. The goal was to fly 200 miles in less than 2 hours using 1 gallon of fuel per occupant, or the equivalent in electrical energy. This competition was won by a modified Pipistrel Taurus G4 motor glider. The U.S. division of the Slovenian glider manufacturer modified their four-seater Taurus, replacing the internal combustion (IC) engine with a 150 kW electric motor. This aircraft has a unique twin-fuselage design with the engine mounted on the wing between the two fuselages. Even though the goal of the Green Flight Challenge was not to build an electric aircraft, the first and the second winners were AE designs. The second place was taken by an AE aircraft built by e-Genius. It is interesting to note that the HY4 fuel cell aircraft mentioned above has an almost identical design as the Taurus G4.

The next category of aircraft that were targeted for modifications were the smaller one- or two-seater aircraft. The IC engines were replaced by battery packs and electric motors. Some companies even designed and developed such aircraft from scratch. Airbus started a company in Europe called VoltAir to develop electric planes [1.10]. The E-Fan concept aircraft, a two-seater that made its first public appearance in 2014, has two 32 kW electric motors mounted above and behind the wings. It is powered by a bank of lithium-ion batteries giving it about an hour of flight endurance. Its claim to fame was to be the first manned crossing of the English Channel, and this was attempted on July 9, 2015. In a bizarre twist of fate, a French stunt pilot called Hughes Duval, piloting the Cri-Cri Cristaline, the world's smallest twin-engine manned aircraft, beat the E-Fan for this title by mere hours. The only chink in Duval's armor was the fact that the Cristaline's takeoff was assisted by another aircraft. In an even more bizarre coincidence, a team from Pipistrel was ready to attempt the channel crossing earlier that same week with their Alpha Electro plane but was prevented from doing so for various reasons [1.11]. To shore up their innovation credentials, Airbus started a company in Silicon Valley called A^3 (A-cubed) that is working on a vertical takeoff and landing (VTOL) vehicle with electric propulsion called Vahana. This is Airbus's entry into the autonomous personal mobility/air-taxi arena. It now seems that the E-Fan program, which was supposed to make production aircraft, has been discontinued, and the team has moved on to a demonstrator program called the E-Fan X. In this program, a BAe 146 aircraft will have one of its jet engines replaced by a fan electrically powered by a Siemens motor which will get its energy from a Honeywell generator, driven by a Rolls-Royce turboshaft engine installed in the aircraft. The partners in this venture, Airbus, Siemens, and Rolls-Royce, are also working on a larger HE concept commercial aircraft called the eThrust.

The mainstay of many electric aircraft programs are lithium-ion rechargeable, that is, secondary, batteries. There is at least one company using primary (non-rechargeable) batteries as an emergency backup, but the main power source is still secondary batteries. Lithium-ion batteries are much lighter than those made with older chemistries, with energy densities around 220 Wh/kg as compared to 35 Wh/kg for lead-acid or 50 Wh/kg for nickel cadmium (Ni-Cd), the other two popular aircraft battery technologies. And the cost of these batteries has been falling rapidly in recent times, dropping from $1000/kWh in 2010 to around $200/kWh in 2017. Add to this the dramatic increase (up to three times) in the operating life of these batteries—a parameter that can be controlled to some extent by the way the battery is used—and electric propulsion technologies start to make economic sense. This does not mean that all the other aspects of aviation, such as development cost, certification, etc., become easy, but at least the prohibitive costs associated with propulsion are coming down.

Many challenges still remain, and it is obvious that, for the time being, an AE architecture will not be feasible except for very small aircraft. To conceptualize the challenges being faced, the Airbus E-Fan has two 32 kW electric motors. A Boeing 747, seating some 400-600 passengers, needs an estimated 94 MW for takeoff and about 40 MW to cruise. That is 94,000 kW or about 1500 times as much power as the E-Fan. Taking this calculation a little further (and it must be understood that this is all very "back-of-the-envelope"), the total weight of the battery system, the power electronics, and the motors for an aircraft the size of a 747-8 to fly 62 min would be about 375 tons, allowing for an additional 30 min of reserve capacity for emergency situations per regulations. This assumes that power electronics for these power levels have a specific power of 15 kW/kg and motors and generators have a specific power of 10 kW/kg. These are very generous assumptions because current technology cannot deliver these numbers. The maximum takeoff weight (MTOW) for this aircraft is about 450 tons, which means that the Boeing B747-8 will be able to carry its design payload of about 75 tons only if all the batteries can fit into the space left over, which is not at all evident. A flight of 82 min would not be able to carry any extra payload, because all the weight would be used up by the batteries! Incidentally, the energy stored in the batteries for the roughly one-hour flight is the same as in some 1.1 million of today's average laptop batteries having a capacity of 60 Wh.

This brings us to 2018. According to the American Helicopter Society (AHS), there are more than 70 eVTOL programs around the world [1.12]. Many of these are small startups but larger companies are getting into the act as well. Airbus is developing the Vahana VTOL aircraft in Silicon Valley; Boeing recently bought Aurora Flight Sciences, which is developing the LightningStrike VTOL aircraft which, until recently, was funded by DARPA. Aurora flight-tested a scaled, autonomous, AE version of this aircraft in 2017. Aurora is working with Uber as well to develop an aircraft for their Uber Elevate

FIGURE 1.4 Lilium VTOL full-scale prototype flight test.

Courtesy of electrek.co

Network [1.13]. Many of the smaller aircraft are being developed for the urban air-mobility market and are VTOL designs.

Some, like the Lilium (Figure 1.4), are equipped with swiveling propulsors, while others have two sets of propulsors, one for lift and the other for forward motion. The Kitty Hawk Cora (Figure 1.5), which was unveiled in New Zealand in March 2018, is a good example of this design. The Aurora LightningStrike will be a much bigger version of the Lilium design.

While many aspects of this industry have yet to show economic viability, there is overwhelming enthusiasm for this field with many new entrants. It is easy to find skeptics among mature aerospace engineers, mainly about the practicality of these designs. However, it is clear that electric propulsion is here to stay. As advances in technology make these systems more efficient, compact, and intelligent, electric aircraft will become more practical, and we will see entirely new businesses like air-taxi services take off at local airports in urban areas. In the longer run, with increased pressure from governmental clean air regulations, HE systems will become more economically feasible as well and may displace the current generation of aircraft in the commercial aviation marketplace.

FIGURE 1.5 Kitty Hawk Cora.

Courtesy of cora.aero

1.6 Way Forward

To make electric aircraft a reality, technical advances in energy storage, highly efficient and dense power electronics, compact generators and motors, and novel cooling strategies are needed. Researchers are even looking at high-temperature superconductors to further reduce weight. To make all of this come to fruition, all components need to be designed so that all systems including the entire propulsion systems can be certified to be safe enough to carry passengers.

Regardless of how disruptive the design of novel aircraft architectures will be, solving the size, weight, and power equation becomes the outstanding challenge in electric aircraft R&D. More than the ability to innovate, what is key is the capability to deliver mature innovations needed by the aviation sector, where safety and reliability are paramount and regulations are very strict.

Lastly, it must make *economic sense* to invest in a completely new paradigm for air travel. The aircraft of the future is clearly going to depend a lot more on electricity than the current generation and will be designed to leave a smaller environmental footprint.

1.7 Book Structure

The book is structured in such a way that the reader can get an overview of what the electric aircraft is all about and what challenges loom ahead. To achieve this goal it follows a step-by-step approach starting out with the basics.

Chapter 2 captures the stakes and challenges encountered with the electric aircraft paradigm. Orders of magnitude of power levels as well as the functional breakdown of system architectures are brought up. One section sheds light on the technologies in play.

Chapters 3-5 cover various aspects of systems electrification and provide in-depth details on both recent aircraft and research work for future applications. After a brief overview of the ME engine, Chapter 3 focuses on power generation and distribution, exploring the technology shifts there. The evolution of conventional electric networks and architectures towards high voltage is also underpinned. Chapter 4 follows up on systems electrification and describes how traditional pneumatic systems such as environmental control and wing ice protection can be replaced with more efficient electrical solutions. This chapter also deals with enabling technologies such as motors and power electronics. As for Chapter 5, it thoroughly describes the shift from hydraulic to electric actuation, addressing pros and cons as well as challenges. A variety of electric actuation systems ranging from flight controls to braking are shown as they exist on ME aircraft in service today. Research topics and demonstrators are also covered.

Chapter 6 is dedicated to electric propulsion options. Here explanatory block diagrams accompany detailed description and performance analysis of HE and electric propulsion architectures. Novel distributed propulsion and its benefits are analyzed as well. Last but not the least, enabling technologies for electric propulsion such as motors, motor controllers, energy storage, etc. are addressed and their key performance metrics are matched against timelines. A road map for target applications is also presented.

Chapters 7 and 8 provide descriptions of a variety of aircraft applications using electric and HE propulsion architectures. Theoretical aircraft designs and demonstrators across all aircraft segments, from general aviation to wide-body aircraft, are put into perspective. While discussions on aircraft powered by electric propulsion and aircraft-based electric taxiing come under Chapter 7, applications of HE propulsion are part of Chapter 8.

Chapter 9 deals with topics linked to maintainability and operations of electric aircraft. The discussion cuts across maintainability, ground, airport, and in-flight operations. Some open views on future autonomous travel are also shared.

Finally, Chapter 10 addresses the performance and business value of electric and HE aircraft across various segments. Operating costs, fuel price impacts, weight impacts, and efficiency gains through fuel burn reduction are all presented in detail for these aircraft, along with metrics measuring how greener and quieter they are.

References

[1.1] Sinnett, M., "787 No-Bleed Systems—Saving Fuel and Enhancing Operational Efficiencies," *AERO Magazine* Q4, 2007, Boeing.

[1.2] Wikipedia Contributors, "Gaston Tissandier," https://en.wikipedia.org/wiki/Gaston_Tissandier, retrieved January 2018.

[1.3] Tissandier, G., "*La Navigation Aérienne L'aviation Et La Direction Des Aérostats Dans Les Temps Anciens Et Modernes*," 1886 (Translated by Pascal Thalin).

[1.4] Garrison, P., "The Electric Airplane," *Air & Space Magazine*, August 2009.

[1.5] BBC News, "Solar Impulse Completes Historic Round-the-World Trip," July 2016, http://www.bbc.com/news/science-environment-36890563, retrieved February 2018.

[1.6] Urs, M. and Vezzini, A., "Electric Flight—History—State of the Art and First Applications," *Proceedings of the EVS28 International Electric Vehicle Symposium and Exhibition*, South Korea, May 2015.

[1.7] Wikipedia Contributors, "Electric Aircraft," https://en.wikipedia.org/wiki/Electric_aircraft, retrieved February 2018.

[1.8] Hanlon, M., "First Manned Fuel-Cell Flight," New Atlas, April 2008, https://newatlas.com/first-manned-fuel-cell-plane-flight/9117/, retrieved March 2018.

[1.9] Ridden, P., "Hydrogen Fuel Cell Four-Seater Passenger Plane Takes to the Air," New Atlas, September 2016, https://newatlas.com/hy4-hydrogen-fuel-cell-passenger-plane-test-flight/45687/, retrieved March 2018.

[1.10] Rajamani, R., *Electric Flight Technology: The Unfolding of a New Future*, (Warrendale: SAE International, 2018), ISBN 978-0-7680-8469-6.

[1.11] Tovey, A. and Yeomans, J., "French Pilot in Tiny One-Seater Beats Airbus to First Electric Cross-Channel Flight," *The Telegraph*, July 2015, http://www.telegraph.co.uk/finance/newsbysector/industry/engineering/11729838/Airbus-E-Fan-2.0-set-to-recreate-aviation-history-with-cross-Channel-flight-only-this-time-its-electric.html, retrieved November 2017.

[1.12] The Vertical Flight Technical Society, "The Electric VTOL News," http://evtol.news/aircraft/, accessed May 9, 2018.

[1.13] Kucinski, W., "Aurora Flight Sciences Partners with Uber in Contested Airspace," SAE Article 15397, April 2017, http://articles.sae.org/15397/, retrieved February 2018.

[1.14] JN Passieux. http://jn.passieux.free.fr/html/MbE1.php, accessed May 9, 2018.

CHAPTER 2

The Electric Aircraft Paradigm

Pascal Thalin
Chair and Member - SAE Electric Aircraft Steering Group

2.1 Scope and Stakes

For the electric aircraft design, an aircraft manufacturer has to take into account various requirements and expectations from the customer side. This flow-down from the airlines takes into account passenger expectations.

Baseline requirements of cost efficiency and environmental friendliness from the airlines can be split into the following:

- Fuel efficiency and range performance
  (fuel consumption savings = lower operating costs and air pollution)
- Reduced cost of operations and ownership
- Enhanced reliability and maintainability
- Available and dependable aircraft
  (lower downtime: more passenger comfort and operating revenues)
- Going green: reduced gas emissions and carbon footprint
- Reduced noise (perceived by passengers and communities)
- Sustainable design, manufacturing, and operations

FIGURE 2.1 Comfort and service-centric flight experience [2.1].

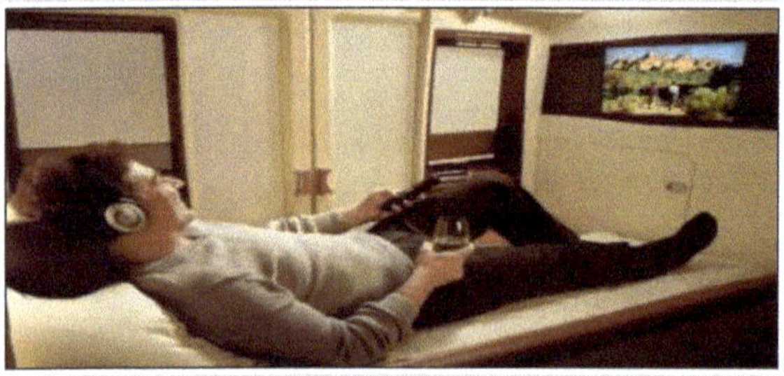

Passenger expectations for better flight experience are shown below (**Figure 2.1**):

- Comfortable cabin air conditioning and pressure level
- Better seating arrangements
- Low noise at airport and on flight
- Punctuality
- Onboard services (connectivity, inflight entertainment, etc.)

2.2 Route to Electric Aircraft

The design of an electric aircraft has to undergo a disruptive approach in order to achieve better overall performance, weight and fuel burn being major drivers. In conventional aircraft, on the non-propulsive side to begin with, some systems operate thanks to hydraulic or pneumatic power derived from the propulsion system. But this comes with a drawback in efficiency. Electrifying these systems as well as the power supplied to them can bring weight savings and efficiency gains resulting in fuel efficiency gains. This leads to decreased engine power offtake, allowing for optimizations in turn leading to lesser fuel burn. Systems where this has been possible recently are shown below:

- Air conditioning
- Engine start
- Airframe ice protection
- Actuation

From a power perspective it is worthwhile comparing the orders of magnitude involved in conventional pneumatic and hydraulic systems on one hand and their electrified versions on the other. **Figure 2.2** shows the power levels in play in a Falcon 2000 business jet taken as an example. This aircraft is fitted with two aft-mounted turbofan engines.

At aircraft level, for both propulsion and systems operation, the engines feature a maximum power capacity of 5000 kW extracted from fuel. This is only a tiny fraction

FIGURE 2.2 Falcon 2000 power levels [2.2, 2.3].

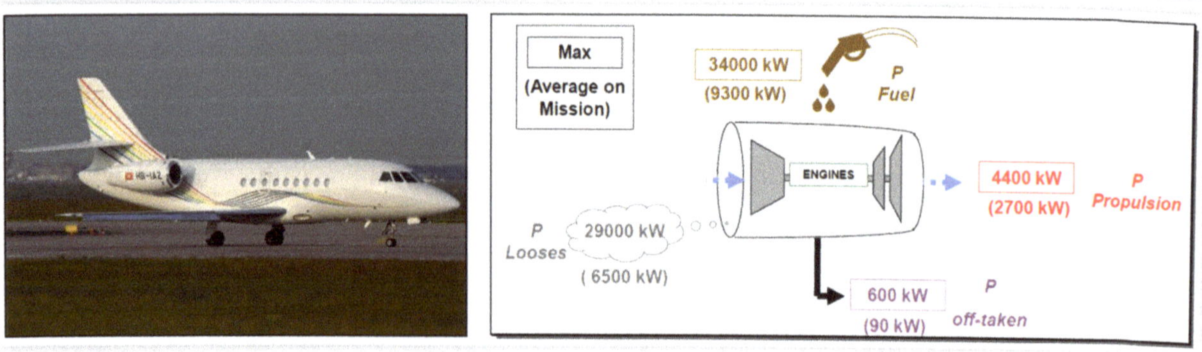

(around 15%) of the maximum power capability (34,000 kW) of jet fuel carried on board. The proportion goes up to around 30% during a typical mission, but is still far off the full fuel capability. The reason behind this is that the power capacity to which engines are designed takes into account design margins and also the power needed to offset the huge losses during the conversion from fuel power to power utilized for aircraft propulsion and other systems. In other words, the engine efficiency also comes into play. Therefore, while producing the lift and thrust necessary to carry and move the aircraft safely through air, improving engine, system, and structural efficiencies is at the forefront of fuel burn reduction at the aircraft level.

Likewise, at the engine level the maximum power offtake capacity for supplying power to non-propulsive systems amounts to a significant 14% (600 kW) of the max propulsion needs (4400 kW), whereas only a tiny fraction of 3% (90 kW) is used on average during a flight. The huge difference is once more due to design margins that have to be taken into account while offsetting losses. On one hand, these losses are attributable to the low efficiency of the extraction and generation of pneumatic and hydraulic power and, on the other, the low efficiency of systems during operation. It is straightforward to see low efficiency creates losses and can lead to engine oversizing. In addition, unfortunately, power offtake capacity has to cover cumulatively the margins of each individual system taken separately, whether pneumatic, hydraulic, or electric.

Hence the electrification paradigm, wherein the cumulative approach leaves way to more optimized monolithic design margins taken into account all at once, without compromising overall performance. Moreover, owing to the far better efficiency of electric systems replacing pneumatic and hydraulic ones, oversizing of engine power offtake capacity for systems operations can be avoided. Electrification of systems helps unify and optimize design margins, and also avoids excessive useless provisions just to compensate losses.

Continuing on the Falcon 2000 business jet example, **Figure 2.3** illustrates the above analysis. It shows that electrification helps significantly cut back from sevenfold to just twofold the ratio of engine power offtake to the power really utilized by the systems.

The overall idea of moving conventionally powered systems to electrical ones holds well. On the flip side, increased electrical power needs on an aircraft require stepping up electrical power generation and distribution capacity on board. This in turn leads to

FIGURE 2.3 Electrification benefits on engine power offtake [2.3].

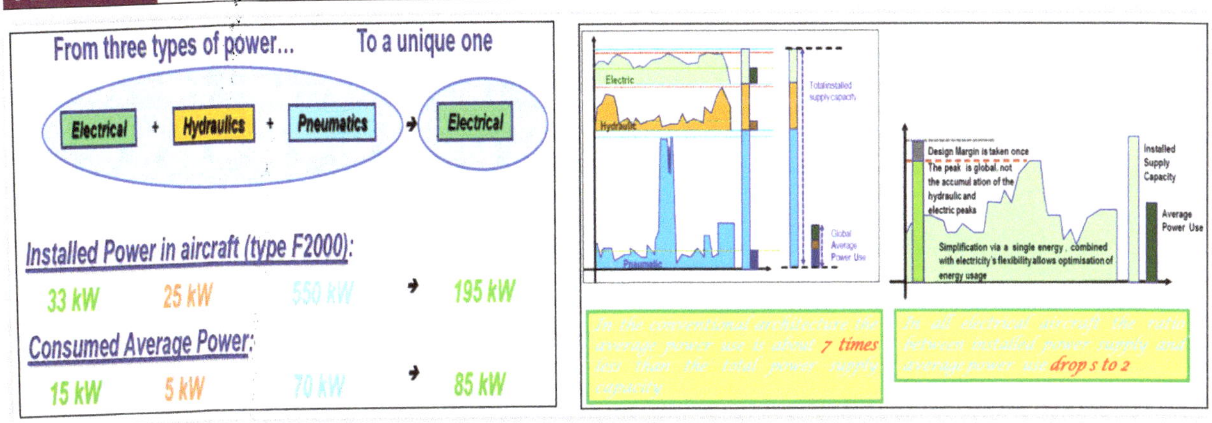

additional and larger power conversion equipment and distribution such as cabling and protective devices. The flow and control of electrical power down to a given electrified end system (load) is only possible via a power electronic drive or power supply allocated to that system. Therefore, using state-of-the-art technology for large-scale system electrification, due to growth of power needs and the multiplication of additional equipment, can quickly lead to additional weight which may undermine the other benefits of electrification.

Rethinking electrical system architectures and developing lighter technologies for airframe structure and systems has become the foremost priority in order to alleviate penalties from potential weight increase during electrification. Therefore, an overall design approach is necessary in order to prevent the weight issue from compromising the valuable benefits and efficiency gains brought by electrification.

One of those instances is the move towards high-voltage electrical networks. Switching to variable frequency of the electrical power generation, as opposed to the legacy fixed frequency network, also eases overall design. Furthermore, novel electrical architectures in combination with power-dense solutions allow significant scope for optimization of both power demand and utilization. Implementation of lighter electrical architectures within lighter composite airframe structures takes the overall aircraft weight optimization to the next level, notwithstanding the constraints specifically linked to the integration of electrical systems in composite airframes. A recent track record for such a move towards more electrical systems is shown in the following aircraft programs taken as examples:

FIGURE 2.4 Airbus A380 [2.1].

FIGURE 2.5 Boeing 787 [2.1].

FIGURE 2.6 Airbus A350 [2.1].

- Airbus A380 (largest passenger aircraft) (**Figure 2.4**)
  - First variable frequency generation, solid-state power control-based distribution
- Boeing 787 Dreamliner (wide-body aircraft) (**Figure 2.5**)
  - Innovative architecture, variable frequency, and high-voltage electrical network compatible with composite airframe
  - Intensive electrification: air conditioning (ECS), engine and auxiliary power unit (APU) start, braking, ice protection (anti-icing/de-icing), new battery technology (lithium-ion)
- Airbus A350 XWB (wide-body aircraft) (**Figure 2.6**)
  - Variable frequency and high-voltage electrical network compatible with composite airframe
  - Electrical APU start
  - New battery technology (lithium-ion)

Electrically powering non-propulsive systems, thereby forgoing non-efficient pneumatic and hydraulic legacy power alternatives, winds up over time in a gradual increase in

FIGURE 2.7 Evolution of onboard electrical power for systems [2.1].

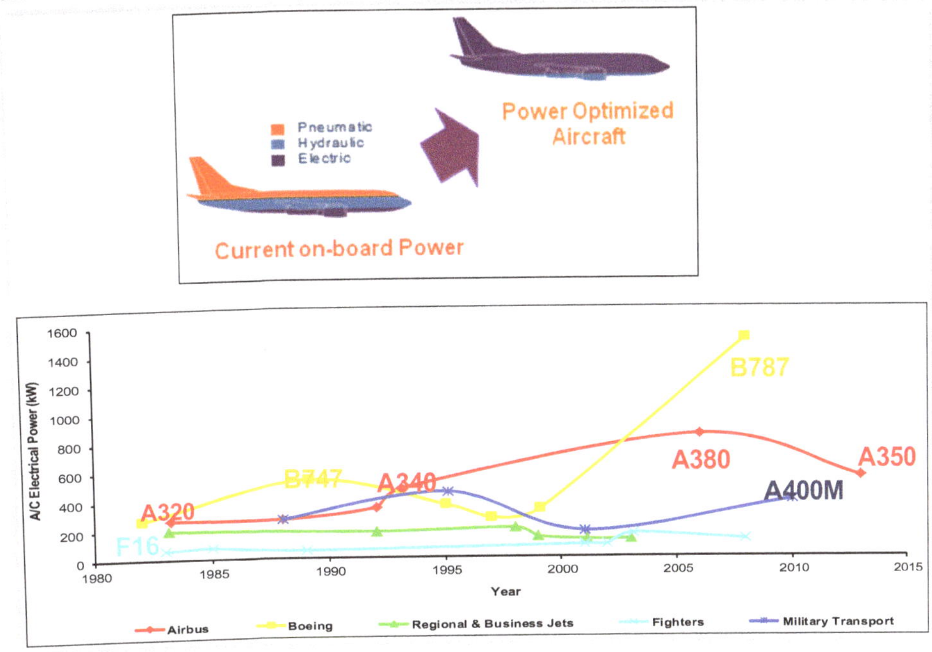

electrical power on board as shown in Figure 2.7. Obviously, this depends on the degree of electrification and the size and the nature of the aircraft segment considered.

Figure 2.7 doesn't take into account the electrical power needs related to hybrid-electric and fully electric propulsion. This will be discussed in Chapter 6.

2.3 Electrical Chain Breakdown

Concurrent breakthroughs in propulsion, non-propulsive systems, and composite airframes contribute to the success of the electric aircraft. Figure 2.8 features the functional breakdown of the electric chain. For the sake of clarity, various functions that can benefit from their electrification are shown.

In fact, according to the ATA100 standard devised by the Air Transport Association of America[1] (ATA), all aircraft components, from head to tail, are categorized into ATA chapters and subchapters. This subdivision into ATA chapters is common to various aerospace industry players such as aircraft manufacturers, suppliers, airlines, and maintenance, repair and overhaul (MRO) companies. Therefore, in Figure 2.8, blocks representing candidate functions for electrification also contain their denominated ATA chapter number.

Typically, the conventional aircraft systems that use pneumatics and hydraulics are air conditioning, airframe ice protection, flight controls, landing gears, and brakes. Figure 2.8 illustrates that a lot of electric optimization is possible by replacing these conventional pneumatic and hydraulic systems with electrically powered systems.

[1] Renamed Airlines for America (A4A).

FIGURE 2.8 Functional breakdown of the electrical chain [2.1].

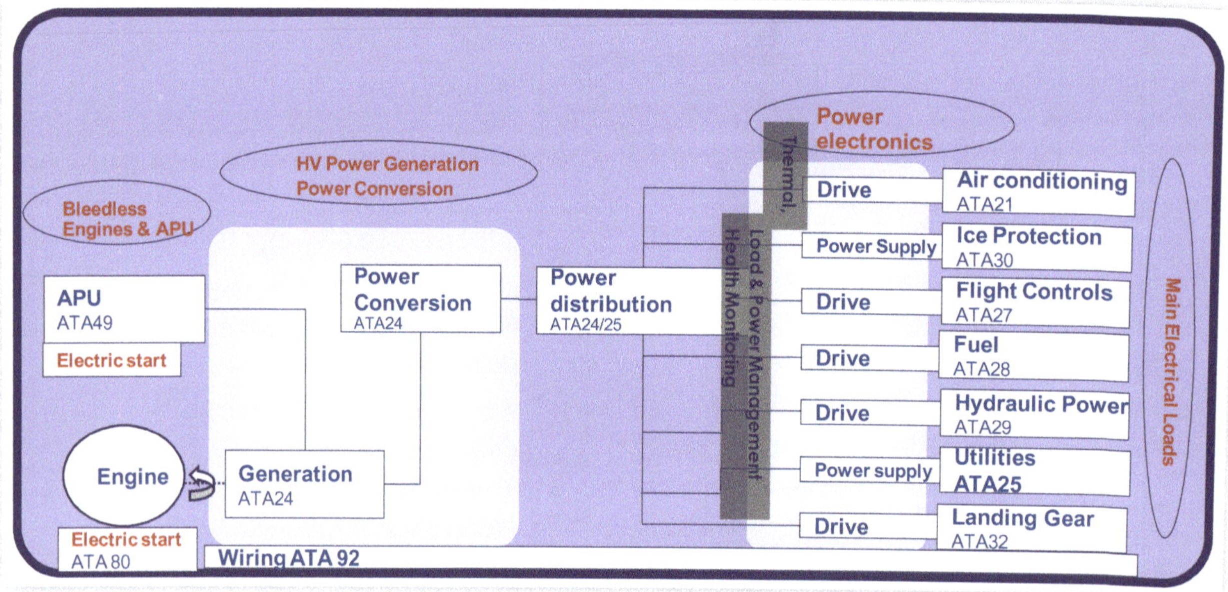

Air conditioning is a good example. On conventional aircraft, hot and pressurized air from the engines called "bleed air" is transformed and used to ensure air conditioning and pressurization. Hot bleed air is also the conventional means for airframe ice protection: hot air is simply blown over the wings and engine air inlets via exhaust holes.

In the past years, however, manufacturers have moved to "bleedless" engines on what has come to be a high point in the development of "more electric" aircraft. These engines still use jet fuel for propulsion but don't provide hot air as a by-product, but *electricity*—which, instead, can be used to power systems having undergone "electrification", including air conditioning and ice protection. In such a bleedless system, an "electric blanket" is built directly into the wings, which heats the wings to keep them free of ice.

Electrification of propulsion systems can also bring a variety of benefits to the electric aircraft. In this case, systems built around (e.g., thrust reverser) or dedicated to engine operation (fuel pump) undergo electrification in what has come to be called the "more electric" engine.

Down the line, comparable to the evolution in the automotive industry, energy storage alternatives could partially or totally replace jet fuel carried on board. This will usher in hybrid-electric and electric propulsion solutions capable of delivering thrust power from electricity. This evolution is a paradigm shift improving aircraft performance and environmental footprint, while departing from aircraft forms and shapes we are used to see in the skies.

2.4 Technology Stakes

In the electrification process, the surge in electrical power capacity on the aircraft has to be met with power generation techniques outside the beaten paths, in order to keep weight impacts at bay. Therefore, instead of simply scaling up traditional solutions, the focus of research and development concerns the following disruptive concepts and technologies:

- High-Voltage Alternating Current (HVAC) power source:
  - High speed, **V**ariable **F**requency **G**enerators with engine **S**tarting capability (VFSG)
- High-Voltage Direct Current (HVDC) power source derived from either of the following which also comes with engine starting feature:
  - Power converters connected to the above HVAC sources
  - Novel machine topologies with HVDC outputs

These starter-generators are conventionally stand-alone and are attached to a so-called gearbox located within the powerplant. The gearbox ensures the mechanical coupling between the rotating engine core and all system-related power sources, such as electrical power generators or hydraulic pumps. Depending on the aircraft platform, whenever a high degree of integration is required, departing from their conventional standalone installation, the above power sources could be embedded into their respective propulsion systems or auxiliary power sources.

The power converters used for HVDC electrical power generation could remain "passive," when using only transformers and diode-based rectifiers. They are classified as "active" when they implement semiconductor-based power switching devices commonly called "power electronics" which understandably require control electronics to operate.

In order to take into account the various power ratings and voltage capabilities of the loads to be supplied, the HVDC power outputs have to be in turn converted to lower voltage levels. Dedicated converters ensure this "power conversion" function on an aircraft and once again could be either "passive" or "active."

The distribution of electrical energy from the sources down to the loads is ensured by the electrical network comprising optimized wiring and power distribution bays, which require the following technology bricks:

- High-voltage cabling (resistant to "partial discharge" occurring at altitude...)
- Cabling fault detection (arc and ground faults, short/open…)
- Semiconductor-based **S**olid-**S**tate **P**ower **C**ontrollers (SSPC) for protection and power dispatch to loads
- Load and power management functions for enhanced power availability

Depending on architectures, the power distribution bays could be either centrally located in one section of the aircraft or distributed all over the aircraft, even in areas with uncontrolled environments.

The "load" end of the aircraft electric chain requires implementation of electric components which could be basically schematized as either motors, in the case of actuation systems, or resistances in the case of heating systems. The following motor technologies are needed for the electric aircraft:

- AC and DC brushless fault-tolerant machine topologies
- High speed, high power, and reversible motors
- Light, low-loss alloys (nanocomposite and high elastic-limit steel)

Moreover, Figure 2.8 shows that these loads interface with power distribution through either motor drives or power supplies.

Motors require variable speed drives in order to operate as required by their respective aircraft functions. Similar to power supplies, these drives comprise power switches and controls and are commonly identified as "power electronics" or power converters.

As discussed above, intensive electrification of systems leads to scaled-up electrical power generation needs on board and different voltage levels. The flow of that power down to end systems implies distributing the power through scaled-up distribution capabilities in the first place. Secondly, before reaching the motors and resistances at the far end of the electrical chain, the last stage of power transmission is only possible through the power electronics (or power supplies) at the interface of each system gone electrical.

Challenges in the field of power electronics used for power conversion, motor drive, and power supplies for the electric aircraft are listed below:

- Fault tolerance, high reliability, and availability
- High power density
- High efficiency and high temperature capability
- Optimized cooling and thermal management
- Installation in unconventional areas (e.g., unpressurized area, explosive atmosphere)
- Maintainability

Let us now focus on the power electronics. From an aircraft manufacturer perspective, it is a no brainer that overall systems weight cannot be optimized if each system uses its own power electronics, especially when they are used transiently during a flight. It can be easily understood that transient system operation results in fuel penalties owing to dead weight carried by the aircraft. An integrated modular approach to power electronics could be the right answer.

In this approach, power electronics are designed as rackable generic modules similar in principle to what is already in service in the avionics domain. A given module could be configured to address any one of the aircraft functions requiring power electronics, namely, power conversion, motor drive, or power supply. During aircraft operation, the module could also be reconfigured on demand and switched from one system to another, thereby reducing downtime and dead weight inherent to transient operation of dedicated power electronics.

Functional availability is also improved thanks to built-in redundancy. In fact, automatic reconfiguration in the event of a module failure is achieved by simply switching off the failed module and switching on another available module. Additionally, the modular design allows power scalability in that multiple modules can be paralleled to increase delivered power meeting the requirements of some large systems.

Hence, the modular approach is not only unavoidable, but a necessity in order to undercut weight penalties. A demonstration of this concept (**Figure 2.9**) was performed in the frame of Clean Sky, a major European collaborative research project. On an aircraft such a power center would be located in between the power sources and the electrified system loads. As discussed above, it incorporates the power rectifier generating the HVDC bus from the sources, and the multiple power electronic modules called up by loads downstream.

Real-time surveillance and analysis of aircraft functions can help contribute to overall availability and maintainability improvement. This is possible thanks to the diagnostic and prognostic capabilities brought by Health Monitoring. On one hand, upon occurrence of abnormal conditions, the "diagnostic" feature singles out the causes so that timely corrective actions are applied ensuring continued operation. On the other hand, for a given monitored system, the "prognostic" feature involves sensing selected parameters and comparing them to values output by that system's lifetime model stored in memory. The potential failure of system components can therefore be anticipated in this way, allowing preplanning of appropriate maintenance tasks thereby avoiding interruption of flight operations.

The advent of electric propulsion is all but tied to a yet-to-come performance enhancement of energy storage alternatives to jet fuel. Even the shift towards hybrid-electric propulsion relies heavily on the capability of batteries to deliver a major step increase in specific energy in the forthcoming years.

FIGURE 2.9 Electrical power distribution center—short-range more electric aircraft [2.4].

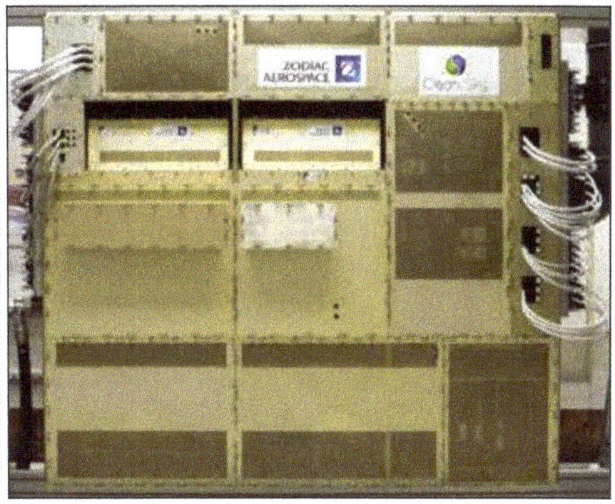

2.5 Conclusion

In summary, the two major challenges in hybrid-electric and electric propulsion reside in the following areas:

- Batteries with high specific energy
- Motors handling high voltage with high power density

Upending conventional fixed-wing aircraft design, electric propulsion may also bring a suite of opportunities to drastically improve aircraft efficiency and safety thanks to tighter propulsion-airframe integration. This includes distributed electric propulsion wherein multiple electrically powered propulsors are distributed all over the wing.

The electric aircraft is a paradigm shift enhancing aircraft performance and economics. With the help of novel architectures and optimized weight, the electric aircraft can bring strong added value to aircraft manufacturers and airlines by improving efficiency while reducing costs over time and the environmental footprint.

References

[2.1] https://sunjet-project.eu/sites/default/files/Thales AES - Thalin.pdf, accessed April 29, 2018.

[2.2] https://fr.wikipedia.org/wiki/Dassault_Falcon_2000, accessed November 10, 2017.

[2.3] Stoufflet, B., "Towards an All Electrical Falcon," Dassault Aviation, *The More Electrical Aircraft—Achievements and Perspective for the Future—ICAS Workshop*, Cape Town, South Africa, September 2, 2013, http://www.icas.org/media/pdf/Workshops/2013/Towards All Electrical Aircraft Stoufflet.pdf, accessed July 12, 2017.

[2.4] Skyline Magazine Issue 15—Show time Demonstrators in the Sky, March 2015, http://www.cleansky.eu/sites/default/files/inline-files/skyline15_march_2015.pdf, accessed April 29, 2018.

CHAPTER 3

Electrification of Aircraft Systems—Part I

Power Generation and Distribution, Electrical Networks and Architectures

Pascal Thalin
Chair and Member - SAE Electric Aircraft Steering Group

3.1 Conventional Aircraft and Engine Systems

Conventional aircraft and engines achieve their functionalities thanks to various systems dedicated to their operation, namely, aircraft systems and engine systems. In order to supply these systems with energy, a flying aircraft has to be autonomous in producing its own energy needs. On conventional aircraft, this is done by extracting energy from fuel onboard. Transformation from fuel to the energy supplied to systems is mainly performed by the engines, on top of their principal function of producing thrust. Energy storage is generally used as backup sources for systems' operation and is called up during starting sequences for small engines and APUs. APUs are auxiliary power units delivering both bleed air and electricity. Even though APUs are able to cater to energy needs, they are generally considered as "non-essential" from a certification standpoint, and their normal utilization is confined to ground and low altitude operations.

FIGURE 3.1 Energy sources and systems—conventional large commercial aircraft [3.1, 3.2].

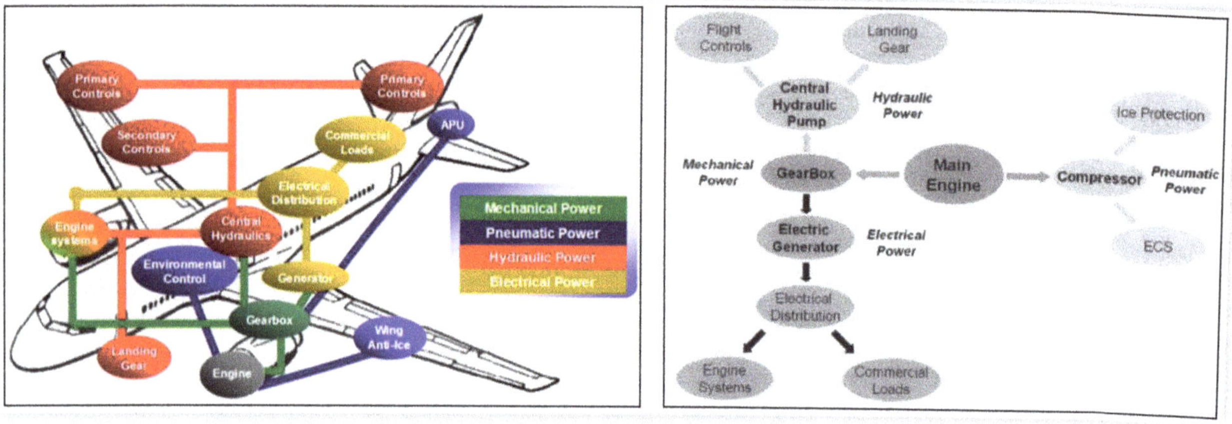

To begin with, let us consider how and in what form the energy for systems operation is extracted from the engines on a conventional large commercial aircraft [3.1, 3.2]. As shown in **Figure 3.1**, four types of energy sources are required to power all the systems.

In conventional engines, secondary power (SP) is the power distributed around the engine. Aircraft systems using this power are called secondary power systems (SPS) and not used for propulsion (this being the primary power). On most aircraft SP is distributed in four forms:

- Electric (avionics, lights, instruments, entertainment, etc.)
- Hydraulic (primary and secondary flight controls, landing gears, brakes, steering, doors, etc.)
- Bleed air or pneumatic (environmental control system (ECS), cabin pressurization, engine cowl and wing ice protection, engine starting, etc.)
- Mechanical (engine systems: fuel pump, oil pump, etc.)

For a twin-engine long-range aircraft, **Figure 3.2** shows the typical engine power breakdown between thrust and SP systems, their functions and the type of energy used to power them [3.3, 3.4].

FIGURE 3.2 Engine power breakdown (thrust and SP)—twin-engine long-range aircraft (A330) [3.3, 3.4].

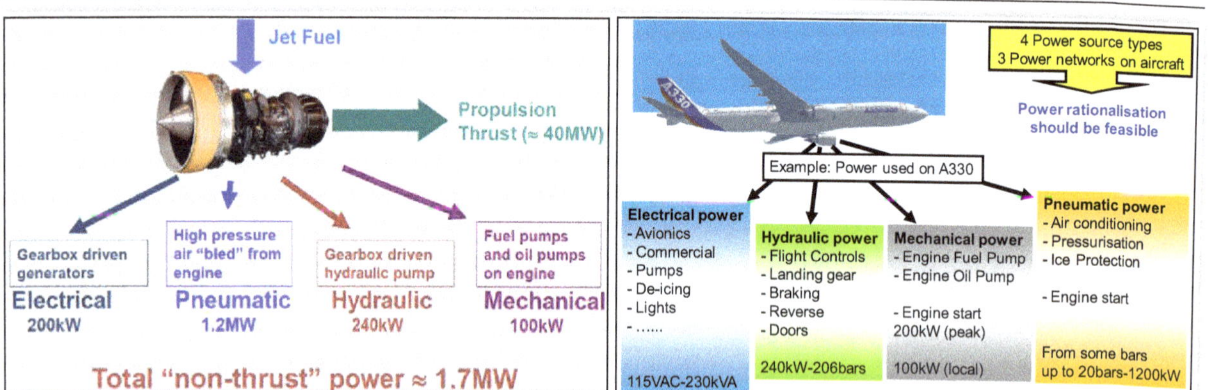

FIGURE 3.3a Gearbox location and architecture [3.5, 3.6].

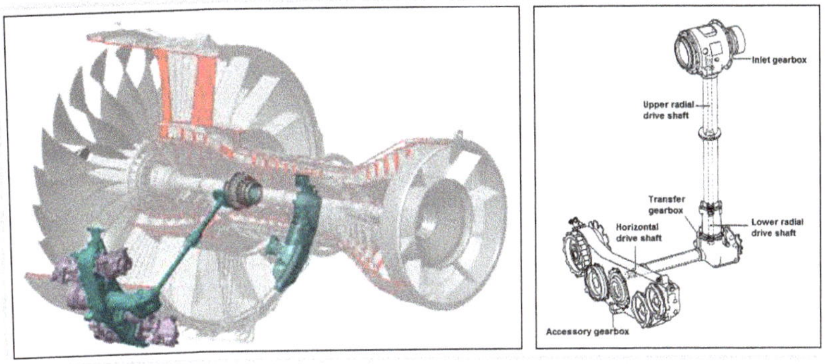

Past and current turbofan engines are capable of supplying electric and hydraulic power through generator(s) and hydraulic pump(s) driven by the engine accessory gearbox (AGB). As shown in **Figure 3.3a**, shaft power is extracted from the engine's high-pressure (HP) shaft to drive the gearbox on which the SP components are mounted in addition to the engine's fuel and oil pumps (**Figure 3.3b**). Mechanical torque is transferred from the engine HP shaft to the AGB. Hence, the AGB enables accessories to be connected to the engine HP shaft, allowing in this way also the possibility of electrically starting the engine thanks to a starter generator. Apart from the gearbox, hot, pressurized air is extracted from the engine's compressors thereby providing pneumatic power.

But there is a downside to such extraction and distribution of SP systems, because on top of reducing maximum thrust capability, it affects the design and performance of the engine by increasing the following factors:

- Complexity and weight
- Maintenance and operational costs
- Aircraft drag (profile and induced)
- Turbine entry temperature (TET)
- Specific fuel consumption (SFC)

FIGURE 3.3b Location of SP components on gearbox [3.7].

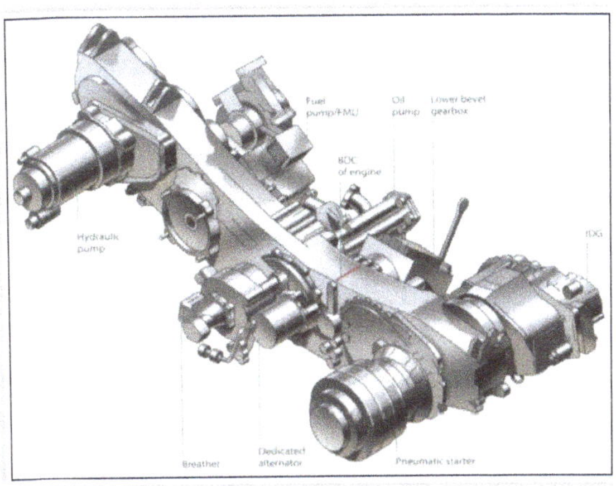

Bleeding air from the engine is largely inefficient and a large amount of power is wasted. Taking bleed air from the compressor requires an increase in fuel flow and hence temperature. When shaft power is extracted the turbine has to deliver extra work at a given speed and this can occur by increasing TET.

Engines on a twin-engine aircraft are oversized to accommodate takeoff conditions with one engine inoperative. During cruise, the engines operate at a significantly lower power setting, and as a result the power extracted in the form of shaft power and bleed air constitutes a larger percentage. The reduction of shaft power and bleed air reduces the overall specific fuel consumption.

Within an overall aircraft design optimization, the adoption of all-electric technologies has the goal to reduce the operating empty weight of the aircraft and leads to reductions in fuel load, required thrust, maximum take-off weight, etc. Reductions in maximum aircraft takeoff weight would reduce the wing area, which in turn could allow for reduced wingspan and increased aspect ratio, thus improving the aerodynamic efficiency of the aircraft. Reductions in thrust requirements and TET could reduce jet velocity and noise as well as NOx emissions. Reduced SFC, aircraft mass and improved aerodynamic efficiency would lead to reduced fuel consumption, thus CO_2 emissions. But today, more electric aircraft (MEA) tend to be heavier instead of lighter, at least for less bleed.

Therefore, an all-out systems electrification is still far from being a reality because of the huge challenges that systems design and enabling technologies are confronted with.

3.2 More Electric Engine (MEE) Systems

In the more electric engine (MEE) [3.8], a majority of engine systems would be electrically powered. An MEE takes the engine control elements of the engine, such as fuel pumping, oil pumping and engine actuation that are conventionally powered through mechanical means and converts them to electrically powered operation. The idea that a MEA necessitates the use of a MEE to power it is not true. However, both MEE and MEA have significant but different effects on the engine.

On an MEE, components/functions that may go directly electrical, or indirectly via the electrification of aircraft systems, are as follows:

- Variable stator vanes (VSV)
- Variable bleed valves (VBV)
- Transient bleed valves (TBV)
- Fuel pump
- Oil pump and scavenge system
- Active clearance control
- Blade pitch variation
- Thrust reverser actuation

- Variable area fuel nozzle actuation
- Nacelle anti-ice
- Engine starting system
- Bleed extraction for engine use

MEE should not be confused with electric engine or propulsion which refers to the electrification of the thrust-producing function. While the A380, the Boeing 787, and the A350 XWB are considered as MEA, they are not fitted with MEEs. Nevertheless, electrification of engine control functions is being considered in ongoing research programs. However, the benefits and issues related to the implementation of these functions are still under investigation. Consequently there is not currently an MEE in service. Differing from the engine controls electrification performed in an MEE, the following nacelle (or cowl)-related aircraft systems may go electrical on clean sheet designs of conventional aircraft, even though they are not categorized as MEAs:

- Thrust reverser actuation (e.g., short-range COMAC C919)
- Nacelle anti-ice

3.3 More Electric Aircraft (MEA) Systems

In the MEA, a certain number of SP systems get to be electrically powered, thus moving away from conventional types of energy. However, to a certain extent, present forms of hydraulic, mechanical, or pneumatic energy extraction may be retained as well. Dedicated or localized hydraulic systems may be incorporated together with the extraction of bleed air. Hydraulically powered loads may be retained to a certain extent while others are electrified. On the hydraulic power generation side, some pumps may be electrically driven (electric motor pump, or EMP) as opposed to a conventional system that mostly utilizes pumps mechanically driven by the engine at the AGB output (engine-driven pump or EDP). This applies to recent MEA long-range aircraft programs such as the Airbus A380, A350 XWB, and the Boeing 787 with varying levels of electrification.

On an MEA, if all systems were to be electrified, only one form of energy (electrical) powering both aircraft and engine systems would be derived from engines via large generators. Nevertheless, in this case, as opposed to a long-range MEA with larger engines, smaller short-range MEA engines with lesser ground clearance may obviously run into integration issues with powerful large-sized starter generators, despite the simplification ensured by the removal of bleed components. Research work has looked into options like integrating such machines into the engine core allowing for AGB removal, for instance, or integrating them mechanically into the AGB thereby avoiding drag-producing nacelle protuberances caused by their installation on the AGB.

Compared to the engine power breakdown shown in Figure 3.2 for a conventional configuration, the proportion of electric power in play on an all-electric systems

FIGURE 3.4 Bleedless engines—twin-engine long-range all-electric-systems aircraft [3.4, 3.9].

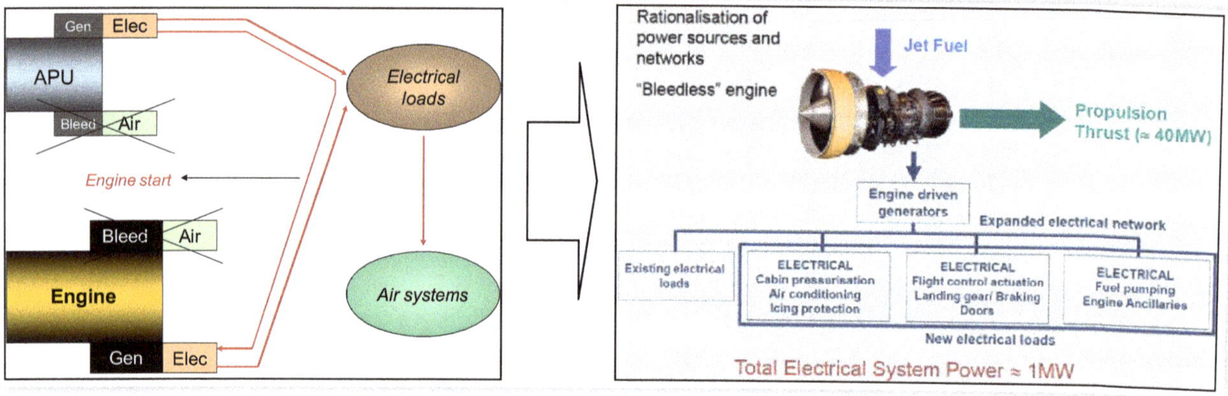

long-range MEA goes up to a much higher level (**Figure 3.4**) [3.9]. As can be seen, on such an aircraft, all four conventional types of energy are substituted by electrical energy alone, thereby lowering overall power needs thanks to efficiency gains brought by electricity. Adoption of a single form of SP system and implementation of all- or more electric technologies offers a number of distinct advantages over conventional SP systems:

- Improved aircraft systems utilization and implementation of more efficient power units
- Reduced engine core size and an increase in engine bypass ratio, overall pressure ratio and the turbine inlet temperature, thus improving engine performance which leads to a more efficient aircraft

Consequently, the removal of cumbersome and heavy bleed systems on engines and the APU allows for power optimization with full-scale system electrification. In the absence of bleed air, engine starting is no longer achieved using an ATS, and the start torque is provided by the engine-mounted starter generator operating in motor mode drawing electrical power from the APU.

Large bleedless engines have enabled the advent of the more electric Boeing 787, in service since 2011. **Figure 3.5** provides a comparison of electrical power levels of conventional aircraft and long-range MEAs [3.9, 3.10]. Value shown for the Boeing 787 includes 450 kW of electrical power generated by the all-electric APU. Also portrayed is the extent of systems electrification on the Boeing 787.

The Boeing 787 bleedless engines are electrically started by powerful starter generators mounted on the AGB which thereafter switch to the generation of main electrical power. Like on conventional aircraft, the ram air turbine (RAT), a wind turbine deployed under emergency conditions, pressurizes a hydraulic circuit which in turn generates electricity via an electrical generator. In fact, the Boeing 787 uses a hybrid RAT that includes both a hydraulic pump and an electric generator.

While hydraulic power generation is partially achieved using EMPs, landing gear and flight control loads have remained hydraulic, though some flight control actuators have gone electrical. Even though the Boeing 787 is not an all-electric systems aircraft,

FIGURE 3.5 Systems electrical power (conventional aircraft vs. MEA) [3.9] and Boeing 787 systems electrification [3.10].

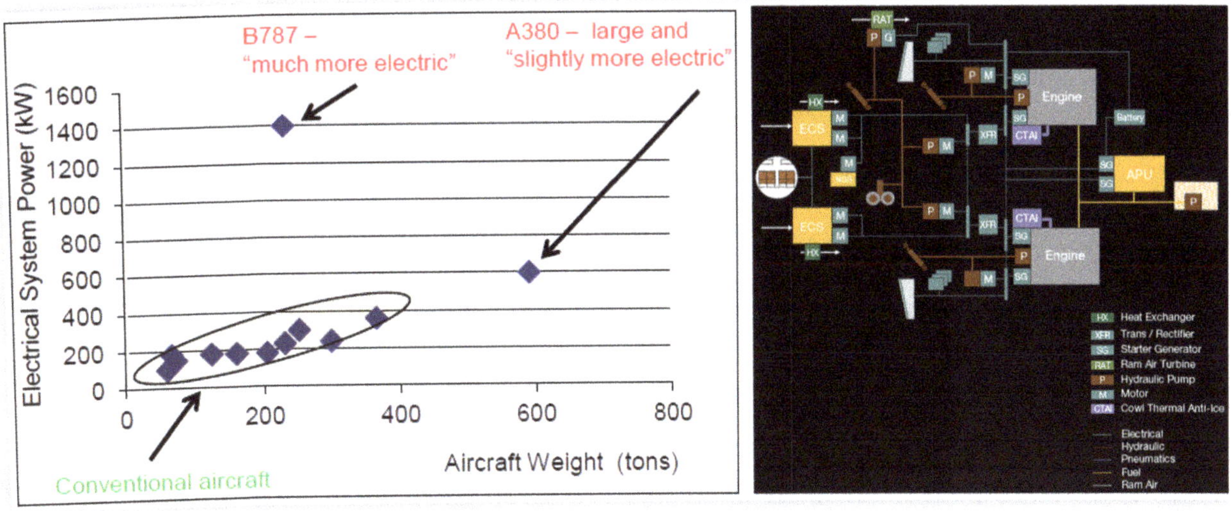

it features extensive electrification of systems. Previously pneumatic large loads such as ECS and wing anti-ice protection are fully electrical whereas significantly smaller loads such as engine cowl anti-ice protection have stayed pneumatic as before.

3.3.1 Electrical Power Generation and Distribution

Small aircraft (e.g., helicopters and business jets) in general do not have an alternating current (AC) network at all, and their low-voltage direct-current (DC) network is directly generated by an engine-mounted DC starter generator which also ensures engine starting. But even on conventional medium to large aircraft, long wiring distances are commonly encountered due to aircraft size. Cabling weight is a major contributor to overall aircraft weight. Weight stakes are surprisingly high like the 292 miles of wiring found on an Airbus 380, for example, weighing nearly 6 tons. Therefore, minimizing currents flowing in the cables for a given power rating would allow smaller gauge wiring to be installed thereby leading to weight reductions. Currents can be lowered by simply increasing the supply voltage, the reason why 115 VAC networks are found on medium to large conventional aircraft rather than the 28V DC networks found on smaller aircraft. When working with AC networks, power required for supplying DC loads is derived from the AC networks thanks to power conversion equipment. Following this approach, on recent large MEA, AC voltage levels have been pushed even higher, like the 230 VAC networks found on the A350 and the Boeing 787.

On the electrical generation side, legacy aircraft such as the Airbus A320, A330, A340; the Boeing 747 and 777; or the Embraer E-Jets use engine AGB-mounted integrated drive generators (IDG) supplying 115 VAC busbars. As shown in **Figure 3.6** [3.11], an IDG comprises a constant-speed drive (CSD) driving an AC electrical generator inside the same housing. In order to produce the proper voltage at a constant AC frequency, usually three-phase 115 VAC at 400 Hz, the generator needs to spin at a constant specific

FIGURE 3.6 VSCF [3.2] and IDG-based constant-frequency power generation principles [3.12] + generators installed on a General Electric GE90 engine (Boeing 777) [3.13].

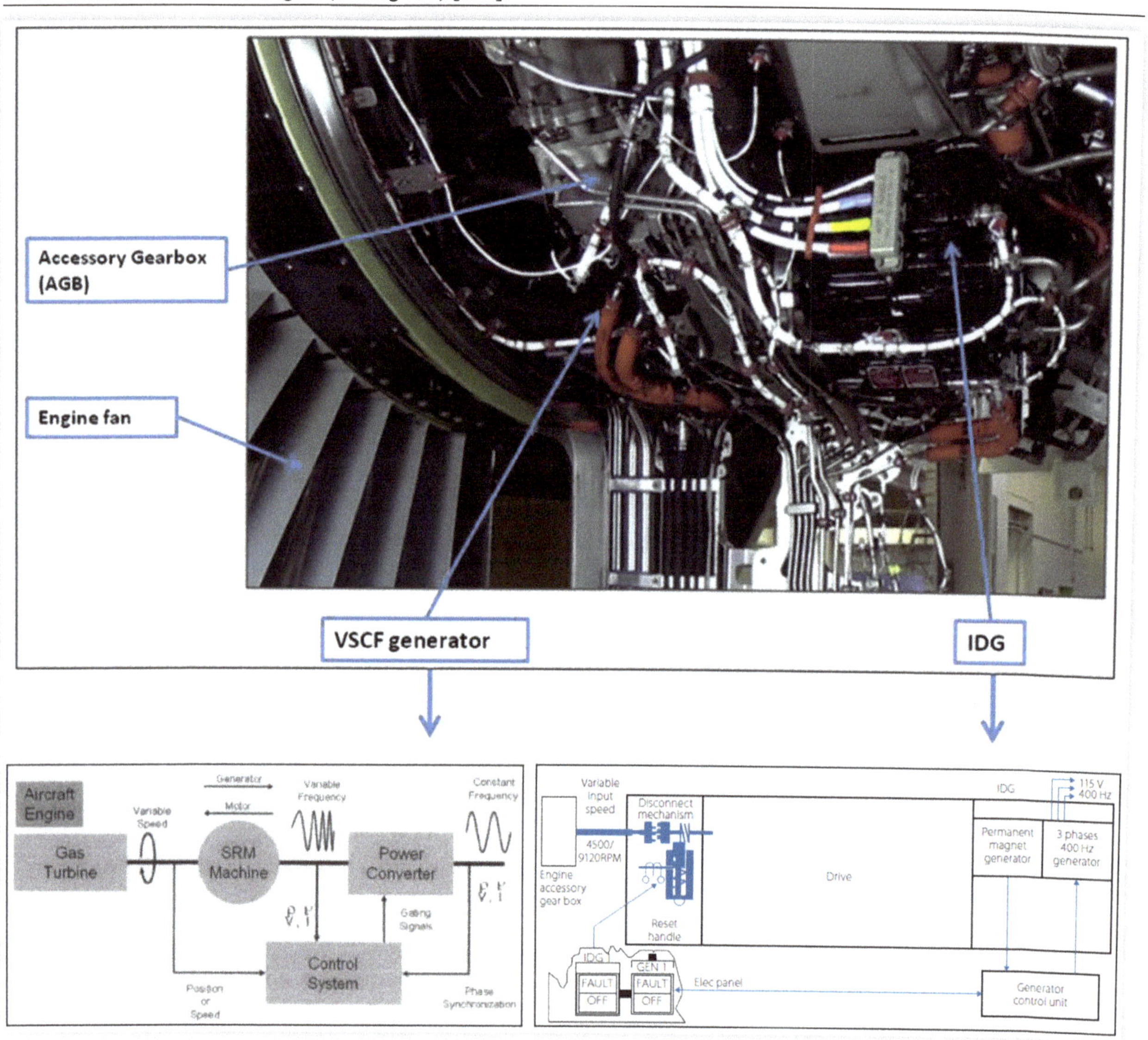

speed (typically 6000 RPM for air-cooled generators, up to 24,000 RPM for oil-cooled ones). Since the jet engine AGB speed varies from idle to full power, this creates the need for the CSD. The CSD takes the variable speed output of the AGB and hydro-mechanically produces a constant speed output that is used to drive the generator.

Therefore, IDGs output a constant-frequency (400 Hz) voltage which is supplied to the onboard power distribution centers in charge of dispatching electrical energy to various loads. The IDG is electronically controlled by a GCU (generator control unit). In case of an IDG overheat or oil loss condition, pilot action can activate a mechanical disconnect device within the IDG, resulting in its mechanical decoupling from the AGB (Figure 3.6) [3.12].

The CSD part of the IDG calls for high-precision parts in a complex integration setup that only advanced know-how and manufacturing workmanship can achieve. As a result, those demanding capabilities are available only from a single source with patent protection supplying a majority of legacy jet aircraft including their reengined versions.

The IDG is still the most commonly used fixed-frequency AC generator. However, operating experiences under the new requirements of lower cost, increased reliability, easier maintenance, and higher operating speed and temperatures have shown that a replacement for the CSD using a static power converter comprising power electronics has obvious advantages [3.2]. The resulting system shown is known as variable-speed constant-frequency (VSCF) system, a typical block diagram of which is shown in Figure 3.6. In the generating mode, the main engine AGB drives the machine, providing electrical power at variable frequency which is transformed into constant frequency by the power converter. The Boeing 777 uses such VSCF generators for supplying backup power.

In order to achieve weight and cost-effective operations on new aircraft, more reliable electrical power generation solutions offering the possibility of procuring them from multiple suppliers. Therefore, recent aircraft developments chose to go for a variable frequency power supply for aircraft loads, wherein the CSD concentrating all the complexity is totally suppressed, allowing for better maintainability, more compact engine installation, and better cost of ownership. The move to large variable-frequency generators (VFG) on the A380, at the cost of requiring added power electronics for many loads, marked the paradigm shift from fixed frequency to variable-frequency (VF) AC power generation. This trend has now gone mainstream on large commercial clean-sheet aircraft developments (Figure 3.7). Recent programs include the A350, Boeing 787, CSeries, C919, and MC-21.

This opened up new avenues in the integration of multiple functions such as the combination of engine start and VF power generation functions within the same housing. This gave birth to AC starter generators going by the name of variable-frequency starter generators (VFSG) as implemented on the Boeing 787 (Figure 3.8).

The gradual increase in the degree of electrification can also be characterized by the number of electric circuits versus the number of hydraulic circuits found in the architecture of a given aircraft. The increase in the number of electric circuits indicates an increase in electrification.

FIGURE 3.7 Variable-frequency power generation principle [3.3] + 150 kVA/115 VAC A380 VFG [3.14].

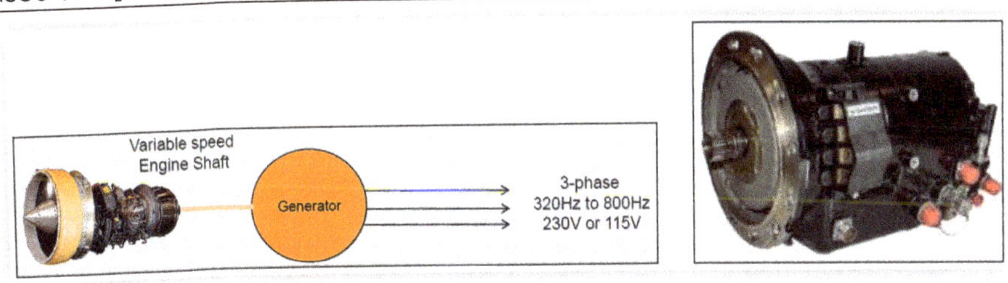

FIGURE 3.8 250 kVA/230 VAC variable-frequency starter generator (VFSG) [3.15] + installation on a Boeing 787 bleedless engine [3.16].

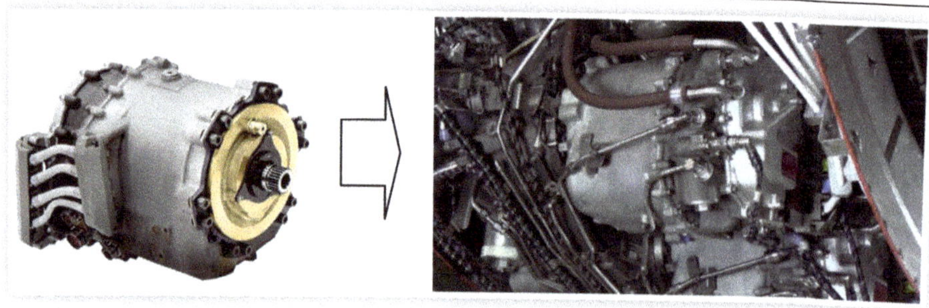

Legacy aircraft generally comprise three hydraulic circuits ensuring power supply for actuation in an architecture dubbed "3H." A typical 3H architecture for a legacy twin-engine aircraft, based on the Airbus A330, comprising three hydraulic circuits as well as AC and DC networks is presented in **Figure 3.9** [3.17].

Electric networks for legacy short and long range generally have a primary fixed frequency AC network powered by IDGs and another network, derived from the primary AC network via transformer-rectifier units (TRU), supplying low DC voltage. For instance, on the Airbus A330 aircraft, the AC power is about 300 kVA. Batteries are used for aircraft power on and come into play during emergency situations. The CSM/G is a constant-speed motor generator which supplies emergency electric power to the AC bus derived from hydraulic power. Depending on the fault condition, this hydraulic power originates from either engine-driven pumps or the RAT, a wind turbine deployed in

FIGURE 3.9 Typical 3H architecture—twin-engine legacy aircraft [3.17].

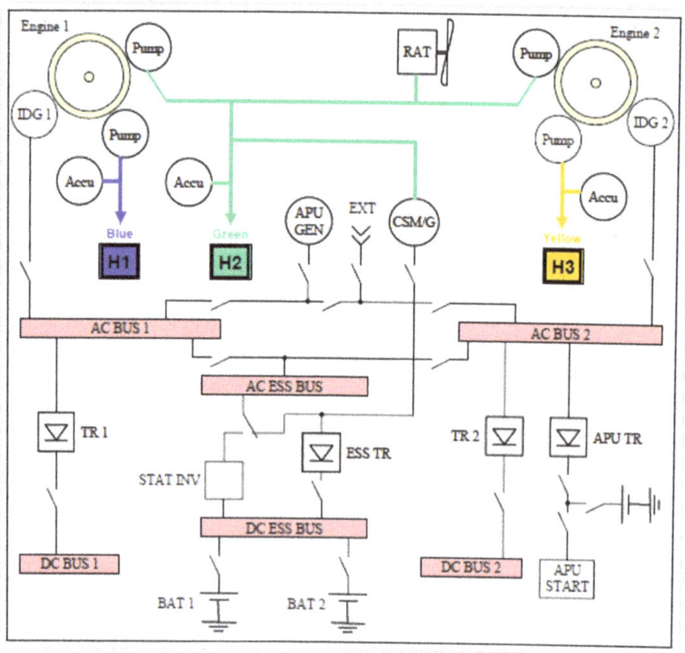

emergency situations. In fact, in the extremely rare case of total engine power loss (TEFO, or total engine flameout), the RAT deploys into the airstream and produces power for the cockpit and flight controls to safely land the aircraft. The CSM/G power range can go from a few kVAs up to about 10 kVA.

The APU is an auxiliary power unit which in fact is a turboshaft engine producing pneumatic power and driving an AC electrical generator with fixed frequency (400 Hz) outputs. It is normally used on ground and produces pneumatic power for starting the engines and also for ensuring cabin air conditioning. Generally the APU generator has about the same power rating as the IDG. Contrary to common belief it is not an emergency power source, so much so that an aircraft may be dispatched with the APU inoperative. It is not designed to be started at very high altitudes, and power availability gets more and more limited with altitude.

Now let us now consider a "2H+2E" architecture for a twin-engine MEA (**Figure 3.10**). This architecture comprises only two hydraulic circuits along with the AC and DC electrical networks. In fact, the third hydraulic circuit usually powered by engine-driven hydraulic pumps on legacy aircraft is replaced here by an electric circuit. The following electrification and new technology shifts are also considered [3.17]:

- Flight control actuation powered by electricity (~30% of flight controls)
- Emergency electrical power generation using an electrical RAT
- VF electrical power generation for the AC network
- Solid-state power controllers (SSPC) in lieu of traditional circuit breakers (CBs)

This 2H-2E type of architecture has been scaled up on the Airbus A380 and applied to the A350 XWB. Moreover, the latter aircraft's AC electrical network voltage at 230 V

FIGURE 3.10 2H-2E architecture—twin-engine MEA [3.17].

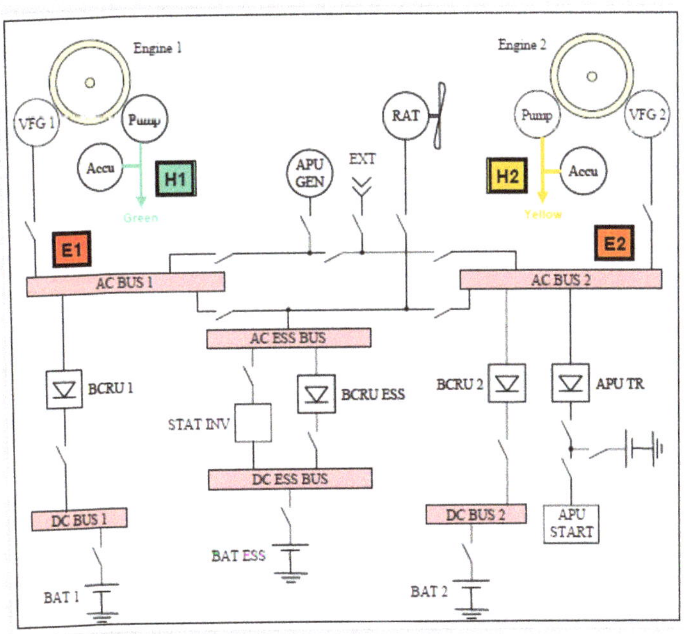

FIGURE 3.11 AC and DC network components and location on the Airbus A380 [3.12].

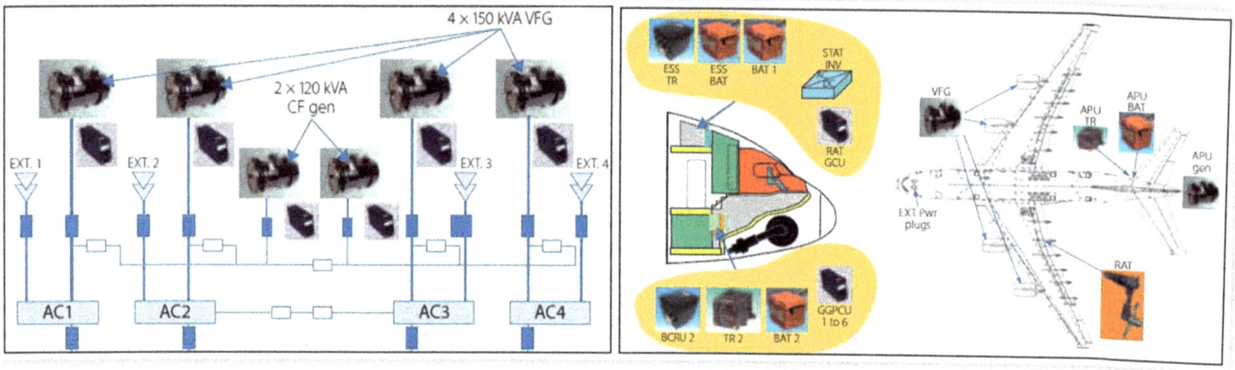

is double the level implemented on conventional aircraft and on the A380. Powerful loads are supplied by the AC network whereas small loads are supplied by the DC network, which is in turn is derived from the AC network.

The DC network on the A380 has a NBPT (no-break power transfer) capability which ensures uninterrupted DC power supply thanks to batteries connected to DC buses. The TRUs, traditionally ensuring AC to DC network conversion, are replaced by BCRUs (battery charger rectifier unit), which supply the DC network while also managing battery charge.

Figure 3.11 shows the AC network on the A380. Also shown are the locations on the aircraft for VFGs, APU generators, their control units (GGPCUs), DC network components, as well as the RAT emergency generator.

The interconnection between power sources, voltage networks, and loads is achieved via distribution power centers installed in the aircraft. They gather AC and DC contactors, cables, electrical busbars, devices ensuring circuit protection (e.g., against short circuits) and control electronics. The primary electrical power distribution center (PEPDC) interfaces with power sources, large loads, and the SP distribution. Electrical network management (ENM), electrical load management (ELM), and circuit breaker monitoring (CBM) applications are part of the power distribution perimeter. On the A380, the ELM and CBM functions are hosted on IMA (integrated modular avionics) modules instead of previous stand-alone computers.

In conventional aircraft, electromechanical CBs are mainstays for electric circuit protection. They are either manually commanded CBs featuring push buttons or remotely controlled (RCCB), thanks to combination with power contactors. SSPCs with better configurability and control features made their first appearance on the A380 for current ratings not exceeding 15 A. Ever since, thanks to technology advances, SSPCs have become key protective devices to reckon with when designing electrical power distribution whatever the size and type of aircraft. SSPCs ensure protection/switching features when distributing power to both AC and DC loads. Their flexibility allows for eased load reconfigurations and user-friendly status display, monitoring, and reset. Circuit boards comprising SSPCs are less voluminous and enable chip-like

FIGURE 3.12 Electromechanical circuit breakers replaced by SSPCs [3.18, 3.21, 3.19, 3.20].

integration on modular circuit boards that are directly insertable in power centers (**Figure 3.12**).

The A380 electrical distribution sub-assemblies and their location are displayed in **Figure 3.13**. SSPCs are deployed in the secondary distribution comprising SEPDCs and SPDBs in charge of cabin and cargo loads.

Electrical power on the A350 is generated onboard by the following power sources (**Figure 3.14**) [3.23]:

- Four main VF 230 VAC 100 kVA generators (two per engine)
- One auxiliary 400 Hz 230 VAC 150 kVA APU starter generator
- One emergency electrical RAT-driven 50 kVA AC generator

Also based on the 2H-2E architecture, the A350 has the following electrical networks [3.24]:

- 230 and 115 V AC networks
- 28 V DC network

Each network has a normal and an emergency electrical distribution. Two electrical power distribution centers (EPDC) manage the power distribution for normal and

FIGURE 3.13 Electrical power distribution block diagram and location on the Airbus A380 [3.22].

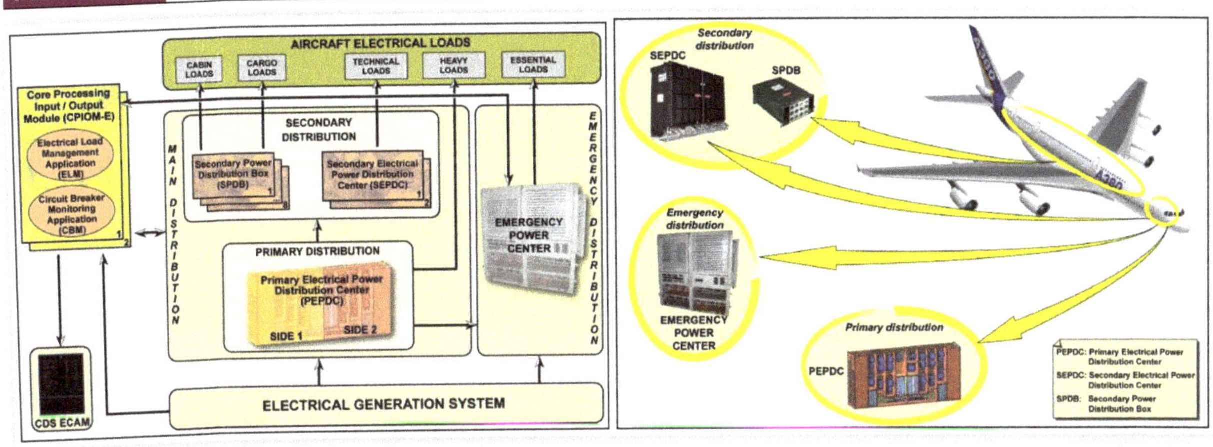

FIGURE 3.14 Electrical power distribution centers and AC power sources on the Airbus A350 [3.24, 3.25].

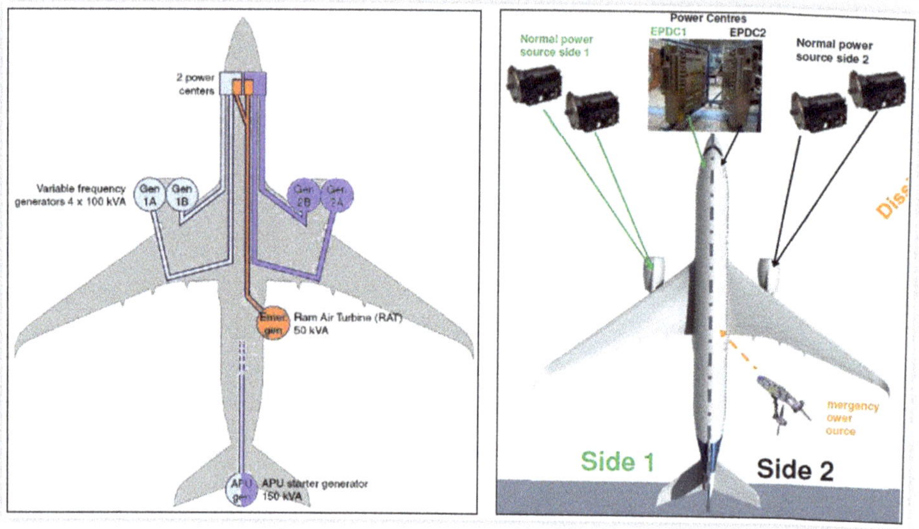

emergency operation. Then the autotransformer units (ATUs) convert the 230 VAC into 115 VAC to supply the 115 VAC network. During ground operations, when the generators and the APU are not operating, the ATU helps converting the 115 VAC that comes from the ground power unit (GPU) into 230 VAC.

The APU drives an auxiliary AC starter generator. On ground, the APU can power the entire electrical network for normal operations. In flight if one or more main generators fail, the APU generator can take over the failed main generator(s). On ground, it is possible to connect two GPUs to the aircraft through two external power connectors. The GPU supplies 115 VAC and a normal power of 90 kVA.

The 230 VAC network supplies the 28 VDC network via four TRUs. As part of a redundant system, four identical lithium-ion batteries are connected to the 28 VDC network in order to:

- Ensure the NBPT function
- Provide standby DC power
- Provide DC power on ground if AC power is not available

Two out of the four batteries can provide temporary supply in an emergency electrical configuration.

The electrical network supplies the aircraft systems with three voltages via the following networks:

- 230 VAC network (normal and emergency): large power consumers (e.g., fans, compressors, pumps)
- 115 VAC network (normal and emergency): commercial loads such as galley equipment
- 28 VDC network (normal and emergency): DC loads

On the Boeing 787, as described in [3.10], bleed air is only used for engine cowl ice protection and pressurization of hydraulic reservoirs. The electrified functions are wing deicing protection, engine starting, driving the high-capacity hydraulic pumps, and powering the cabin ECS (which includes cabin pressurization). The engine starting is performed by the VFGs in initial "motor" mode, after which they revert to "generation" mode. The motor controller used for the engine start is diverted to the ECS. Here are two examples of how on the equipment side and the controls side it is possible to host two different functions provided they are not running in parallel but in a sequential fashion.

In the no-bleed architecture, electrically driven compressors provide the cabin pressurization function, with fresh air brought onboard via dedicated cabin air inlets. This approach is significantly more efficient than the traditional bleed system because it avoids excessive energy extraction from engines with the associated energy waste by pre-coolers and modulating valves. There is no need to regulate down the supplied compressed air. Instead, the compressed air is produced by adjustable speed motor compressors at the required pressure without significant energy waste. That results in significant improvements in engine fuel consumption.

The Boeing 787 uses an electrical system that is a hybrid voltage system consisting of the following voltage types: 235 VAC, 115 VAC, 28 VDC, and ±270 VDC. The 115 VAC and 28 VDC voltage types are traditional, while the 235 VAC and the ±270 VDC voltage types are the consequence of the no-bleed electrical architecture that results in a greatly expanded electrical system generating twice as much electricity as previous Boeing airplane models. The system includes six generators—two per engine and two per APU—operating at 235 VAC for reduced generator feeder weight. The system also includes ground power receptacles for airplane servicing on the ground without the use of the APU.

The generators are directly connected to the engine gearboxes and therefore operate at a variable frequency (360-800 Hz) proportional to the engine speed.

The electrical system features two electrical/electronics (E/E) bays, one forward and one aft, as well as a number of remote power distribution units (RPDU) for supporting airplane electrical equipment. The system saves weight by reducing the size of power feeders. A limited number of 235 VAC electrical equipment is supplied from the aft E/E bay, while the majority of airplane electrical equipment, being either 115 VAC or 28 VDC, are supported by the forward E/E bay and RPDUs as shown schematically in **Figure 3.15**. The RPDUs are largely based on SSPCs. The ±270 VDC system is supplied by four autotransformer rectifier units (ATRUs) that convert 235 VAC power to ±270 VDC. The ±270 VDC system supports a handful of large-rated adjustable speed motors required for the no-bleed architecture. These include cabin pressurization compressor motors, ram air fan motors, the nitrogen-generation-system compressor used for fuel-tank inerting, and large hydraulic pump motors.

The system, as shown in Figure 3.15, features two forward 115 VAC external power receptacles to service the airplane on the ground without the APU and two aft 115 VAC external power receptacles for maintenance activities that require running the large-rated adjustable speed motors.

There are four VFSGs on the Boeing 787, two mounted on each main engine and connected through the AGB. The VFSGs perform two primary functions: electric starting of the main engines and, once the engine is started, providing electric power to the airplane. The frequency of the VFSG AC output varies with the speed of the engine.

FIGURE 3.15 Boeing 787 remote power distribution (vs. traditional centralized distribution) [3.10].

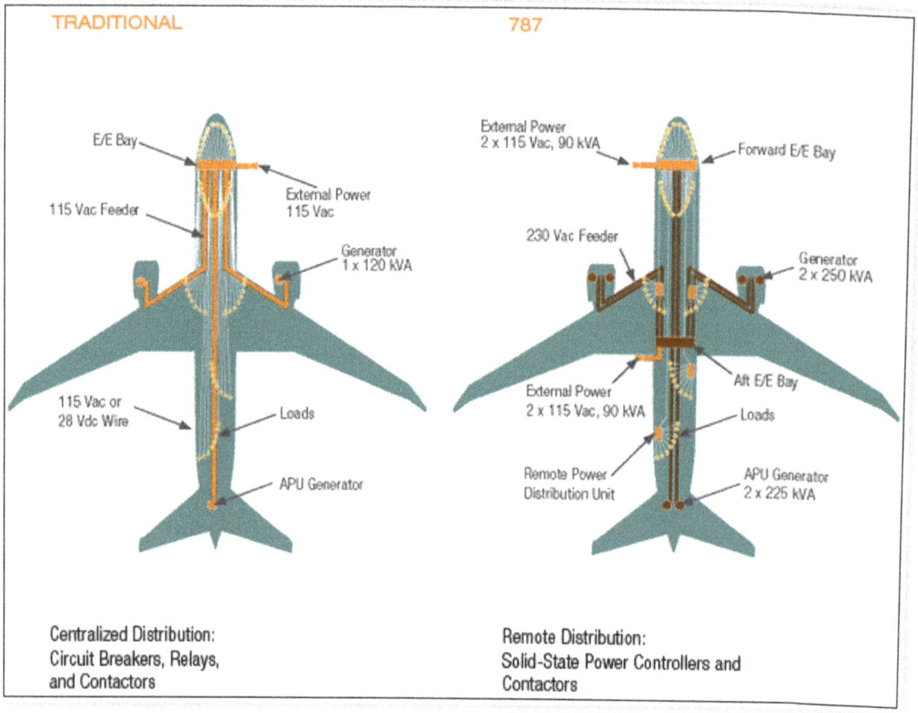

There are six GCUs on the Boeing 787, one for each generator (four main engine generators and two APU generators). Each GCU controls power from its respective generator and reconfigures the power system to ensure airplane capability is maintained if an engine or generator fails. Each GCU also provides voltage regulation and protection of the respective generator.

Primary power panels house the engine generator GCUs and contactors used to control and distribute the power from the engine VFSGs to the rest of the airplane. There are two primary power panels located in the aft electronics bay, each receiving power from the two generators on each engine.

Spoiler electromechanical actuators (SEMAs) control two of the seven spoiler pairs on the wing surfaces and provide roll control, air/ground speedbrake, and droop capabilities similar to the hydraulic actuators used on the remaining spoiler surfaces. The associated electronic motor control unit (EMCU) controls the operation of the SEMA. This is the first use of an electromechanical actuator on the primary flight control surface of a production civil transport airplane or military aircraft. Hydraulic actuators are used to position all of the primary airplane control surfaces (ailerons, flaperons, elevators, and rudders).

The VFSG, GCU, and primary power panels are all part of the airplane electrical power generation and start system (EPGSS) as illustrated in Figure 3.16. These components comprise the equipment necessary to generate and distribute the VF electrical power used on the airplane.

FIGURE 3.16 Boeing 787 electric power generation and start system (EPGSS) and electrical architecture [3.26, 3.27].

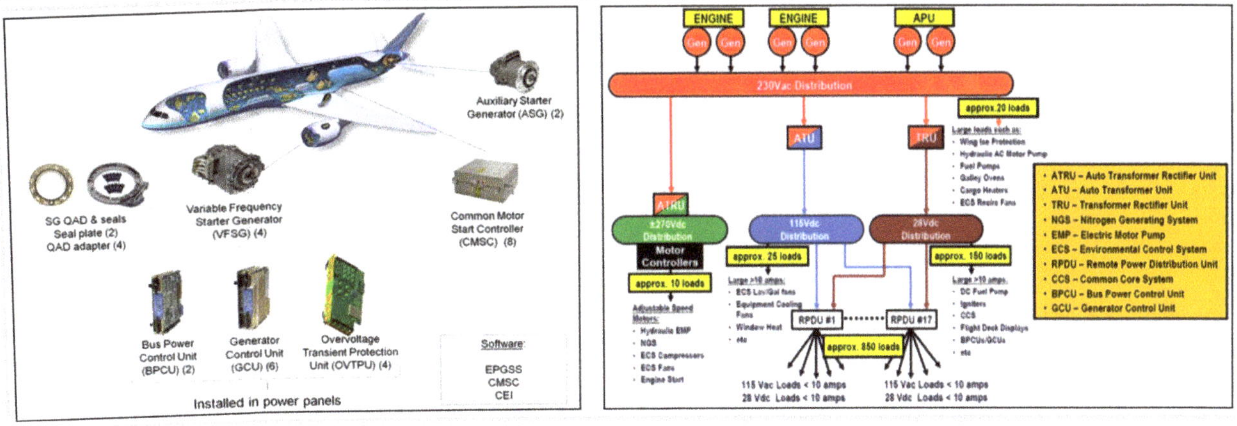

One of the advantages of the no-bleed electrical systems architecture is the greater efficiency gained in terms of reduced fuel burn—the Boeing 787 systems architecture accounts for predicted fuel savings of about 3%. The Boeing 787 also offers operators operational efficiencies due to the advantages of electrical systems compared to pneumatic systems in terms of weight and reduced life cycle costs.

Let us now compare the electrical load analysis (ELA) of a twin-engine conventional long-range aircraft (Boeing 777) and the more electric Boeing 787. Figure 3.17 shows for the 777 the evolution of electrical loads during flight on the one hand [3.28] and on the other the load distribution at cruise for the "one-engine-off" failure case [3.31].

Let us consider the ELA for the Boeing 787. For operation during weather conditions requiring maximum deicing power, the amount of power drawn from the engine generators is relatively constant over the major portions of the flight. At lower altitudes, greater power is used for ice protection and hydraulics, while at higher altitude, more power is used for the ECS and cabin pressurization. The two effects offset, resulting in a relatively constant power output from the engine generators. There are brief periods when the power drops noticeably. For example, when landing gear is retracted, the ice protection

FIGURE 3.17 Electrical load analysis—conventional long-range aircraft (Boeing 777).

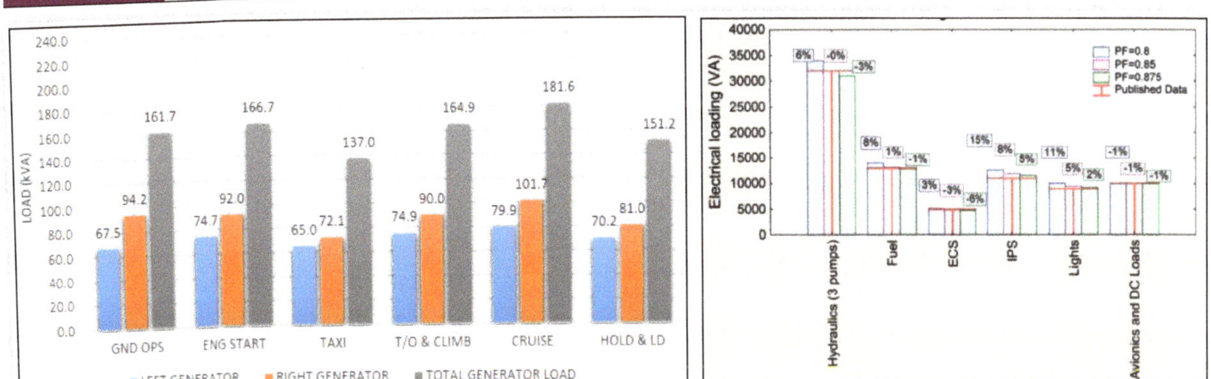

Total aircraft loads [3.28]

Load distribution at cruise (one engine out) [3.31]

FIGURE 3.18 Electrical load analysis—more electric long-range aircraft (Boeing 787) [3.29].

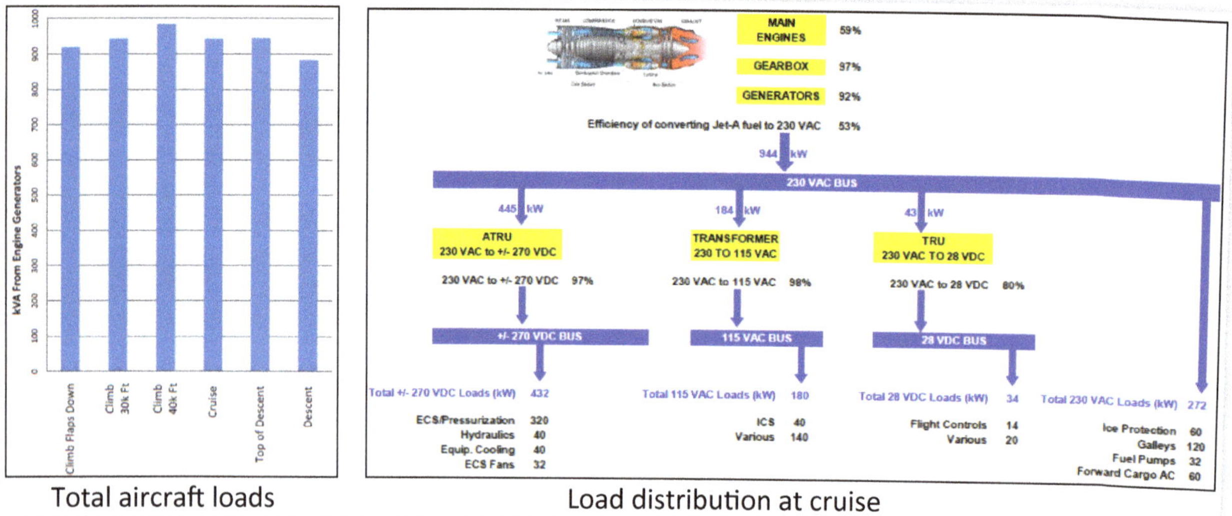

is turned off, briefly resulting in a dip in the load on the engine generator. However, these periods are brief and unlikely to have a noticeable impact on fuel consumption for electrical generation. The variation in engine generator loads for major portions of the flight are shown in **Figure 3.18** [3.29].

The Boeing 787 electrical generation and conversion efficiencies are significantly higher than earlier non-MEA. The improvements in efficiency are primarily due to the use of VFGs and advances in electronics that allow much higher power conversion efficiencies. For example, the efficiency actually measured at the AC output of the generator is 52% for the Boeing 787 compared to 44% for the Boeing 777.

The Boeing 787 has engine-mounted generators that produce a VF 230 VAC output. About 30% of the generated power is used directly, but to satisfy the largest loads, the AC power is converted to ±270 VDC in an ATRU with 97% efficiency. Power is also converted to 115 VAC 400 Hz in a 98% efficient transformer and to 28 VDC in an 80% efficient TRU. The electrical system as currently configured for engine generators with loads specified at cruise condition is shown in Figure 3.18.

As the aerospace industry is edging towards the electric aircraft, there is a growing demand for electrical power to supply onboard systems. Resorting to VF AC networks has shown that power generation scale-up is possible while improving efficiency and reliability. Furthermore, implementing high-voltage DC (HVDC) networks instead of AC ones would allow for full systems electrification. There are reasons why 270 VDC, or a higher 540 VDC, networks are gaining focus in research studies for all-electric systems aircraft. Actually, static converters extensively used with actuators enable HVDC utilization. Moreover, with HVDC networks, input rectifier stages of a majority of large electrical loads could be deleted, bringing weight reduction benefits [3.4].

Nevertheless, with either AC or DC networks, the increase in voltage levels gives rise to the corona effect impacting cablings and windings. In fact, the localized electric field near a conductor can be sufficiently concentrated to ionize air close to the conductors. This can result in a partial discharge of electrical energy called a corona discharge, or corona, which may degrade cabling or winding insulation if special care is not taken during the design phase. Aviation applications take an additional hit due to the fact that the corona effect has an inverse relationship with air density and hence increases with altitude. Therefore, when working with HVDC or HVAC networks, special precautions are necessary in the sizing and design of cabling and windings (motors, generators, or transformers) in order to alleviate the drawback from the corona effect. This corona effect may probably also result in periodic maintenance checks to make sure the cabling is not degraded over time.

Figure 3.19 shows one side of the HVDC architecture studied for a twin-engine aircraft within the European MOET research project [3.30]. The buck-boost converter unit (BBCU) shown allows for HVDC to be converted to LVDC (28 VDC) and vice versa. The ensuing Clean Sky project has focused on voltages of 270 VDC when addressing helicopter, business jet and regional aircraft and 540 VDC for short-range aircraft. Figure 3.19 also provides orders of magnitude of electrical power requirements regarding a short-range aircraft with HVDC networks supplying systems gone fully electrical [3.3].

Considering the fact that all loads are not powered at the same time, the load analysis in [3.31] provides a comparison between a conventional 180-passenger twin-engine short-range aircraft and its version with all-electric systems (Figures 3.20 and 3.21).

Bearing in mind that the loads shown in Figure 3.21 have to be borne by a single generator in case of engine or generator failure, generator sizing will have to take this into account on top of design criteria and margins addressing power requirements under various weather conditions. Moreover, power ratings for acceptable 5-minute and 5-second generator overloads have a direct impact on generator size and weight. Therefore, from this perspective, there is room for generator optimizations depending on how transient loads are managed. In research, normal and overload power ratings are undergoing various adjustments depending on whether one or two electrical generators are considered, whether they are similar in technology, etc. In order to strike the best balance between weight and performance, refined load requirements, power management strategies, and efficiency, improvements with new technology are being continuously fed into design iterations. In this way, many research projects are aiming to solve

FIGURE 3.19 HVDC architecture (MOET project) and power requirements—short-range aircraft [3.30, 3.3].

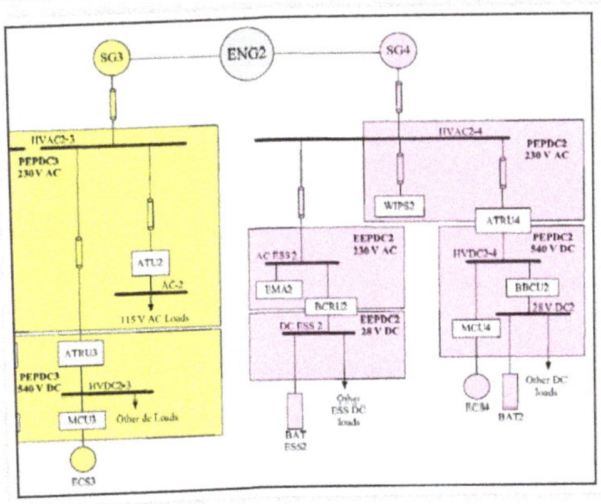

Power User	Comments	Typical Power level
Air Conditioning	ECS	4 x 70kW+
Flight Controls	Primary and secondary	3 kW to 40kW often short duration at high loads
Fuel pumps		about 10kW
Wing Ice Protection	Thermal mats or similar	250kW+
Landing Gear	Retraction, steering and braking	25kW to 70kW short duration
Engine starting	May be used for additional applications	200kW+ short duration

FIGURE 3.20 Electrical load analysis-conventional short—range aircraft (180 passengers) [3.31].

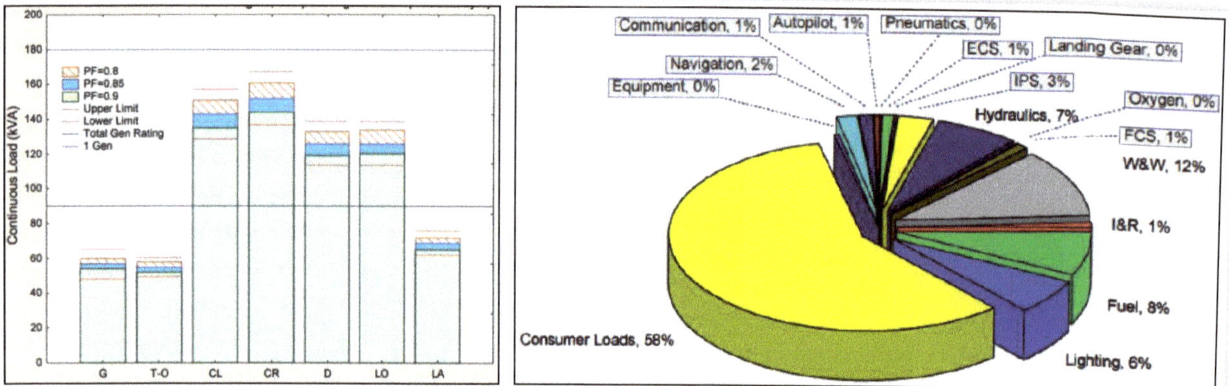

Total aircraft loads   Load distribution at cruise

FIGURE 3.21 Electrical load analysis—all-electric systems short-range aircraft (180 passengers) [3.31].

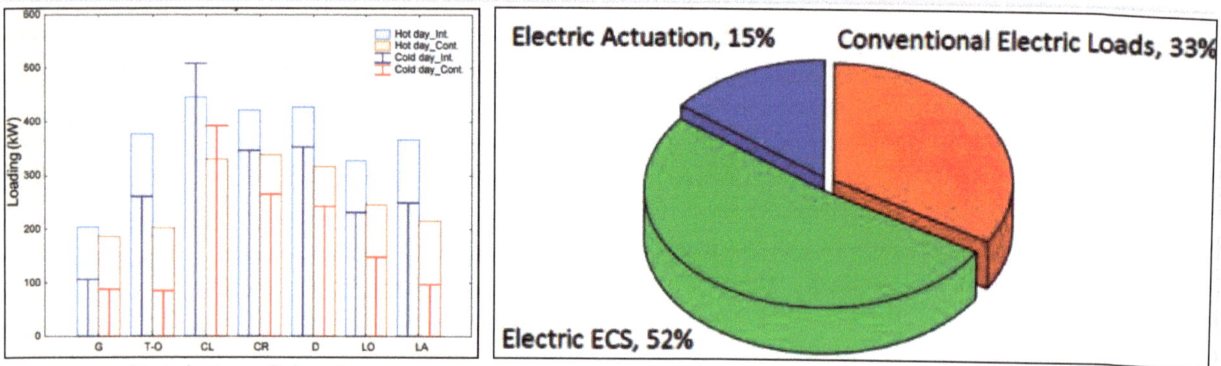

Total aircraft loads   Intermittent load distribution at cruise

the power generation equation for not only short-range aircraft but also other aircraft segments with fully electrified systems.

As discussed previously, on longer-range MEA the initial step has been to move towards variable frequency (Airbus A380), and then the doubling of the AC voltage was introduced (Airbus 350 and Boeing 787). Additionally, even though the Boeing 787 features a HVDC network to power large electrical loads, it still has to deal with AC networks from which, by the way, the HVDC is derived. But AC voltage which is also distributed to other loads has to comply with stringent power quality and stability requirements linked to the operation of electrified loads connected downstream to the HVDC bus. What that compliance means is that the design of generators becomes so constrained that the best power-to-weight ratio may not be achieved due to operational speed limitations. Hence, if AC network power quality constraints were to be alleviated but nonetheless taken into account at the HVDC network level, resorting to high-speed AC generators producing HVDC voltage through rectifier units (RUs) rather than ATRUs would enable sizeable weight reductions (**Figure 3.22**) [3.32].

FIGURE 3.22 Weight-optimized HVDC generation (high-speed AC generator + rectifier unit) (adapted from [3.30]).

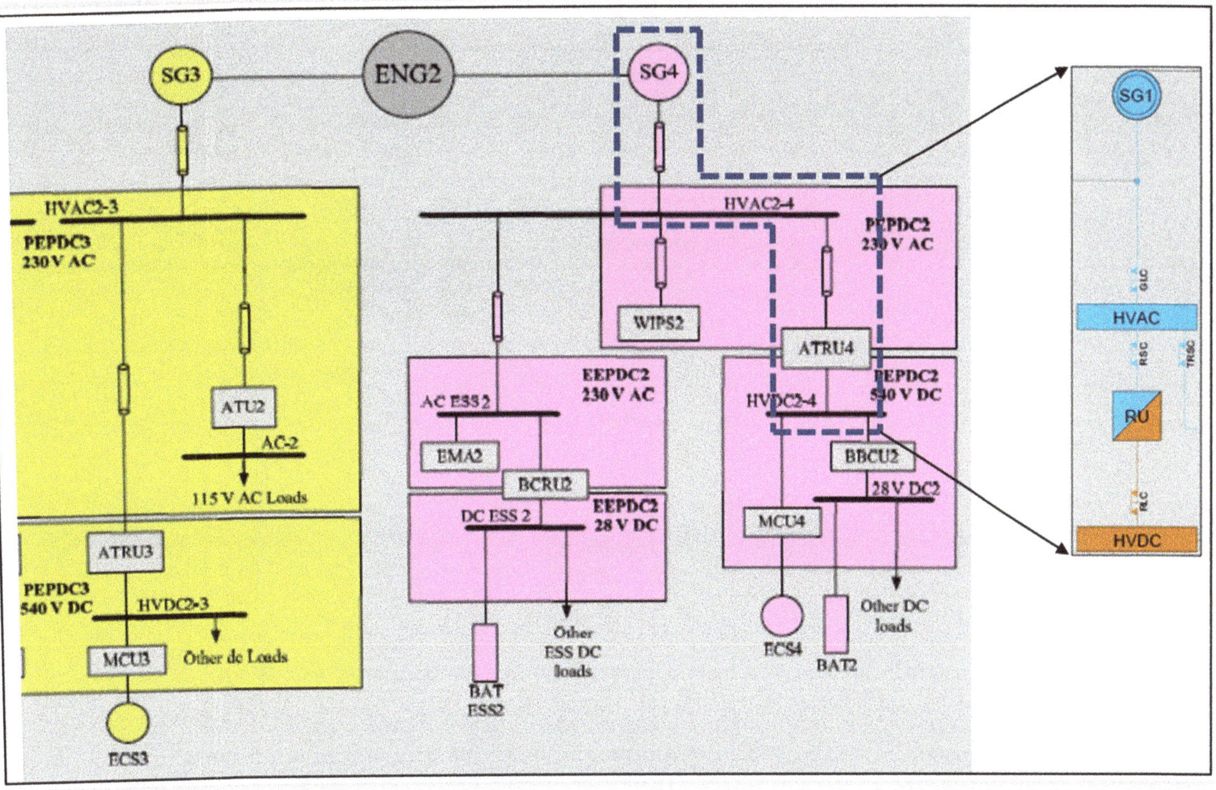

Figure 3.23 provides an order of magnitude of weight reductions achievable with high-speed generators. In fact, a 60% increase in generator speed translates to a 40% reduction in generator weight [3.33]. Moreover, the replacement of ATRUs by RUs, helps removing cumbersome and weighty ATRU transformer windings. Nevertheless, knowing that ATRUs are extremely efficient (97%) and reliable, whether RUs should be active using power electronics or passive with diodes only is an open question.

Bearing in mind the high levels of power to be dealt with when moving from an aircraft with conventional systems towards an all-electric-systems aircraft, solutions offering such significant weight reductions become compelling. Starting out with the electrical network architecture itself, significant weight increase is a core problem accompanying systems electrification. Therefore, working directly from HVDC buses instead of AC buses and distributing them around the aircraft constitutes a weight-optimized approach for the design of electrical networks faced with total electrification of systems.

With regard to engine integration of power generators, the following options may offer optimizations:

- Integrated starter generator and gearbox (oil sharing and integrated design)
- Integral starter generators with gearboxes removed

FIGURE 3.23 Electrical power generation weight reduction with high-speed generators [3.33].

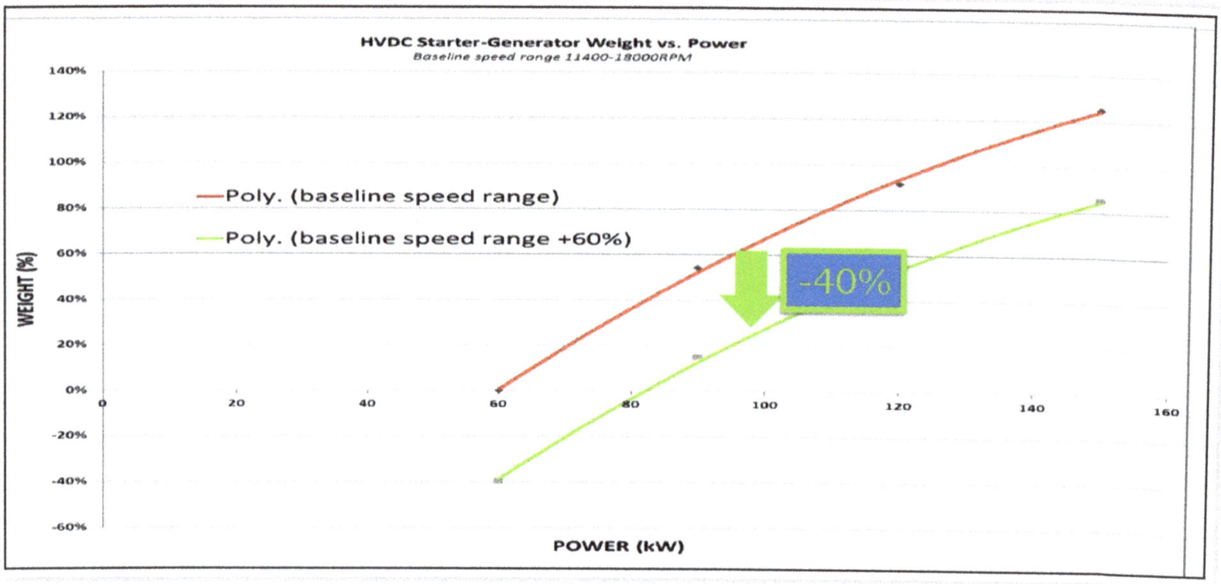

FIGURE 3.24 HVDC electrical architecture—twin-engine aircraft with all-electric systems [3.34].

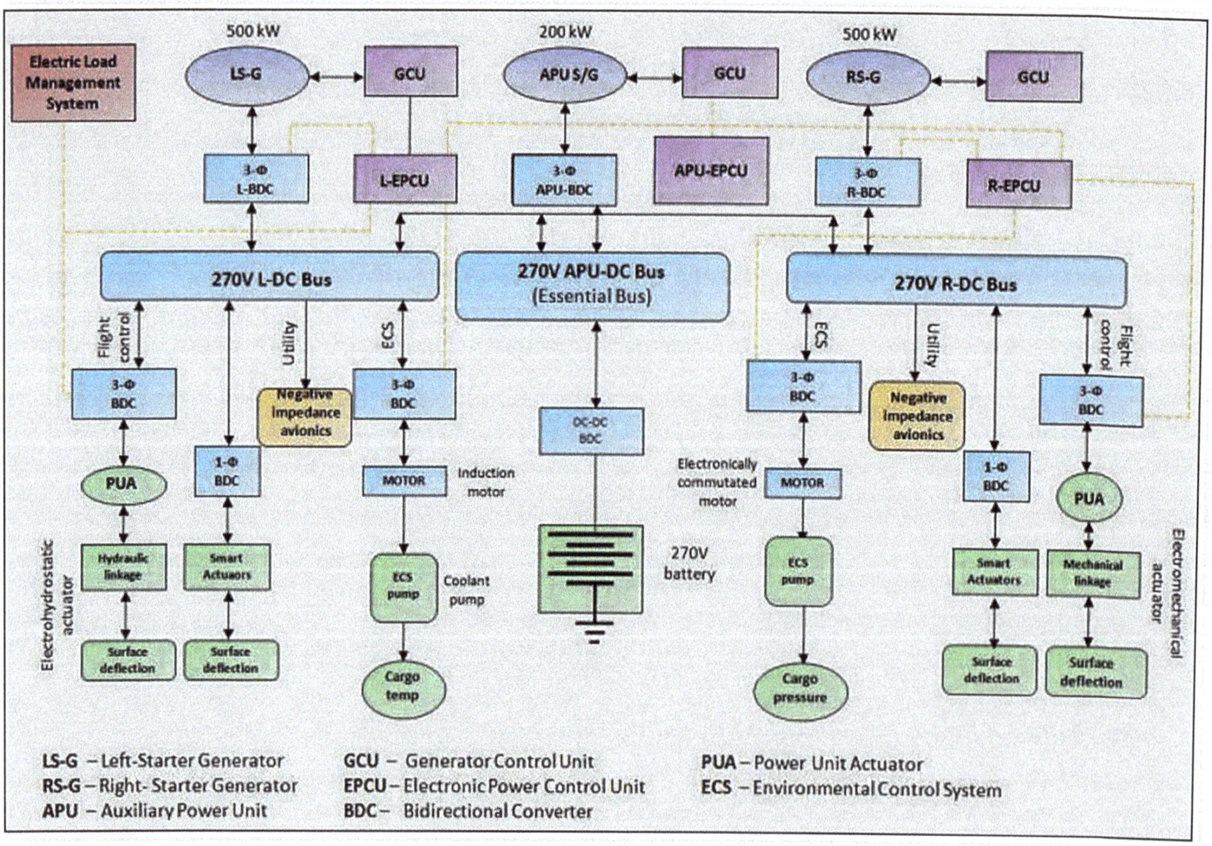

Giant strides in power electronic topologies and switching device technologies may allow for further flexibility to be built into the electrical network. One solution consists of implementing bidirectional active converters instead of RUs for handling normal AC to HVDC conversion and also supporting reverse operation, that is, converting HVDC voltage back to AC voltage whenever necessary. This new bidirectional functionality allows for better operational flexibility on the aircraft and thereby contributes to enhanced "availability" of electrical power onboard. Remember the bidirectional converter (BDC) approach was already introduced in the MOET electrical architecture (Figure 3.19) between the HVDC and LVDC networks. An HVDC architecture example for an aircraft with all-electric systems wherein power conversion is based on BDCs is shown in **Figure 3.24**. Of course, the 270 VDC voltage level shown for the HVDC buses may be higher depending on the aircraft considered. The same comment applies to generator power ratings as well.

References

[3.1] Faleiro, L., Herzog, J., Schievelbusch, B., and Seung, T., "Integrated Equipment Systems for a More Electric Aircraft—Hydraulics and Pneumatics," Liebherr-Aerospace, *24th International Congress of the Aeronautical Sciences (ICAS 2004)*, Yokohama, Japan, August 29-September 3, 2004.

[3.2] Rosero, J.A., Ortega, J.A., Aldabas, E., and Romeral, L., "Moving Towards a More Electric Aircraft," *IEEE A&E Systems Magazine*, March 2007.

[3.3] Wheeler, P., "The More Electric Aircraft—Why Aerospace Needs Power Electronics," University of Nottingham, UK, May 2009, http://www.lboro.ac.uk/microsites/research/iemrc/Events%20write%20up/Power%20Electronics%2014.05.09/More_Electric_Aircraft_000.pdf, accessed January 28, 2016.

[3.4] Langlois, O., Foch, E., Roboam, X., and Piquet, H., "EEA 2004—From the More Electric Aircraft to the All Electric Aircraft: State of the Art & Prospective," Airbus, LEEI, March 18, 2004, Powernet/LEEI, https://www.j3ea.org/articles/j3ea/olm/2005/06/j3ea2005601/j3ea2005601-olm.pdf, accessed March 14, 2018.

[3.5] https://www.safran-transmission-systems.com/company, accessed April 29, 2018.

[3.6] Vaschetto, S., "Design and Development of Permanent Magnet Synchronous Machines Shaft-Line Embedded in Aeronautic Engines," Ph.D. thesis, Politecnico di Torino, Italy, 2011, http://hdl.handle.net/11583/2615701.

[3.7] Rolls-Royce, *The Jet Engine*, 5th ed., (John Wiley & Sons on behalf of Rolls-Royce plc., 2015) ISBN 9781119065999.

[3.8] McLoughlin, A., "Engine Powerplant Electrical Systems, More Electric Aircraft Forum," *European Power Electronics and Applications Conference (EPE 2009)*, Barcelona, Spain, September 8-10, 2009.

[3.9] Wheeler, P., "Power Electronics—The Enabling Technology for the More Electric Aircraft," University of Nottingham (UK), *IQPC 1st International More Electric Aircraft Conference*, Hamburg, Germany, November 25-27, 2013.

[3.10] Sinnett, M., "787 No-Bleed Systems—Saving Fuel and Enhancing Operational Efficiencies," *AERO Magazine* Q4, 2007, published by Boeing.

[3.11] Wikipedia, "Constant Speed Drive," https://en.wikipedia.org/wiki/Constant_speed_drive, accessed April 30, 2018.

[3.12] Félice, E. and Oreal, P.-Y., "Vers un avion tout électrique," Lycée Martin Luther King (France) & Lycée Evariste-Galois (France), *technologie Magazine* issue 149, Canopé Editions, April 2007.

[3.13] Image extract from "GE90—Bug Removal & Installation, GE Aviation Maintenance Minute," YouTube video uploaded by GE Aviation on September 27, 2016, https://www.youtube.com/watch?v=rmr9c5B_gcQ, accessed March 24, 2018.

[3.14] https://www.safran-electrical-power.com/fr/systemes-electriques/nos-systemes-de-generation-electrique, accessed March 17, 2018.

[3.15] Clark, S.F., "787 Propulsion System," *AERO Magazine* Q3, 2012, published by Boeing.

[3.16] Photo extract from "GEnx-1B—VFSG Alignment Tool—GE Aviation Maintenance Minute," YouTube video uploaded by GE Aviation on September 8, 2016, https://www.youtube.com/watch?v=0rOpjIkY6_k, accessed March 15, 2018.

[3.17] Langlois, O., Foch, E., Roboam, X., and Piquet, H., "De l'avion plus électrique à l'avion tout électrique : état de l'art et prospective sur les réseaux de bord," Airbus, LEEI, J3eA, Journal sur l'enseignement des sciences et technologies de l'information et des systèmes, Volume 4, Hors-Série 1, 1 (2005), doi:http://dx.doi.org/10.1051/bib-j3ea:2005601.

[3.18] http://cdn.crouzet-aerospace.com/assets/library/Crouzet-Aerospace_Catalog_Circuit-breakers.pdf, accessed April 30, 2018.

[3.19] https://www.utcaerospacesystems.com/product-gallery/power-distribution-and-management-systems/, accessed May 1, 2018.

[3.20] http://www.esterline.com/powersystems/ProductRange/RelayPanelsSSPC.aspx, accessed May 1, 2018.

[3.21] https://www.wirtschaftsforum.de/fileadmin/wirtschaftsforum/unternehmen/697405_hs-elektronik-systeme-gmbh/programmierbar_wartungsfrei_steckbar_power_distribution_modul_DEFAULT_2_1.jpg, accessed May 1, 2018.

[3.22] A380 Technical Training Manual—Maintenance Course—T1 & T2 (RR/Metric)—Level I—ATA24 Electrical Power, Airbus, March 15, 2006, https://fr.scribd.com/document/226098143/A380-LEVEL-I-ATA-24-Electrical-Power, accessed March 24 & 25, 2018.

[3.23] Johnson, F.L., "A350 Technology," Airbus, Aviation Technical Education Council (ATEC) Conference, Orlando, FL, USA, April 21, 2009, http://www.vaughn.edu/assets/downloads/ATEC-2009-4.pdf, accessed March 15, 2018.

[3.24] Airbus, "A350-900 Flight Deck and Systems Briefing for Pilots Issue 02," Airbus, September 2011, http://www.smartcockpit.com/docs/a350-900-flight-deck-and-systems-briefing-for-pilots.pdf, accessed March 15, 2018.

[3.25] Roemelt, S., "Electrical Systems Engineering & Integration in AIRBUS," Airbus, *ICAS Biennial Workshop 2015*, Kraków, Poland, August 31, 2015, http://www.icas.org/media/pdf/Workshops/2015/ICAS_Workshop_Roemelt.pdf, accessed March 15, 2018.

[3.26] Boeing, "Boeing 787–8 Critical Systems Review Team Report," Boeing, March 19, 2014, https://www.faa.gov/about/plans_reports/media/787_Report_Final.pdf, accessed March 15, 2018.

[3.27] Nelson T., "787 Systems and Performance," Boeing, 2005, http://myhres.com/Boeing-787-Systems-and-Performance.pdf, accessed March 16, 2018.

[3.28] Liska-Guzman, J. and Stuckless, J., "B777 RAVE Electrical Load Analysis (ELA) Amendment," Trillium Aerospace, Document No: AV071016910, July 10, 2016, http://pubhtml5.com/uhqv/aanc/basic, accessed March 16, 2018.

[3.29] Whyatt, G.A. and Chick, L.A., "Electrical Generation for More-Electric Aircraft using Solid Oxide Fuel Cells," prepared by the Pacific Northwest National Laboratory for the

US Department of Energy, April 2012, https://www.energy.gov/eere/fuelcells/downloads/electrical-generation-more-electric-aircraft-using-solid-oxide-fuel-cells, accessed March 16, 2018.

[3.30] Wu, T., Bozhko, S.V., and Asher, G.M., "High Speed Modeling Approach of Aircraft Electrical Power Systems under both Normal and Abnormal Scenarios," GE Global Research, University of Nottingham, 2010, doi:10.1109/ISIE.2010.5637279, https://www.researchgate.net/publication/251965555_High_speed_modeling_approach_of_aircraft_electrical_power_systems_under_both_normal_and_abnormal_scenarios, accessed March 15, 2018.

[3.31] Seresinhe, R. and Lawson, C., "Electrical Load-Sizing Methodology to Aid Conceptual and Preliminary Design of Large Commercial Aircraft," Cranfield University (UK), *Proceedings of the IMechE Part G: Journal of Aerospace Engineering*, 229, no. 3 (March 2014): 445–466, doi: 10.1177/0954410014534638, http://journals.sagepub.com/doi/abs/10.1177/0954410014534638, accessed March 16, 2018.

[3.32] Roboam, X., Sareni, B., and De Andrade, A., "More Electricity in the Air: Toward Optimized Electrical Networks Embedded in More-Electrical Aircraft," *IEEE Industrial Electronics Magazine* 6, no. 4 (2012): 6-17, ISSN 1932-4529, doi:10.1109/MIE.2012.2221355, http://oatao.univ-toulouse.fr/9269/, accessed May 1, 2018.

[3.33] Thalin, P., "Game-Changing Power System Solutions for MEA," Thales, *IQPC 1st International More Electric Aircraft Conference*, Hamburg, Germany, November 25-27, 2013.

[3.34] Setlak, L. and Ruda E., "Review, Analysis and Simulation of Advanced Technology Solutions of Selected Components in Power Electronics Systems (PES) of More Electric Aircraft," *World Academy of Science, Engineering and Technology International Journal of Computer and Systems Engineering* 9, no. 10 (2015), http://scholar.waset.org/1307-6892/10002777, accessed October 18, 2018.

CHAPTER 4

Electrification of Aircraft Systems—Part II

Replacement of Pneumatics, Enabling Technologies

Pascal Thalin
Chair and Member - SAE Electric Aircraft Steering Group

4.1 Pneumatic Power Generation

Before getting into the details of how traditionally pneumatic Environmental Control System (ECS) may undergo electrification, to begin with, let us consider engine functions and components involved in the pneumatic energy extraction from conventional engines.

Bleed air produced by gas turbine engines is compressed air that is taken from the compressor stage of those engines, which is upstream of the fuel-burning sections. As presented in [4.1], on a turbofan engine, the amount of air which is tapped from engine compressor stages for cooling and other purposes can exceed 20% of the core flow. At that level, it has a marked influence on the performance of the engine. A turbofan engine has the need for compressed or "bleed" air for many functions:

- Cooling of engine parts, static or rotating
- Clearance control of rotating parts
- Sealing
- De-icing of the engine's front parts and the aircraft's wing and empennage leading edges

FIGURE 4.1a Classical bleed architecture [4.1].

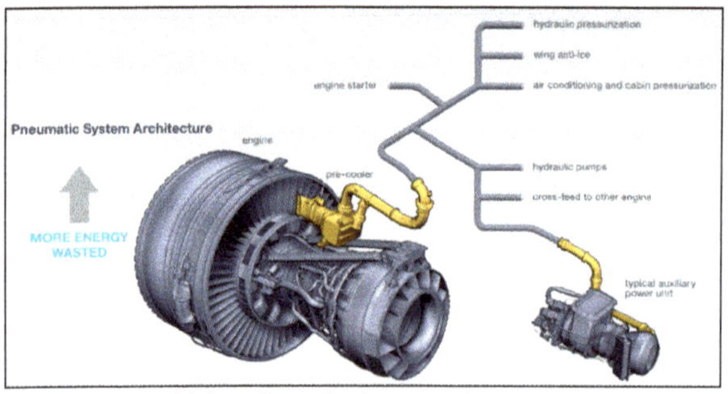

mixed bleed and electrical energy

FIGURE 4.1b More electric bleedless architecture [4.1].

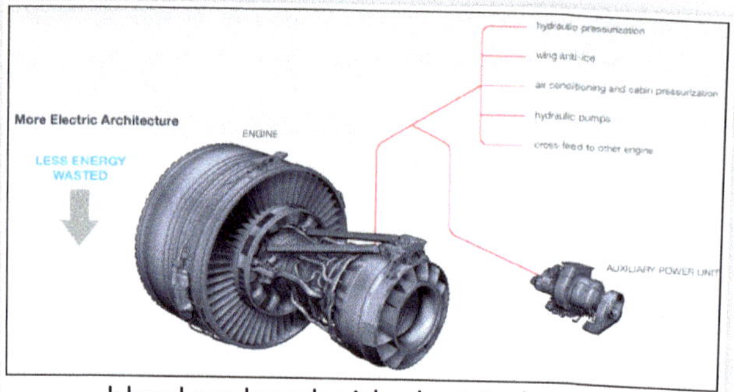

bleed replaced with electrical energy

- Provide the air source for the aircraft's air conditioning system, cabin pressurization
- Starting a second engine via its air starter (cross-start)
- Operation of pneumatic actuators and air-driven motors
- Pressurization of the hydraulic reservoir and waste and water storage tanks

The air is always tapped at the earliest possible place in the compression chain which has the appropriate pressure level. The least amount of energy is then spent on the air and it has the lowest possible temperature.

Bleeding air lowers the performance of the compressor and hence the engine. When bleed air is tapped for cabin air conditioning and de-icing, it increases the fuel consumption of the engine by up to 5%.

The bleed air tapped from the high-pressure compressor has too high of a temperature to be routed around the aircraft. It is therefore pre-cooled in a pre-cooler which uses the fan's airstream for cooling. This bleed air gets cooled and regulated to 200°C–250°C and 275 kPa/40 psi in the engine pylon's pre-cooler. It is then routed to users around the aircraft (**Figure 4.1a**).

While bleed air is a simple way to get compressed air to the aircraft's air conditioning system or heat to de-ice the wings, it requires cooling away energy in the pylon pre-cooler, and it means routing 250°C air around the aircraft. Any leaks can destroy electrical cabling or even weaken aircraft structures. Therefore leak detection systems are required and different shutoff or rerouting valves need to be operated by the pilots should a leak occur.

In order to decrease the amount of energy wasted due to the inefficiency of bleed power extraction, bleed components (bleed air plumbing, pre-cooler, etc.) may be removed from the engines. Fitted with such world's first bleedless engines, the more electric Boeing 787 does not use bleed air for aircraft systems except for nacelle anti-ice. On this aircraft, the removed bleed power is replaced with more efficient electrical power (**Figure 4.1b**).

4.2 Environmental Control System

As discussed in [4.2], the task of an ECS is to control air temperature, to pressurize relevant aircraft compartments, to provide sufficient ventilation and fresh air to passengers, to control the level of humidity within acceptable limits in the cabin, and to remove pollutants. The core system where air is conditioned is the so-called "packs."

4.2.1 Conventional ECS

FIGURE 4.2 Simplified typical engine bleed air system [4.2].

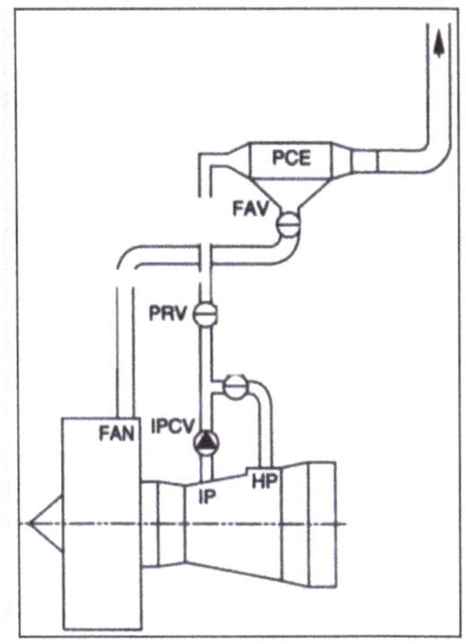

Currently the ECS packs of large civil aircraft use air which is bled from the engine compressor to provide conditioned air (flow, pressure, temperature) to the cabin (**Figure 4.2**). As the thrust produced by the engines depends on aircraft flight phase, the pressure of the engine bleed air varies such that it is necessary to use two different ports: the Low Pressure (LP) or Intermediate Pressure (IP) port which is used during most of the flight, and the High Pressure (HP) port which is used when the engine is operating at low thrust (especially in landing, hold, and descent conditions).

The choice of the bleed port (LP/IP or HP, controlled by valves LPCV/IPCV and HPV) for the supply of the ECS (but also the Wing Ice Protection System—WIPS) is driven by the available air pressure and is controlled by complex laws.

The pressure of the air delivered is limited by the Pressure Regulating Valve (PRV) and its temperature by the Pre-Cooler (PCE) which uses engine fan air to cool down the air coming from the core of the engine. The Fan Air Valve (FAV) modulates the coolant flow so that the temperature of the air leaving the PCE doesn't exceed 200°C.

Bleeding air from the engine has an impact on engine fuel consumption. This depends mainly on air mass flow, air pressure, and engine thrust. The ECS has to cope with all energy level supply situations coming from the engine, regardless of air temperature (low or high) and air pressure (low or high). This means that the ECS has to be designed for the worst case (low pressure and high temperature), but that when flying conditions are not so stringent (most of the time), the ECS supply system is oversized, and the ECS is supplied by high levels of energy that it doesn't need (this occurs particularly during the takeoff and climb phases): the engine provides air with a high pressure which is reduced (wasted) by the PRV and Flow Control Valve (FCV) at the inlet of the packs.

Martinez [4.3] indicates that in order to satisfy the air conditioning needs in flight and on ground, that is, cabin air renovation flow of 5 L/(s·pax) at 75-100 kPa and 22°C, different systems can be used. Here are a few examples:

- Until 1950, only heating was available (from engine heat recovery, electrical heaters, or burners).
- A vapor-cycle system like in a car's air conditioning (best with heat-pump capability), supplemented with a separate ventilation system, mostly used in small aircraft.
- An air cycle system that provides both heating/cooling and ventilation. This is the standard in medium and large aircraft because of its compactness and reliability, in spite of its poor energy efficiency (it draws typically 1 kW/pax at full capacity, some 350 kW for a 400-passenger aircraft, about 75% of all non-propulsive energy consumption on board). The ECS packs are typically located close to the wing box section in the belly fairing, in the vicinity of the center fuel tank (**Figure 4.3**).

FIGURE 4.3 Location of ECS packs—twin-engine aircraft (two packs) [4.3].

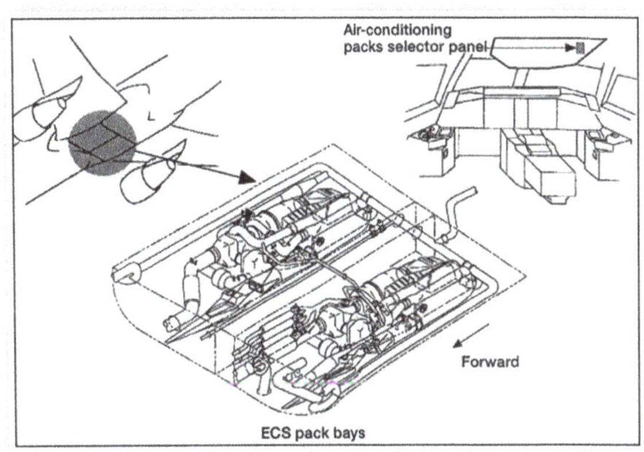

The state of the art is provided below:

- All-in-one solution used in most aircraft: the air cycle machine (ACM) pack (but heating/refrigeration efficiency is ~50%). A separate electrically driven ACM would be more versatile and efficient (~70%).
- Vapor-compression heat pump (but no pressurization, no ventilation) used in many small aircraft.

On ground, instead of using the aircraft's APU bleed air to feed the ECS, high-efficiency ground refrigeration systems are used in main airports. A novel ground ECS service is based on water chillers with ice storage (ice is made at night, and used on the spot to avoid large refrigerators).

Let us go through the operating principle of an ACM. Operating principles of the typical air cycle conditioner pack are described below:

- Air is bled from the main engine compressors through a FCV at around 250 kPa. It is very hot (>200°C) because of the adiabatic compression heating. As pressure and temperature of bleeding depends on compressor stage and spinning rate, this pressure is regulated by having two or three bleedings at different stages and control valves.
- The hot bled air always requires cooling before entering the cabin, but a simple heat exchanger (HE) with outside air is not efficient (a big HE is needed to cool that amount of air from 200°C to 20°C, particularly at low altitudes). To guarantee that bled air can be safely piped from the engine through the fuselage to the ACM and de-icing system, a HE (pre-cooler) in the engine cools bled air from the compressor (at >200°C) to some 180°C (**Figure 4.4**).
- The ACM is based on an inverse Brayton refrigeration cycle. Pre-cooled air from the bleed system is further cooled in a so-called primary HE down to some 110°C, and then enters a compressor with pressure ratio around 1.8 that raises the air to 210°C, and a second HE lowers the temperature again to some 100°C. Pressure losses in these compact HEs are high (20–40 kPa). Afterwards, air passes through a turbine and exits at about 5°C, to be mixed with some hotter bleed air at around 100°C to get the 10°C–35°C (it depends on operation phase) needed to keep the cabin air at around 22°C at all times, accounting for internal heat release (passengers and equipment), outside heat gain and loses, and air recirculation. The thermodynamic process is sketched in Figure 4.4.

FIGURE 4.4 ECS—block diagram and thermodynamic cycle of a pack [4.3].

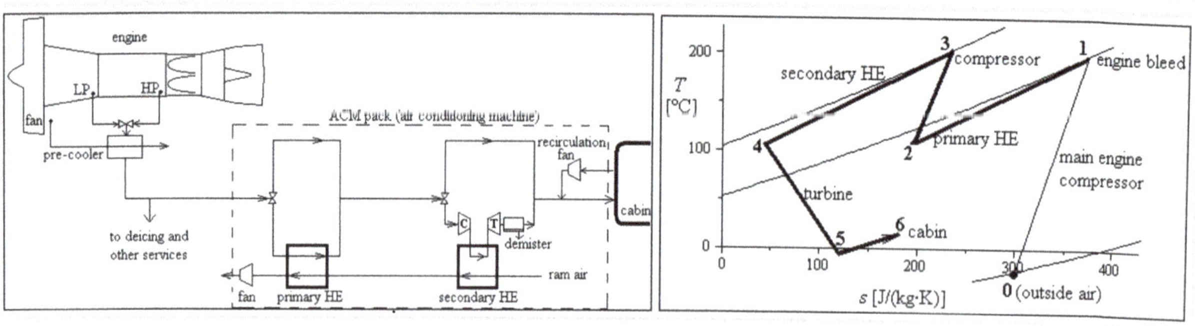

- Additional filters/separators are provided to remove water condensate on ground and low altitudes in humid climates (just after the ACM turbine in most cases, and before the turbine too when very cold cabin air is needed on ground, to avoid the risk of icing inside the turbine). The condensate is injected into the ram air stream to help HE efficiency. The common-axis turbo-compressor is made of aluminum, with air bearings.
- Ram air is used as a heat sink in both HEs, which are of the compact type (cross-flow corrugated aluminum plates). Ram air is captured through an inlet door and a diffuser, and forced by a fan (driven by the ACM turbine) to go through the two HEs mentioned above, an exhaust nozzle, and a louvered exit door. The inlet and outlet doors are linked and driven by the same actuator (one for each pack). These doors remain closed when there is no cooling need (e.g., when the ACM compressor-outlet temperature is <~120°C). Typical mass flow rate of ram air is two or three times greater than bleed air, in each ACM.

For redundancy, the ECS is usually split into an ACM pack per engine (Figure 4.3). The ACM size is dictated by the ventilation requirement of 6 g/s minimum per passenger (e.g., 1.2 kg/s minimum for the ~200-passenger capacity of the A320 aircraft family, ~2 kg/s being the typical design value). The A340 ECS weighs 720 kg, and has four ACMs, each fed with 1 kg/s of air.

On large aircraft, ~1 kg/s of compressed air is bled from each engine (from the 80 to 100 kg/s of primary air used). Each air pack (ACM) weighs ~150 kg. Each ACM takes some 3 kg/s of ram air per pack (plus 0.5 kg/s of post-fan air for the pre-cooler at each engine).

The nominal ECS power requirement depends on the work spent in compressing bleed air, which is 200 kW per pack [4.3].

It is always interesting to see how the real system looks like. The conventional ECS pack of the short-range Airbus A320 supplied by Liebherr is shown in Figure 4.5. The figure also highlights the location on the aircraft of the pack as well as the ram air inlet and outlet ports of its HEs.

Let us now consider the much larger Boeing 747-8 *Intercontinental*, a four-engine jumbo jet capable of transporting 467 passengers (three-class cabin) over a 8000 nm range. It entered into service with newly developed ECS packs. There are four of them on the aircraft, and these are located in the belly fairing, two per side. The bleed air is used for cabin air conditioning and temperature control system (CACTCS), engine anti-ice, wing anti-ice, the hydraulic air-driven pump, the leading-edge flap drive unit,

FIGURE 4.5 Conventional ECS pack and location on aircraft—short-range A320 aircraft.

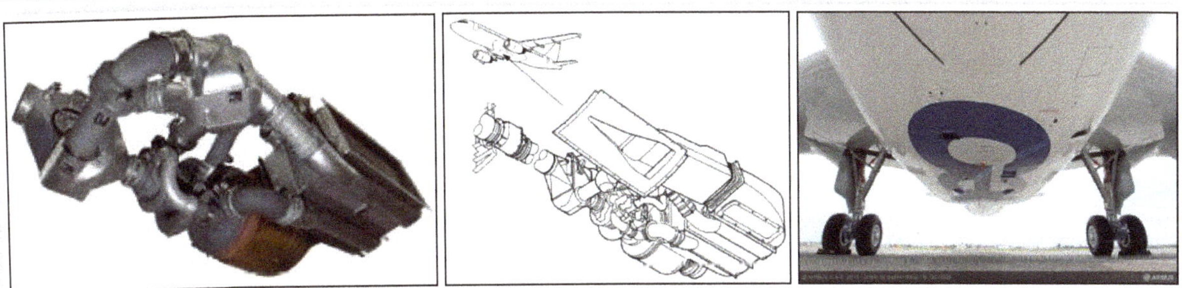

Conventional ECS pack [4.5] | Pack location [4.6] | Pack heat exchanger inlet/outlet ports [4.7]

FIGURE 4.6a Conventional ECS pack—long-range Boeing 747-8 aircraft [4.9, 4.10].

FIGURE 4.6b Boeing 747-8 ECS pack [4.8].

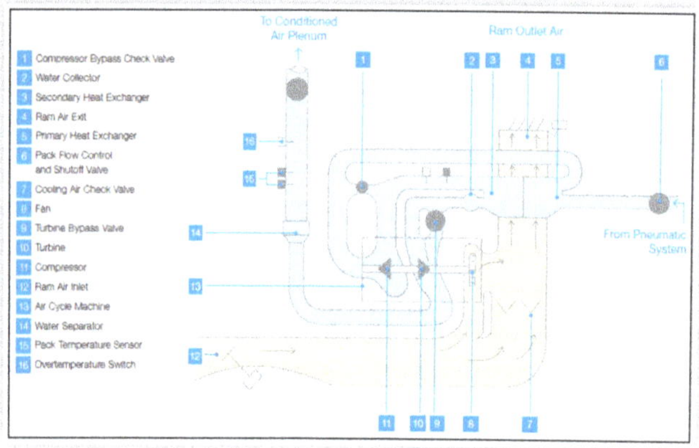

the nitrogen generation system, aft cargo heat, total air temperature probe aspiration, and hydraulic reservoir pressurization. All four engine bleed systems are connected by a common manifold. The pack temperature is modulated using the ram air door actuators (RADAs) and the temperature control valve (TCV). Technological advancements enabled Boeing to make several improvements to the conventional bleed system on the Boeing 747-8 (**Figure 4.6b**) [4.8].

4.2.2 Electric ECS

Conventional ECS design is reliable, but expensive to operate (1 kW/pax). There are also some maintenance problems due to overheating on ground and during ascent/descent (when ram air flow must be forced with a fan, the engine fan or a dedicated one) [4.3].

At aircraft level, the electrification of the ECS is a key enabler of the MEA as it is the biggest steady-state power consumer during aircraft cruise, but could also allow for a lot of fuel savings, due to its consequential adaptability and lower impact on the operation of the engine. **Figure 4.7** shows the approximate power consumption for pressurization and ventilation as well as the cooling of an aircraft with 100 and 350 passengers (valid for a cruise flight at 40,000 ft, hot day conditions). The values of electrical power consumption of a full-electric ECS are directly linked to the number of passengers, as the number of passengers determines the fresh air flow to be introduced into the cabin. Although the values reported in the figure are approximate, they give an order of magnitude of the needed electrical power to be supplied. For an aircraft with 350 passengers, the electrical power needs will be in the range of 400 kW and will therefore largely influence the sizing of the Electrical Power Generation System (EPGS) [4.3].

The benefits of the electrification of the ECS are numerous, and some of them are listed below:

FIGURE 4.7 Full electric ECS power consumption vs. aircraft size [4.2].

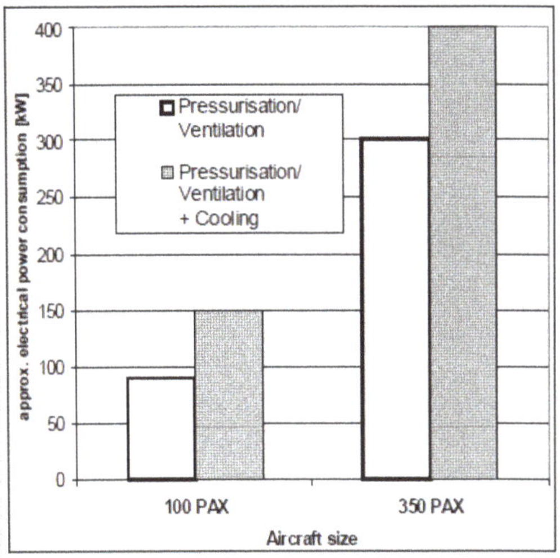

- No direct intervention of the ECS in the thermodynamic cycles of the engine. Due to the fact that conventional bleed air extraction from the engine compressor is avoided, the engine performances are not degraded anymore. An electric ECS will allow the engine compressor to be designed independently from the needs of the pneumatic systems.
- A high-efficiency engine can only be realized with a high bypass ratio (fan flow divided by core flow). Therefore, in a high bypass-ratio engine, the core flow is relatively reduced to a minimum and the percentage of bleed air (determined by the aircraft fresh air needs) related to engine core flow is higher because the mass flow of air required by the ECS is taken from the engine core flow. But the lower the core flow the higher the impact on engine Specific Fuel Consumption (SFC) of bleed air extraction. Moreover, the potential for reduction in core flow allowing for bypass ratio increase may even become limited due to the need to provide the aircraft cabin (and hence the ECS) with a minimum amount.
- Unlike a conventional ECS, where the energy source is the engine compressor, and therefore the power level depends on the flight phase (high power level during the climb and low level during descent), the electric ECS will demand only the power needed to perform its tasks and not waste energy unnecessarily.
- During some flight phases, the minimum thrust produced by the engine can be driven by the ECS, which requires a minimum pressure to fulfill its tasks. This leads to an additional aircraft fuel consumption and prevents the aircraft from flying fuel-efficient mission profiles. An electric ECS would allow the segregation of engine thrust from ECS airflow needs. But, this is somewhat offset by the negative engine operability impacts of providing increased electrical power.

The electric ECS comes with many advantages, but some challenges have to be overcome:

- Due to the high electric power consumed by the ECS, electric generators have to be appropriately sized. The challenge is much steeper for twin-engine aircraft and especially for the "one engine off" failure case, where half of the generators have to deliver more than half of the ECS power consumption for normal operation.
- The deletion of engine bleed air extraction for the ECS will deliver its full positive potential only if the other aircraft system using a large amount of engine bleed air, the WIPS, uses another power source. The pneumatic power of the WIPS has to be replaced by electrical power, and power consumption of this system is significant when icing conditions are encountered.

- Compact, efficient, and low weight ECS call for turbomachinery working at high speeds. These motorized air cycle machines (MACM) have to work at speeds as high as ~50,000 rpm. Air-cooled motors are feasible for the high power MACM application.
- The implementation of the electric ECS may certainly simplify the engine (deletion of bleed ports, bleed valves, and pre-cooler) and its controls, but will make the ECS more complex, as it will have to produce pressurized air by itself. This higher complexity may result in more weight, installation space, and costs. These impacts have to be balanced against the benefits to the aircraft.

As highlighted, engine power offtake drives the whole electric ECS approach, due to many snowball effects, in particular with regard to the sizing of the electrical power generation and distribution. Thus the system perimeter is as shown in **Figure 4.8** comparing the functional diagram of a conventional ECS with the electric ECS [4.4], as studied in the European MOET research project for the short-range aircraft application.

In both cases the pack discharge air from the MACM turbine section is then sent to the compact mixer prior to distribution into the main airplane cabin area. The compact mixer ensures the efficient mixing of outside air delivered by the ACM with recirculated air from the main cabin zones [4.4].

One approach developed in the MOET project deals with power optimization based on the principle that fresh air cooling is from a thermodynamic perspective more "expensive" than recirculated air. The reason is obvious: fresh air has to be compressed before it can be used for cabin supply. One design principle minimizes the fresh air flow to the required certification flow plus margin. The technical solution is to incorporate recirculation of air cooling, to handle the heat loads. Additionally the occurrence of icing conditions in the dedicated pack HEs can be prevented in that way.

Regarding the power offtake, a vapor cycle may be more efficient than an air cycle regarding the function of cooling. Therefore the additional cooling system is designed as a vapor cycle system. Such an approach was studied in the European MOET project. It includes two electric packs based on the air cycle technology as well as two vapor cycle systems. The functions of the packs are the same as for conventional packs, that is, fresh

FIGURE 4.8 ECS functional diagram (conventional vs. electrical) [4.4].

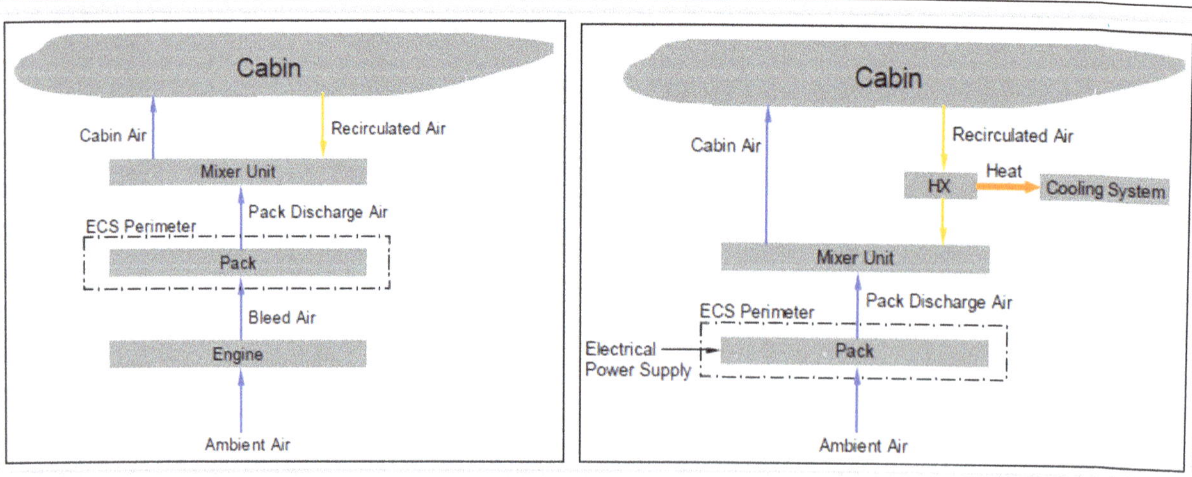

air supply as well as cooling and humidity control. Additionally the electric packs provide pressurization. The function of the vapor cycle in general is cooling and in a very poor manner dehumidification as well. All the systems are supplied by electric energy. The architectures and analysis of such systems are briefly described in [4.4].

The function of the air conditioning pack is to provide conditioned pressurized air to the cabin. The pack has to be designed properly to provide sufficient amount of fresh air to the cabin to comply with ventilation certification requirements. Moreover, supplied air temperature and humidity has to be controlled so as to maintain firstly safe cabin conditions and then sufficient passenger comfort level.

FIGURE 4.9 Block diagram of the electric ECS pack [4.4].

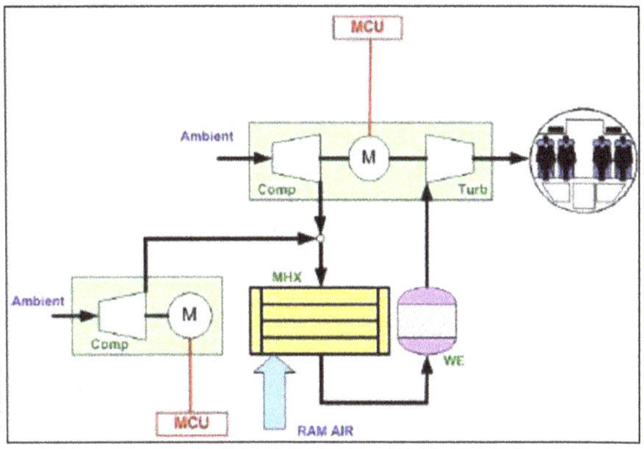

The only available energy for the pack is electrical power as pneumatic power is no longer available. Therefore, the pack is mainly composed of electrical air compressors, which suck air from outside through a dedicated efficient ram air scoop inlet. This air is pressurized through high-pressure-ratio centrifugal compressors (Comp) while the cooling of the hot compressor discharge air is realized via a HE, which uses ram air as a heat sink, and optional expansion cooling in a turbine (Figure 4.9). The outside air will be dehumidified in a high-pressure water extraction loop between the HE and the turbine.

Moreover, the compressor map functional area has to be large enough to comply with ambient air density variation between low and high altitude phases. In any case, efficiency has to be optimized to minimize electrical power offtake from the aircraft engine. Cooling and dehumidification on ground on hot-day conditions need a specific device such as a turbine to provide additional cooling by air expansion and high-pressure water extraction loop. During flight, outside air is cold enough to provide sufficient cooling by ram air main HE ventilation.

In case of failure, the air conditioning system with remaining operating components has to comply with specific certification requirements with respect to cabin ventilation and temperature. This has to be considered for component design. Furthermore, aircraft dispatch objectives imply several component redundancies in order to increase system availability.

A dedicated pack control has to be designed to handle different control modes depending on flow and cooling needs. In order to meet dispatch objectives, controls should also include reconfiguration capabilities in case of component failures. Controls have also to interface with the global aircraft thermal management control, which defines pack cooling needs throughout the aircraft mission. In the frame of the MOET project, Liebherr developed a full-scale demonstrator of an electrical air conditioning pack for a short-range aircraft application.

Main critical components such as motorized turbo machines, motor control units, etc. were developed as part of the demonstrator that is focused on high-performance requirements:

- Compressor stages with wide operating range and pressure ratio of up to 6:1
- Motor control units with power density higher than 4.1 kW/kg

Building upon the MOET project achievements, Airbus, in collaboration with Liebherr, conducted real-condition testing of an electric ECS pack within the European Clean Sky research program (Systems for Green Operations (SGO) platform).

The first flight of the Airbus Flight Lab (A320 MSN1) took place in June 2016. On this test aircraft (**Figure 4.10**) one of the conventional bleed ECS packs was replaced with an electric ECS pack developed by Liebherr (**Figure 4.11**) [4.11].

The electric environmental control system (e-ECS) pack used for the technology demonstration was rated at 50 kW. The flight test campaign covered the complete foreseen envelope and included some additional robustness tests at high altitude. Therefore, the development of the innovative e-ECS in Clean Sky reached Technology Readiness Level 5 (TRL5) by the end of 2016.

FIGURE 4.10 Conventional ECS pack replaced with electric pack—Airbus A320 test aircraft [4.11].

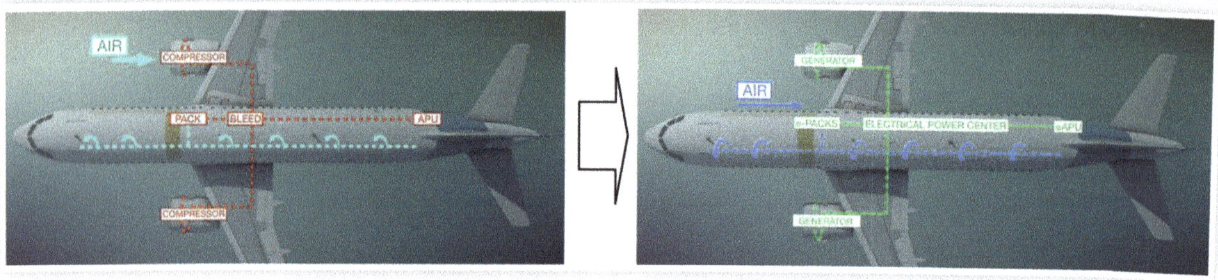

FIGURE 4.11 Electric ECS pack demonstrator + installation—short-range A320 test aircraft.

Electric ECS pack [4.12]

Pack location [4.11]

Pack location [4.11]

Pack compressor ram air inlet port location [4.11]

The Green Regional Aircraft (GRA) was another aircraft platform within the Clean Sky program to test the e-ECS for the regional aircraft segment in cooperation with Leonardo and ATR. The technology was successfully demonstrated in 2016 through a flight test campaign on an ATR72-600 aircraft with one e-ECS pack installed in the aircraft's belly fairing.

In spite of the promising upsides, to date, no short-range or regional aircraft program has selected an electric ECS baseline, all activities being still confined to the research domain. The story is quite different for the long-range aircraft segment.

In recent long-range aircraft developments, the degree of systems electrification has dictated the choice of a conventional ECS or an electric one. While the Airbus 350 has stuck to the conventional pneumatic ECS, the Boeing 787 has resorted to an electric ECS in the frame of extensive systems electrification undertaken for the first time on a large aircraft. UTC Aerospace Systems (UTAS) has a large work share in this electrification process including the ECS. This reflects supply chain consolidation, a departure from previous aircraft programs, enabled by large-scale electrification.

On the twin-engine ~300-seater Boeing 787, which features an electric ECS eliminating engine bleed altogether, cabin pressure is increased to >82 kPa (1800 m cabin altitude). As such, the ECS does not use high pressure air from the engines—bleed air—to drive the two air conditioning packs it is comprised of. Instead, four cabin air compressors (CACs)—radial compressors driven by electric motors—are used within the packs (two CACs per pack). With 110 kW each, their combined power requirement, which roughly compares with the total electrical power consumption of a Boeing 737, is supplied by the electrical generators [4.13]. The CACs run on air bearings at 40–50,000 rpm with a total pressure ratio of 5 (on each pack). They draw ram air (at 20-100 kPa during cruise) and deliver it at 90°C instead of 180°C as with conventional bleed air ECS. This electric ECS adds about 200 kg to the aircraft system weight (and some maintenance costs), but saves some 5% fuel overall. Higher humidity in the passenger cabin is enabled by the use of corrosion-free composites. Ozone is removed from outside air; HEPA filters remove bacteria, viruses, and fungi; and a gaseous filtration system removes odors, irritants, and gaseous contaminants [4.3].

The CACs are powered by common motor starter controllers (CMSCs). Pin-programming allows these standard modules to be also used for other electric drives such as the ram air fan and the motor-driven compressor. The ram air fan supplies air to the ram air duct when the aircraft is on the ground. The motor-driven compressor produces compressed air for the nitrogen generation system, which delivers the gas to the fuel tank inerting system. Therefore, multiplexed CMSCs can be used to power different components on an aircraft at different times, allowing for the complete elimination of extra controllers. On some large commercial aircraft, that may equate to a reduction of five motor controllers at 500 lbs combined [4.14].

Figure 4.12 shows the design comparison of the new all-electric Boeing 787 ECS and the old bleed air ECS. The block diagrams also highlight the energy sources and interconnections with other aircraft systems [4.15].

In Figure 4.12, all the components shown for the Boeing 787, except the engines, are supplied by UTC. The Autotransformer Rectifier Units (ATRUs), symbolized as XFR, are supplied by Thales and have a power rating of 155 kW, giving an idea of the

FIGURE 4.12 Conventional and electric ECS design comparison [4.15].

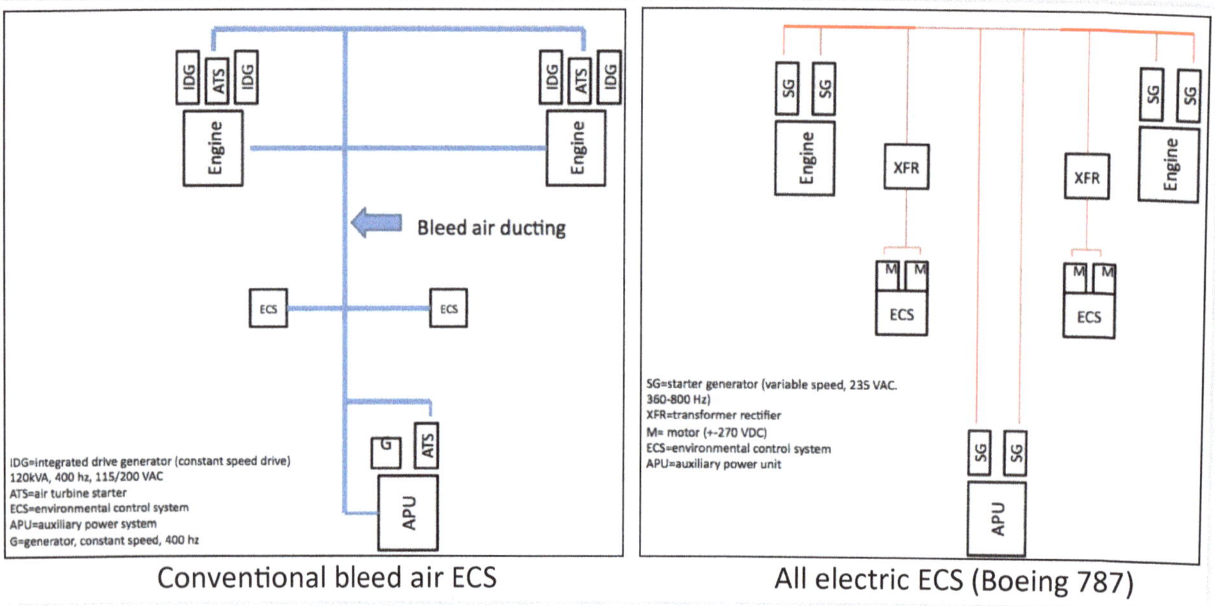

FIGURE 4.13 Boeing 787 electric ECS pack [4.16, 4.17].

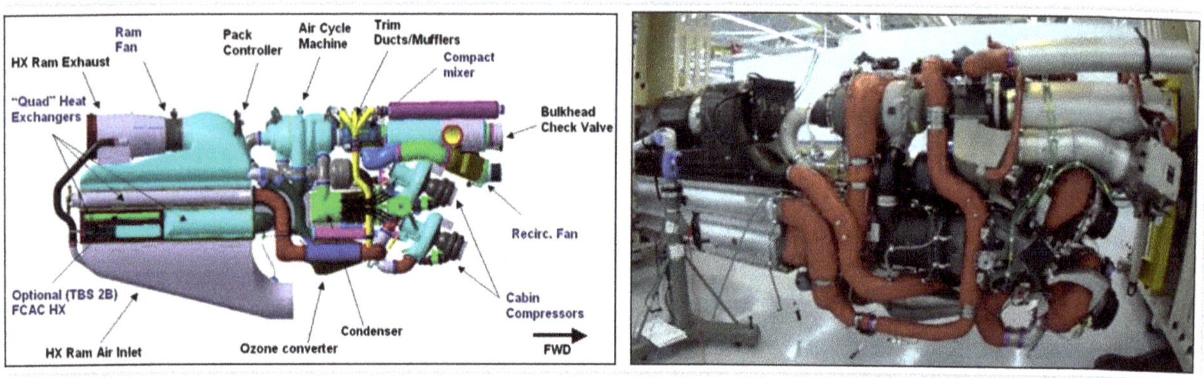

amount of power in play. For reasons of clarity, compressor power electronics, that is, the CMSCs connecting the starter-generators via ATRUs to the ECS compressors (CACs), are not shown.

The Boeing 787 electric ECS pack components are detailed in Figure 4.13. Neither electrical power supply components nor compressor power electronics are shown. Figure 4.14 shows the area on the Boeing 787 where the packs are installed and highlights the locations of the ram air inlets for the ECS HEs and the CACs. It is to be noted that while the HE inlet is necessary for both conventional and electric ECS design configurations, the CAC inlet is required for the electric ECS only.

FIGURE 4.14 Boeing 787 electric ECS: pack and ram air inlet locations [4.18].

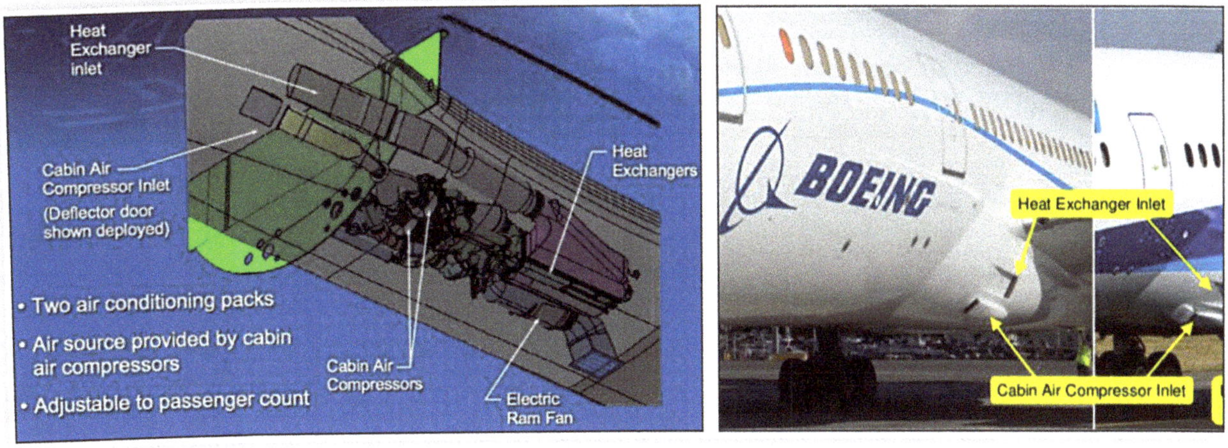

4.3 Wing Ice Protection System

Overall ice protection of the aircraft is necessary in order to ensure flight operations during icing conditions. Ice protection detection/activation and the variety of available techniques and technologies involved are summarized in the block diagram shown in Figure 4.15 [4.19].

In the traditional architecture on large aircraft, hot bleed air is extracted from the airplane bleed system and distributed through the areas of the wing leading edge that need ice protection. For each wing, one valve controls the flow of the bleed air to the wing leading edge, while a "piccolo" duct distributes the heat evenly along the protected area of the wing leading edge. In addition, should ice protection on the leading edge slats be required, a telescopic duct supplies bleed air to the slats in the extended position. The spent bleed air is exhausted through holes in the lower surface of the wing or slat. The Boeing 787 utilizes an electrothermal ice protection scheme for the WIPS, in which several heating blankets are bonded to the interior of the protected slat leading edges. The heating blankets may then be energized simultaneously for anti-icing protection or sequentially for de-icing protection to heat the wing leading edge. This method is significantly more efficient than the traditional system because no excess energy is exhausted. As a result, the required ice protection power usage is approximately half that of pneumatic systems. Moreover, because there are no bleed air exhaust holes, airplane drag and community noise are improved relative to the traditional pneumatic ice protection system.

On the Boeing 787 six electrically powered heater mats are built into each heated slat. Aluminum caps protect the leading edge surface. The heater mats are bonded to the interior surface of the composite slat leading edge skin (Figure 4.16). The function of the WIPS is to provide electrothermal ice protection

FIGURE 4.15 Aircraft ice protection block diagram [4.19].

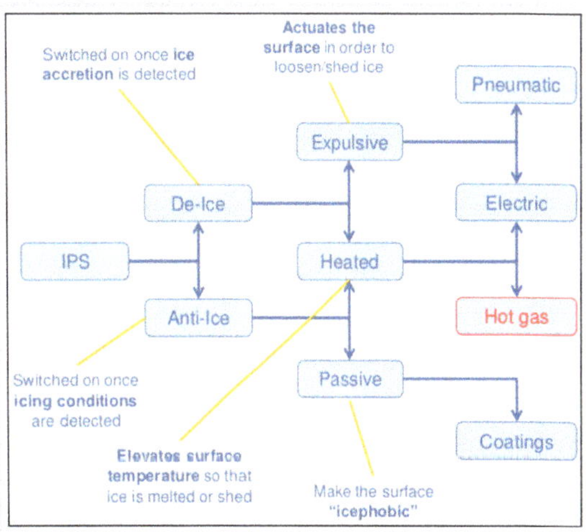

FIGURE 4.16 Location of electrothermal WIPS and slat heater mats—Boeing 787 long-range aircraft [3.27].

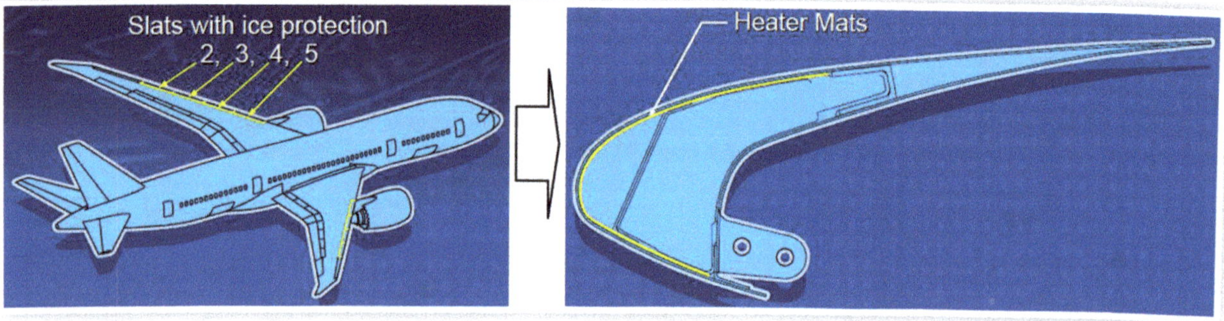

to an area of the aircraft wing leading edge slats. WIPS operates simultaneously both left (port) and right (starboard) wing ice protection to achieve symmetrical wing leading edge ice protection, and so maintains aerodynamic stability.

The Boeing 787 235VAC three-phase power supply is switched by the Wing Icing Protection Control Unit (WIPCU) (Figure 4.17). The WIPCU controls the power to the eight heated slats-four on each wing. There are three types of electronic circuit boards in the WIPCU:

- Power Supply (1 per WIPCU).
- Sequence Controller (1 per WIPCU).
- Zone Controller (24 per WIPCU). The number per IPCU depends on the area of coverage. The Boeing 787 has 24 zone controller cards.

The Power Supply conditions and distributes the dual 28 VDC power to the Zone Controller and Sequence Controller. The Zone Controller controls the power switching to the different heater mat zones. The Sequence Controller maintains the symmetrical ice protection for the aircraft controlling the switching of the Zone Controllers.

Such an ice protection system is designed to operate in either an anti-icing, de-icing, or hybrid mode catering to a wide range of ice protection requirements, from simple, time/voltage, on-off control through to very complex multi-zone, closed loop feedback control. Other applications having adopted this system are [4.20]:

FIGURE 4.17 Wing ice protection control unit—long-range Boeing 787 [4.20].

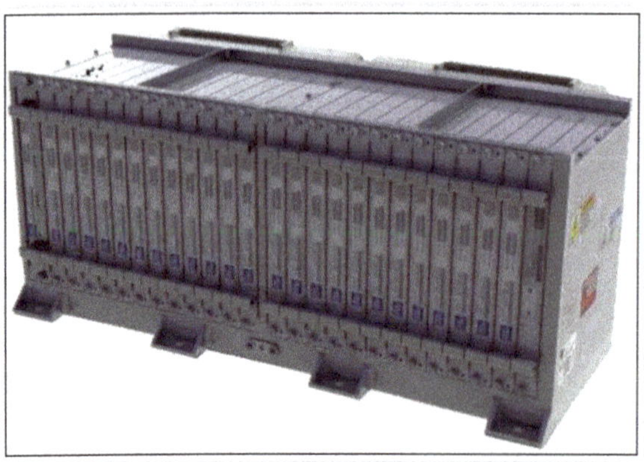

- Pratt & Whitney F135 Engine Inlet (this engine powers the Lockheed Martin F35 fighter)
- Lockheed Martin F35 Lightning II Lift Fan Inlet

4.4 Enabling Technologies

4.4.1 Motors

At the heart of aircraft electrification lies the utilization of electric machines, whether they are used as generators in power generation or motors ensuring motion control, whether that, for instance, means starting an engine, compressing outside air for air conditioning, or actuating flight control surfaces. Choosing the best-fitting machine topology and technology depending on the application is therefore a key decision in the design process. Figure 4.18 shows the variety of machine technologies at the disposal of design engineers [4.21].

Permanent Magnet Alternating Current (PMAC) referred to in Figure 4.18 is actually a generic name covering all brushless PM Synchronous Motor (PMSM) technologies, including brushless DC and brushless AC motors. In essence, brushless DC and brushless AC motors are identical in hardware configuration but fed with different waveforms of supply which can be achieved by modifying control strategy in software. PMACs are characterized with high power density and efficiency, high torque/inertia and torque/volume ratios, and improved reliability.

PM motors can be utilized in several ways with respect to the rotor PM arrangements. Generally, surface-mounted PM motors yield a small rotor diameter with low inertia (thus good dynamic performance), whereas interior permanent magnet motors provide higher per unit inductances and thus field weakening capability. Designs for fault tolerance have generally combined deep PMs and a single-layer concentrated winding topology to minimize mutual inductance.

While it is well accepted that PMAC motors provide higher torque density than reluctance and induction motors, they are less intrinsically fault tolerant. In conventional PM machine topologies, currents can flow in a failed lane, even when the lane is disconnected from the electrical supply, because of the continuing presence of the magnet-induced back electromotive force (emf). Careful parameter selection can make the motor survive such a fault. Appropriate choice of PM motor inductance and current rating ensures that short-circuits at motor terminals caused by a motor controller failure do not negatively impact performance. In short, improved PMAC motors can offer the smallest solution to actuation requirements, but the management of faults is more complex than in SR machines.

From a systems electrification perspective, PM machines are dominating recent aerospace research and development, and their implementation has become widespread on more electric aircraft certified of late. The optimal number of motor phases may differ from application to application due to the varying subsystem requirements and refinement criteria, which in practice would be based on total loss requirements. To improve fault tolerance, it would be ideal to have a multiple of independent single-phase drives or a multiple of independent three-phase drives.

FIGURE 4.18 Machine technology and performance comparison [4.21].

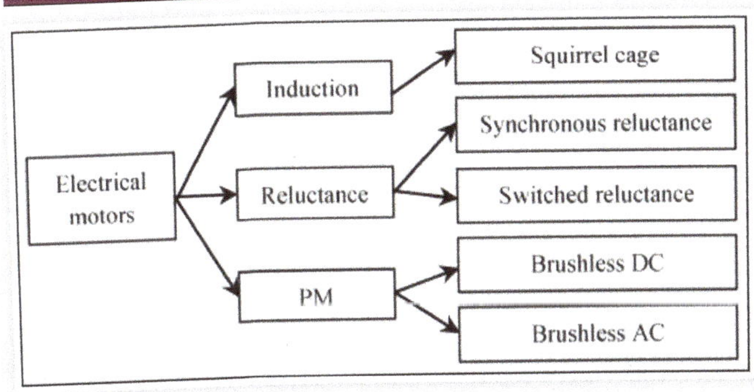

Performance	IM-cage	SR	PMAC
Fault tolerance	low	high	high
Power density	moderate	moderate	high
Robustness	√	√	×
Efficiency	moderate	high	high
Cost	low	low	high
Wide speed range	×	√	√
Open-loop control	√	×	×
Close-loop simplicity	√	√	√
Torque ripple	×	high	low
Acoustic noise	×	moderate	low
Areas of application	actuators	high-temp. engine	fuel pump actuators

FIGURE 4.19 Fault-tolerant PMAC: configuration example and comparison of topologies [4.21].

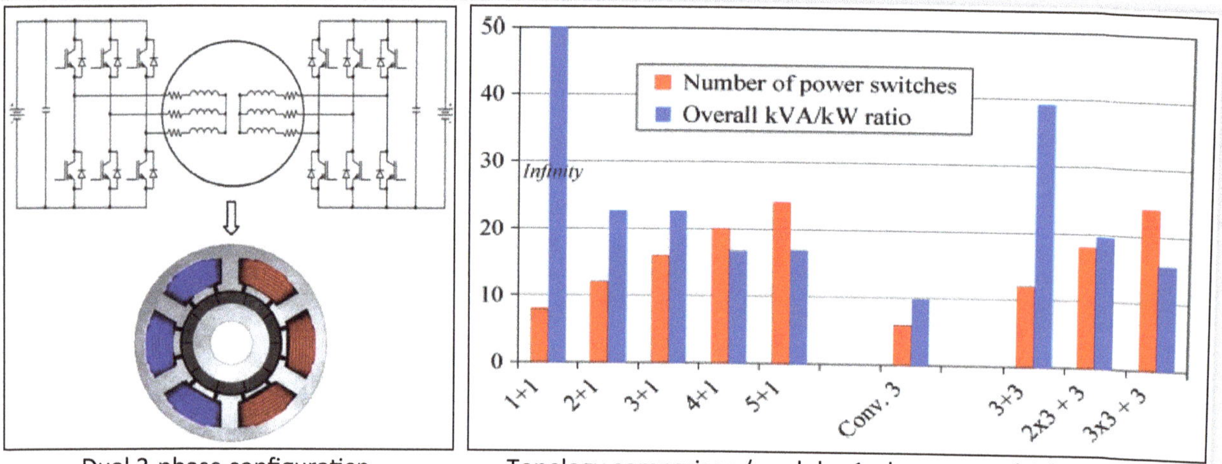

Dual 3-phase configuration

Topology comparison (modular 1-phase vs. multiple 3-phase)

A multichannel three-phase approach offers potential upsides. In this configuration, PM motors with multiple isolated sets of three-phase windings are each driven by independent conventional three-phase power converters. Figure 4.19 shows another comparison of the number of power switches and overall kVA/kW ratio of fault-tolerant topologies for three-phase and modular single-phase systems. The kVA/kW ratio indicates the quantity of power electronic devices for a fixed power output. As the number of modules increases for modular single-phase systems, the overall size of the fault-tolerant drive tends to decrease, but the component count increases. Higher module numbers will show a leveling out of drive size, but an ever-increasing component count. Motors with three-phase modules do not suffer from unbalanced torque dips when one lane fails, but for a given number of lanes, larger numbers of switches are required, and this results in 50% more power devices. It is noted that dual three-phase systems have the highest kVA/kW ratio (thus machine/converter size) for a given power output but use less power switches.

Nonetheless, multiple three-phase (e.g., dual or triple three-phase) systems are still advantageous because of the following:

- They provide a fault-tolerant arrangement with the minimum number of power supplies, processors, and filters, and with a less complicated control scheme.
- They act as a balanced load on the supplies, reducing Total Harmonic Distortion (THD) and ElectroMagnetic Interference (EMI).
- Each lane produces smooth torque, and so, in general, loss of one lane does not introduce ripple or drag torque into the system. Compensation for a single-lane failure simply requires an increase in the torque demand for the other lanes.

4.4.2 Power Electronics

Electrical applications require power electronics whether they are used in active power conversion equipment, switching power between two different types of electrical networks and/or scaling up (or down) voltage levels, or in motor drives where they are part of speed/torque or power control of the electrical loads they are connected to. It is to be noted that motor drives are necessary not only for aircraft and engine systems electrification but also when dealing with electric and hybrid-electric propulsion.

Challenges in the field of power electronics are huge because they are mainstays in an electric aircraft. Therefore, improving their key metrics such as power-to-weight ratio, thermal capability, availability, etc. is paramount to stave off downsides and reap the benefits of electrification.

Power switches are core components of power electronics. The recent development of new Wide Bandgap (WBG) semiconductor technologies (transistors and diodes) allows for a significant increase in efficiency, performance, and power density of adjustable speed electrical power drive systems. Compared to their silicon (Si) versions, these devices, made of Gallium Nitride (GaN) or Silicon Carbide (SiC), offer faster switching speeds with lower losses and are capable of operating at higher temperatures. Hotter temperatures allow for the relaxation of cooling requirements potentially saving weight. Nevertheless, due to their higher switching speed (dv/dt) and voltage overshoot, special care is necessary when WBG devices are used in power converters of an electromechanical chain. Otherwise, EMI issues may lead to additional filtering requirements. Hence, a balance has to be struck between losses and EMI performance by keeping switching speed and overvoltage at reasonable levels. Moreover, susceptibility of inductances and motor windings to high dv/dt is also a matter of investigation [4.22].

For instance, when SiC MOSFETs are used instead of conventional Si IGBTs as power switches in a three-phase inverter, a power module for one inverter leg can be designed as shown in Figure 4.20. Each side of the module comprises two SiC MOSFET dies in parallel (S), one series diode (D1), one freewheeling diode (D2), and one shunt resistor (R) for current sensing. With one such SiC module, switching losses at inverter level could be reduced by more than 60% compared to a Si IGBT-based solution [4.23].

Nevertheless, increased mechanical stress and higher thermal conductivity encountered with SiC are factors to be reckoned with when designing the power module as regards robustness and thermal performance.

FIGURE 4.20 1700 V/40 A SiC inverter leg: schematic and power module (internal view) [4.23].

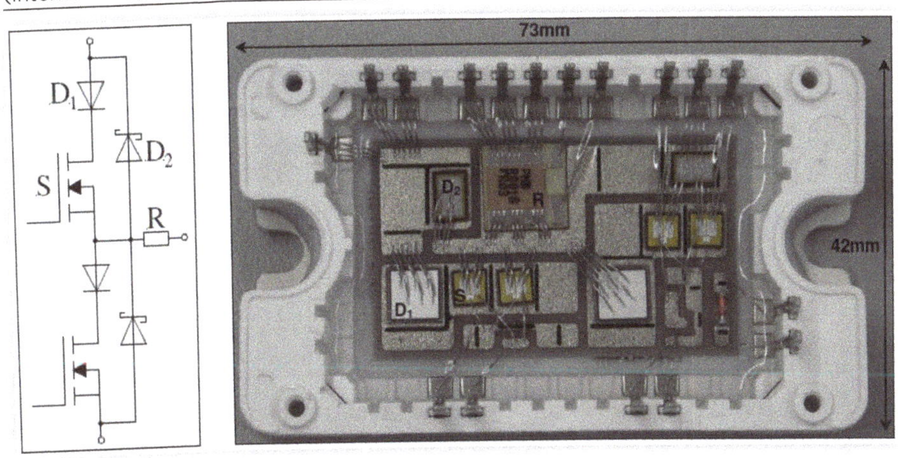

FIGURE 4.21 Three-phase 1200V/100A SiC MOSFET inverter—Intelligent Power Module (IPM) [4.24].

Taking the module integration one step further, the Intelligent Power Module (IPM) concept offers an optimal integration of the gate driver with the power transistors (**Figure 4.21**). The gate driver circuit is optimized taking into account parasitic inductances of the power module while minimizing them when possible [4.24].

Specifically designed for actuation systems such as primary flight controls, landing gear and braking of a more electric aircraft with a 540 VDC supply bus, the Power Core Module (PCM) constitutes a compelling approach in the miniaturization of power electronics [4.25].

In the PCM, the SiC MOSFETs and anti-parallel diodes of the three-phase inverter along with gate drive resistors are all placed on a substrate soldered to the base plate (**Figure 4.22**).

Pushing the envelope in terms of power device and controls integration, the PCM then integrates the power substrate assembly with a controller Printed Circuit Board (PCB) (**Figure 4.23**) which in turn interfaces with a gate driver PCB. The controller board features local power supplies for the logic and circuitry for signal conditioning, communications, control, and signal processing. It interfaces with the driver PCB which

FIGURE 4.22 Three-phase inverter integrated into the power substrate assembly of the PCM [4.25].

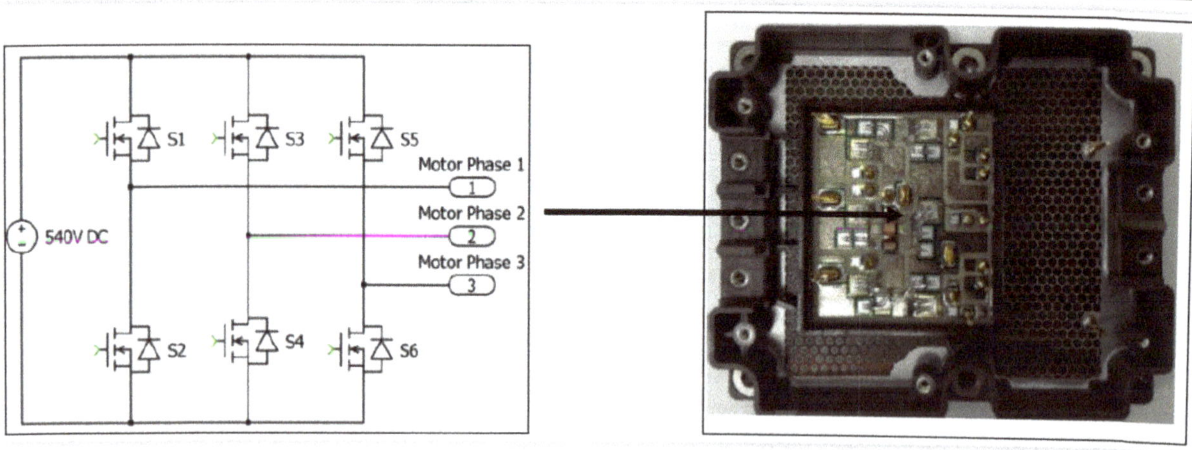

FIGURE 4.23 Three-phase 1200 V/5 kVA SiC inverter—PCM [4.25].

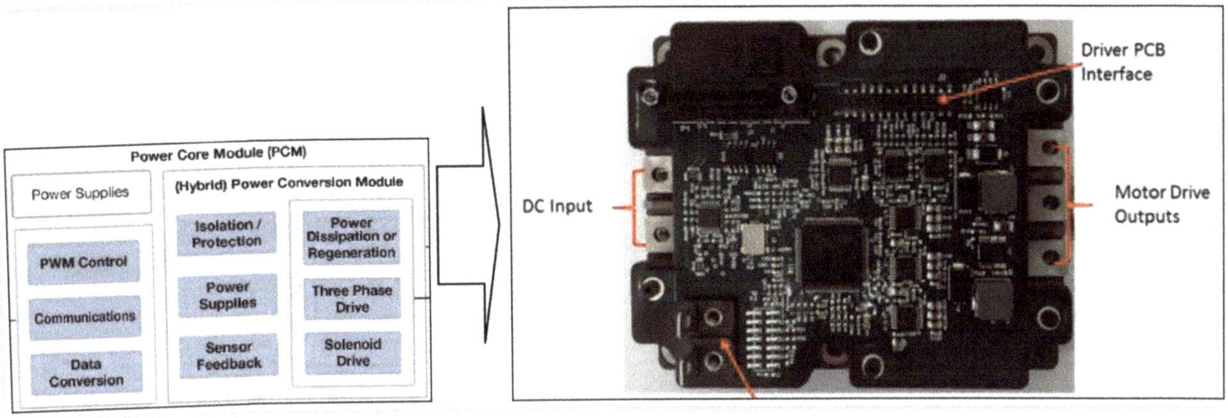

has double-sided component placement and is connected to the substrate assembly through thermal, signal, and power pins.

Such a PCM can be directly used in an actuator motor controller.

Apart from SiC diodes, WBG switching devices have not been able to hit the sweet spot on recently certified more-electric aircraft. Nonetheless, building upon the potential of WBG devices, ongoing research is actively targeting the optimization of power modules on top of other components such as passive components and materials. Increasing the level of integration of power electronics is a major step in boosting the power-to-weight ratio of motor drives, a key design criterion for the electric aircraft.

Figure 4.24 shows the weight breakdown of state-of-the-art silicon-based stand-alone actuation motor controllers on more-electric aircraft. When these controllers are supplied by an AC network and located in unpressurized areas, their power density is around 1.5 kW/kg, whereas this value may go up to 3 kW/kg when considering a DC power supply and installation in pressurized areas [4.26].

Figure 4.25 provides a few examples of motor controllers on certified aircraft and research programs for electrified loads such as flight controls, engine start, ECS, etc.

What can be seen from the weight breakdown of Figure 4.24 is that there is room for optimization allowing for the design of much-needed power dense controllers on an electric aircraft. For instance, removing the AC/DC converter and drawing power directly from DC networks would allow power density improvement. Centralized motor controllers

FIGURE 4.24 Typical weight breakdown of actuation motor controller [4.26].

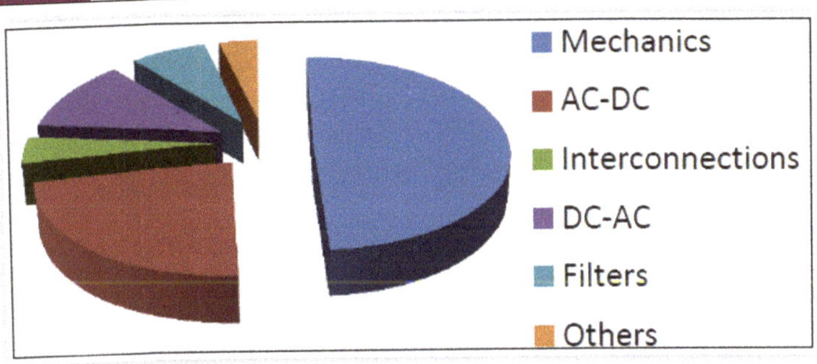

FIGURE 4.25 Electric motor controllers (certified and research programs).

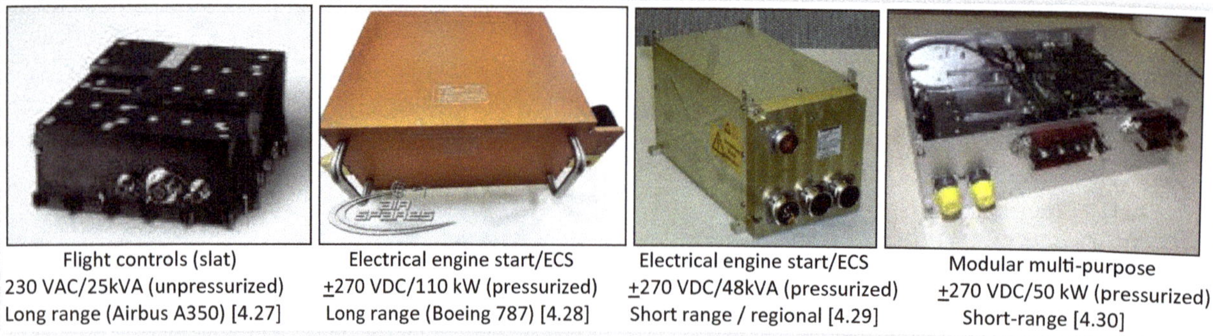

would enable a common cooling technique tailored to address the best trade-off between controller losses and cooling requirements. On the other hand, remote power electronics integrated at the actuator level would also enable high power density values to be reached.

Power electronics demonstrator programs have been able to achieve up to 5 kW/kg of power density in centralized liquid-cooled versions, whereas the power-to-weight target for power electronics working from HVDC networks is aggressively set around 16 kW/kg [4.26].

For the more electric and electric aircraft, highly reliable and efficient power converters for motor drives and power supplies are required. High power density and high temperature solutions are being researched in order to stave off weight penalties due to cooling requirements. Making power electronics generic so that they can be shared, paralleled, and reconfigured, and also host several application software packages, would contribute to the reduction of aircraft dead weight due to controllers functioning during short periods of time only during a flight and stay inoperative otherwise. Such sharing of a modular resource allows for dynamic

FIGURE 4.26 Rationale for IMPE [3.33].

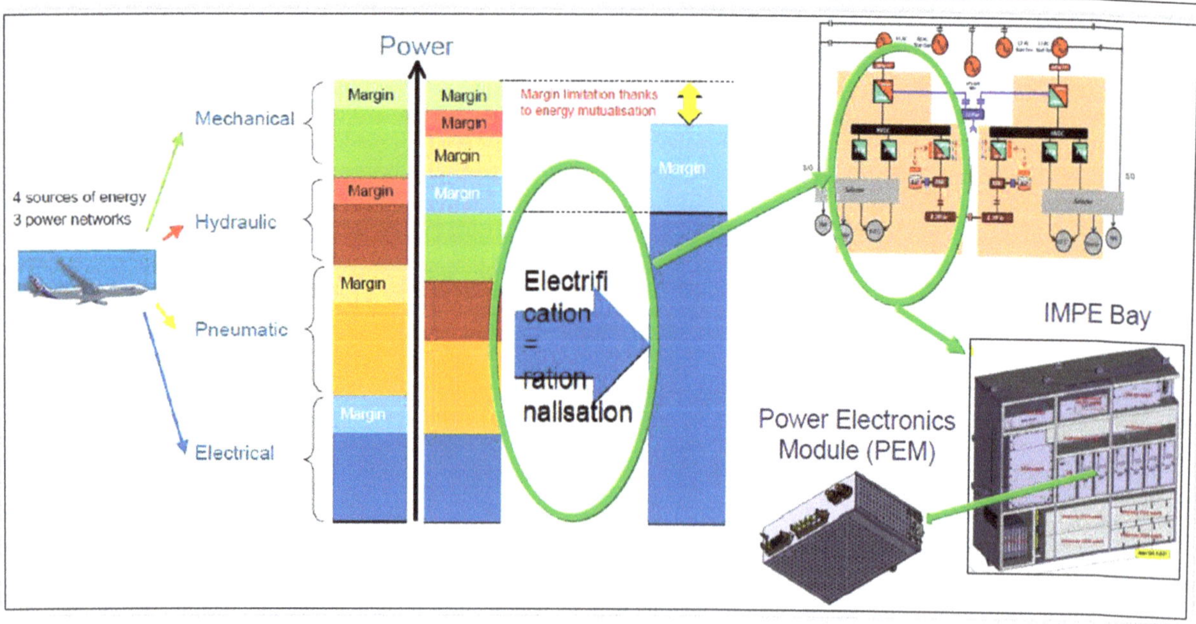

power management across multiple systems and enables improved functional availability.

From an aircraft manufacturer perspective, this kind of Integrated Modular Power Electronics (IMPE) for power conversion and motor drive applications helps increase the effective system power density thanks to the following reckoning [3.33] (Figure 4.26):

- Overall systems weight cannot be optimized if each system uses its own power electronics.
- Transient system operation results in dead-weight penalty over aircraft life.
- Reconfiguration, redundancy, and two-way operation improve functional "availability" and weight optimization.

Power electronic modules (PEMs) integrated in an electronic bay would benefit from overall filtering and cooling mechanisms. Their weight and volume stakes are portrayed in Figure 4.27.

FIGURE 4.27 Power electronics: weight and volume stakes [3.33].

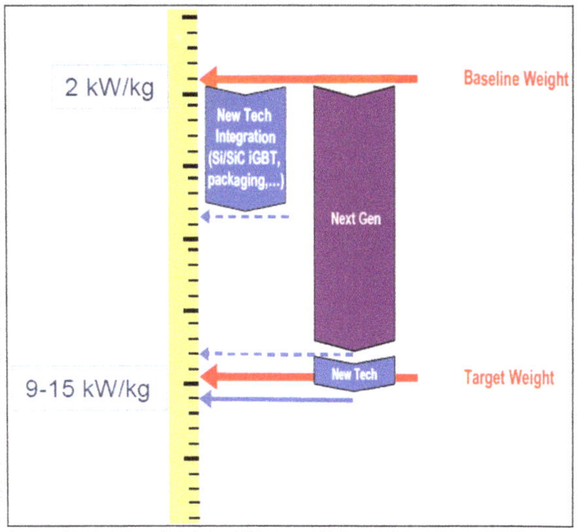

4.5 Conclusions

In summary, from a systems perspective, the challenges of more electric and electric aircraft for aircraft manufacturers are found in the following areas:

- Weight reduction
- Optimized architectures with power management
- Lower manufacturing costs

For airline operators the benefits expected with electrification are:

- Reduced fuel burn and environmental footprint (CO_2, NOx, noise)
- Improved efficiency
- Higher reliability and maintainability
- Increased aircraft availability and dependability
- Lower operational costs

This chapter shows that in spite of power density issues, electrification enables design, manufacturing, and maintenance optimizations. It improves overall system performance and increases availability. Operations can therefore get more flexible, reconfigurable with simplified maintenance operations, while streamlined manufacturing and assembly lines add value to the aircraft manufacturer.

Research is very actively being pursued in the field of aircraft and engine system electrification in conjunction with a host of demonstrations with the goal to improve maturity levels. The following chapter will go into the details of actuation systems that are an integral part of both aircraft and engine systems, whether the aircraft is totally electric or just more electric.

References

[4.1] Fehrm, B., "Bjorn's Corner—Efficient Systems," Leeham News and Comment, February 5, 2016, https://leehamnews.com/2016/02/05/bjorns-corner-efficient-systems/, accessed March 27, 2018.

[4.2] Herzog, J., "Electrification of the Environmental Control System," Liebherr Aerospace, *25th International Congress Of The Aeronautical Sciences (ICAS 2006)*, Hamburg, Germany, September 3-8, 2006.

[4.3] Martinez, I., "Aircraft Environmental Control," September 24, 2013, ETSIAE-UPM, Spain, http://webserver.dmt.upm.es/~isidoro/, accessed March 14, 2018.

[4.4] Galzin, G., Gomez, V., Mevenkamp, C., and Brunswig, H., "Electrical Environmental Control System," Liebherr, Airbus, More Electric Aircraft Forum, *European Power Electronics and Applications Conference (EPE 2009)*, Barcelona, Spain, September 8-10, 2009.

[4.5] https://www.liebherr.com/fr/bel/produits/aerospace-transportation-systems/aerospace/produits-et-solutions/syst%C3%A8me-d-air/syst%C3%A8mes-d-air.html#!/accordion-start-module=air-conditioning-acc-item-start-module, accessed March 17, 2018.

[4.6] Liebherr Aerospace, "A319/A320/A321 Environmental Control System—Familiarization Training," Liebherr Aerospace, Revised: July 30, 2004, http://docplayer.net/21850013-A319-a320-a321-environmental-control-system.html, accessed March 16, 2018.

[4.7] Airbus, "A320neo_details_belly" JPEG image, Airbus, http://www.airbus.com/search.image.html?tags=products-and-solutions%3Acommercial-aircraft/a320-family/a320neo&page=7, accessed March 17, 2018.

[4.8] Brasseur, A., Leppert, W., and Pradille, A., "Inside the 747-8 New Environmental Control System," Boeing, Liebherr, AERO Magazine Q1, 2012, published by Boeing.

[4.9] Majeed, O., "Aircraft Environmental Control Systems—A Presentation of Current Systems and New Developments," Specific Range Solutions, Carleton University AERO 4003 Lecture, November 9, 2010, http://www.srs.aero/wordpress/wp-content/uploads/2010/11/AERO-4003-ECS-Lecture-Final.pdf, accessed March 14, 2018.

[4.10] Boeing, "air_force_one10_960x600" JPEG image, Boeing, https://www.boeing.com/defense/air-force-one/index.page#/gallery, accessed April 30, 2018.

[4.11] Image extract from "Liebherr—First Flight of Clean Sky/Airbus Flight Lab (Long Version)", YouTube video uploaded by Liebherr on July 28, 2016, https://www.youtube.com/watch?v=swB1cp5jRbw, accessed March 15 & 17, 2018.

[4.12] Liebherr, "liebherr-eecs-pack-copyright-liebherr-aerospace.jpg" JPEG image, Liebherr, https://www.liebherr.com/en/are/latest-news/news-press-releases/detail/electrical-environmental-control-system-of-liebherr-successful-during-first-flight-of-clean-sky-airbus-flight-lab.html#lightbox, accessed March 15, 2018.

[4.13] Althaus, A., "Testing for the Dreamliner, Connection," *The Lufthansa Technik Group Magazine*, 6.2017, November and December 2017.

[4.14] Hunt, G. and Mandyck, J., "Green Ideas Born To Fly™: How Weight, Design and Integrated Systems Drive Sustainable Aviation at United Technologies," United Technologies, Green Aviation White Paper 4, June 2017.

[4.15] Bower, G., The All Electric Boeing 787, GM-VOLT: Chevy Volt Electric Car Site, Posted March 30, 2012, http://gm-volt.com/tag/all-electric-boeing-787/, accessed March 16, 2018.

[4.16] https://www.quora.com/How-does-the-Boeing-787-pressurize#, accessed March 17, 2018.

[4.17] Image extract from "UTC Aerospace Systems delivers 100th CACTCS pack shipset for Boeing 787 Dreamliner", YouTube video uploaded by UTCAerospaceSystems on October 8, 2012, https://www.youtube.com/watch?v=FoHzONWixkM, accessed March 16, 2018.

[4.18] Dodt, T., "Introducing the 787," Boeing, ISASI, Salt Lake City, UT, USA, September 12-15, 2011, http://www.ata-divisions.org/S_TD/pdf/other/IntroducingtheB-787.pdf, accessed March 16, 2018.

[4.19] Brooks, A., Cassissa, M., and Halls, S., "GKN Technology—Leading the Way toward More Efficient Aircraft," GKN Aerospace, July 2014, http://studylib.net/doc/10747757/gkn-technology---leading-the-way-toward-more-efficient-ai...#, accessed May 1, 2018.

[4.20] https://www.ultra-pcs.com/app/uploads/2017/06/Brochure-Ice-Protection-Systems.pdf, accessed March 19, 2018.

[4.21] Cao, W., Mecrow, B.C., Atkinson, G. J., Bennett, J.W. et al., "Overview of Electric Motor Technologies Used for More Electric Aircraft (MEA)," Aston University (UK), Newcastle University (UK), *IEEE Transactions on Industrial Electronics* 59, no. 9 (September 2012), https://www.researchgate.net/publication/224254344, accessed May 1, 2018.

[4.22] Dos Santos, V., Cougo, B., Roux, N., Sareni, B. et al., "Trade-Off between Losses and EMI Issues in Three-Phase SiC Inverters for Aircraft Applications," *EMC 2017, International Conference*, Washington, DC, August 2017, https://hal.archives-ouvertes.fr/hal-01588886, Accessed April 4, 2018.

[4.23] Liebig, S., Nuber, M., Engstler, J., Engler, A. et al., "Characterisation and Evaluation of 1700V SiC-MOSFET Modules for Use in an Active Power Filter in Aviation," Liebherr Elekronik and TU Chemnitz, International Conference for Power Electronics, Intelligent Motion, Power Quality and Energy Management (PCIM Europe), Nuremberg, Germany, May 8-10, 2012, https://www.researchgate.net/publication/272684505, accessed March 18, 2018.

[4.24] http://www.cissoid.com/news/sic-intelligent-power-modules.html, accessed March 18, 2018.

[4.25] O'Donnell, S., Debauche, J.-L., Wheeler, P., and Castellazzi, A., "Silicon Carbide MOSFETs in More Electric Aircraft Power Converters: The Performance and Reliability benefits over Silicon IGBTs for a specified Flight Mission Profile," Microsemi and University of Nottingham (UK), *18th European Conference on Power Electronics and Applications*, Karlsruhe, Germany, September 5-9, 2016, http://eprints.nottingham.ac.uk/39658/, accessed April 9, 2018.

[4.26] Engler, A., "Optimizing Power Electronic for More Electrical Aircraft—Liebherr en route to More Electric Aircrafts," Liebherr Elektronik, *IQPC 1st International More Electric Aircraft Conference*, Hamburg, Germany, November 25-27, 2013.

[4.27] https://www.liebherr.com/shared/media/components/documents/control-technology-and-electronics/liebherr-electronics-for-aerospace.pdf, accessed March 18, 2018.

[4.28] http://www.aeroforgetooling.com/boeing-licensed-tooling/k24002-1, accessed March 19, 2018.

[4.29] https://media.springernature.com/original/springer-static/image/art%3A10.1007%2Fs13272-017-0268-x/MediaObjects/13272_2017_268_Fig4_HTML.jpg, accessed May 1, 2018.

[4.30] Bensalah, T. and Thalin, P., "Integrated Modular Power Electronics: Achievements and Challenges," Thales, *MEA 2015 More Electric Aircraft Conference (3AF/SEE)*, Toulouse, France, February 3-5, 2015, OAI: www.see.asso.fr:10638:20104.

CHAPTER 5

Electrification of Aircraft Systems—Part III

Shift from Hydraulic to Electric-Powered Actuation

J.-C. Maré
Institut National des Sciences Appliquées de Toulouse – Institut Clément Ader, Toulouse, France

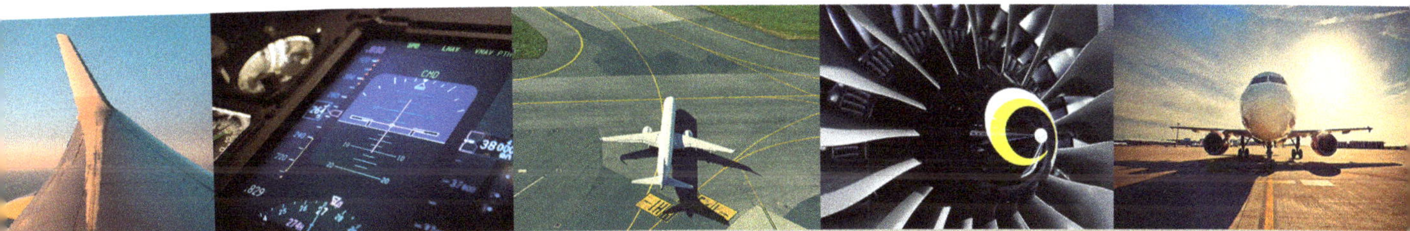

As defined in SAE ARP4386, an actuator is the *component of the actuation system that does work, or dissipates energy, or traps fluid to control, move or hold a load*. In the field of aerospace, the component "actuator" is usually defined as the physical unit that drives a mechanical load, from a centralized source of power.

Any aircraft requires numerous actuators to fulfill its mission. Some of them are aimed at delivering fractional horsepower for non-safety critical functions, for example, seat inclination. This kind of application will not be addressed below because such electrically supplied actuation has been well established for decades, electric motors being used widely with success as the prime power transformer between electrical and mechanical domains. However, numerous actuation functions require high power, in particular high forces at low velocities, and/or are safety-critical, for example, for flight controls or landing gears. For this kind of application, hydraulically supplied actuators—so-called *conventional actuators*—have been extensively employed for more than 60 years. The use of hydraulic power was originally used due to its superiority over electric drives regarding power-to-weight ratio and ease of realizing secondary functions. The tremendous improvement of power electronics and electric machines has recently modified, or even reversed, this well-established situation.

Switching to more electric technologies involves both signal and power levels. Unfortunately, the common wording is not explicit or is ambiguous as "by-wire" is used without distinction. For example, Fly-by-Wire (FbW) only refers in practice to the transmission of signals (or data) by means of electrical wires for flight controls. To make it clear, the present chapter will address *Power-by-Wire* (PbW) wherein wires supply power to the actuator and potentially allow regeneration.

Section 5.1 presents the actuation needs in terms of power transmission and control. Section 5.2 addresses the general considerations that apply to any type of PbW actuator. The following Sections 5.3 through 5.5 focus on specific solutions for more electric actuation. Finally, Section 5.6 presents the remaining challenges and the potential directions on improvement of PbW actuators.

5.1 Actuation Needs for Power Transmission and Control

Depending on the flight phase, actuators may draw hundreds of kW from the centralized power networks. On the Airbus A380, for example, they require up to 350 kW and are supplied by four power networks with a total capability of around 850kW (2 hydraulic networks, 725 kW engine driven pump + 127 kW electric motor pump), and 910 kVA (2 electric networks, 600 kVA variable frequency from engines + 240 kVA fixed frequency from auxiliary power unit + 70 kVA from ram air turbine). In practice, actuation is a major consumer of secondary (*non-propulsive*) power that significantly impacts the aircraft design relative to power generation and distribution levels, as well as its operation. Therefore, addressing the design of actuation systems requires a systems approach, at aircraft and even at airline operational levels, instead of a component view focusing only on the actuator. This is a key aspect to keep in mind. A lot of research and development programs dealing with more electrical actuation intended to simply pick and replace conventional actuators by electrically supplied ones. They all have demonstrated how important it is to first come back to the actuation needs. Indeed, the full benefit of electrical technology cannot materialize if electrically powered actuators are used in an arrangement that was originally designed for conventional actuation. In particular, most of the loads (e.g., flight control surfaces or landing gear legs) are attached to the airframe through a hinge joint that provides a rotational degree of freedom. However, conventional actuators use hydraulic cylinders that are cheap and constitute a reliable means to produce linear motion. Then, power is transformed into rotational motion through dedicated kinematics (generally, a 3 bar mechanism where the cylinder makes the variable length bar). Table 5.1 summarizes the major requirements and constraints that have to be considered when dealing with actuation.

The first lines of this table mainly focus on the capability to develop and control mechanical power to the load. It is interesting to note that the actuators are sized for the

TABLE 5.1 Typical application, requirements, and constraints applying to aerospace actuation (first view)

Applications	
Main applications	Flight controls: primary (PFC), secondary (SFC) Landing gears (LG): extension retraction, doors, locks, steering, braking. Engine: inlet guide vane, thrust reverser, thrust vectoring, intake, and exhaust
Requirements and constraints specific to actuation	
Type of control	End-stop to end-stop: for example, LG extension/retraction, utilities, engine reverse thrust, or closed loop: position (e.g., FC or LG steering) force (e.g., braking), and more rarely velocity or acceleration
Power capability	Max stroke: 10-700 mm Max force: 5-320 kN Max speed: 20-500 mm/s Sized for worst case, but mainly operated in normal case
Type of operation	Continuous (whole flight), e.g., PFC or Transient (few tens of seconds or minutes per flight), e.g., SFC or LG steering or Pulsed (fractional second to seconds), e.g., LG uplock
Response to failure (at actuator level)	Fail-safe response, depending on actuation function. At load level: free, damped, frozen, or at neutral
Dynamics	Bandwidth of pilot command to driven load: 1-15 Hz
Not specific to actuation	
Environment	Temperature, pressure, humidity, vibration, electromagnetic field, e.g., DO160 [5.1]
Service life (at aircraft level)	A few thousand (helicopters, fighters) to 150+ thousand (commercial) flight hours, more than 30 years of total service
Reliability	Failure rates as low as 10^{-9} per flight hour

worst case. Most of them often operate at a very few tens of percent of their rated speed and force on average. This is illustrated in **Figure 5.1** for aileron actuation on a short-range aircraft.

In practice, there are other important needs that require specific power management functions at actuator level. Unfortunately, as they are easily satisfied in conventional actuators, these needs are often forgotten or insufficiently identified although they significantly influence the success of any electric actuation development. This is also a key consideration to keep in mind. The main needs to address are typically the following:

FIGURE 5.1 Example of operating points needed for actuation during a flight [5.2].

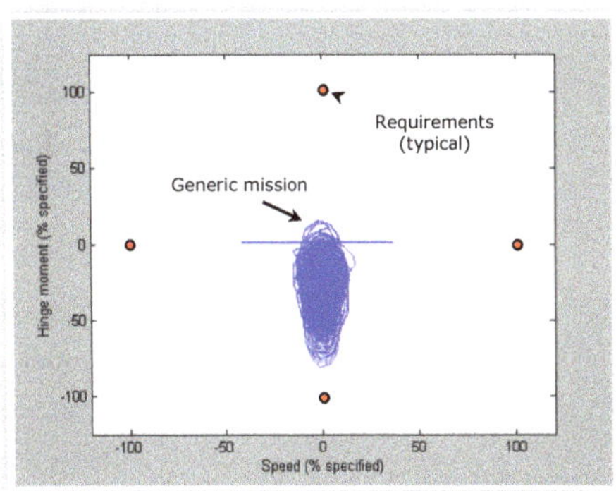

a. Static/dynamic overload protection
   Overload protection aims at avoiding over-stressing the mechanical components and structure due to load inertial effects, backdriving winds on the flight control surface when parked, or in case of a malfunction.

b. Back-drivability/position hold
   It is important to consider the ability of an aiding load to backdrive the actuator(s). Backdriving is generally required for Primary Flight Controls (PFC), and Landing Gear (LG) extension for free

fall or steering in towing mode, or simply for maintenance purposes (e.g., thrust reversers). Position holding is generally required for Secondary Flight Controls (SFC) like slats, flaps, or trimmable horizontal stabilizer (THS). When multiple actuators (or actuator channels) drive a single load in a force summing configuration, actuator (or channel) back-drivability is mandatory to make it fail-passive with respect to the actuation function. Position hold is mandatory for fail freeze or locked response to failure.

c. Cooling/heating
The heat produced by energy losses generated at the actuator level has to be dissipated to avoid excessive heating of actuator parts. Conversely heating may also be required to allow starting and operation under low ambient temperature, in particular by keeping the mechanical efficiency at an acceptable value.

d. Soft end-stops
When moving bodies reach their hard stop at end of stroke, the shock on stop may generate unacceptable mechanical stress. A damping feature is often required to dissipate the bodies' kinetic energy into heat instead of energy of deformation of solids involved in the load path. As a lesson learned from many PbW actuator developments, soft end-stops have to be considered even if bottoming should not occur in active mode.

e. Force equalization
When multiple actuators (or actuator channels) drive a single load in a force summing arrangement, actuators fight each other to position or hold the load. This is generally not perceived at load level although extra power is consumed and internal mechanical loads are increased. Force equalization avoids unacceptable force fighting by balancing the force contribution of each actuator (or channel).

f. Position synchronization
In some applications, independent loads have to be moved with identical position at any time, for example, thrust reverser's transcowls, requiring position synchronization at the level of actuator(s).

g. Damping
The actuator shall dissipate energy to avoid unacceptable vibrations (e.g., flutter or shimmy). Damping may be required in active mode (active damping) or in passive mode only.

5.2 General Considerations for PbW Actuation

This section deals with the electric parts that are common to various designs of PbW actuation. Architecture design is a major phase following the elicitation of needs and their transformation into requirements. However, architectures can be addressed from different points of view versus system levels, engineering activities, phases of life and technological domains, etc., as addressed in the following sub-sections.

5.2.1 From Power by Pipe Towards Hydraulic-Less PbW

When it is intended to make a technological step change, for example, in aerospace actuation, it is a common practice to oppose the natural advantages of the new technology to the inherent drawbacks of the conventional technology. However, the experience acquired for decades with the development of PbW actuators has shown that ignoring the natural advantages of conventional solutions and the issues introduced by the new technology unavoidably leads to unsuccessful outcomes. To avoid this situation, it is of particular importance to identify as early as possible the challenges to be taken into account and where the development effort should be placed. When they are used for power transmission and control in actuation systems, hydraulic and electric technologies have their own natural advantages and drawbacks that are summarized on **Table 5.2** for a wide set of criteria. Gray cells point out the main natural advantages of each technology.

The concepts developed and put into service towards PbW and hydraulic-less actuation have appeared progressively; see **Table 5.3**. They intend to remove the drawbacks of hydraulic technology (hydraulic power networks, control of power by throttling, poor environment-friendliness of hydraulic fluid) while taking the best of electric technology, once it has become sufficiently mature. The upper section of **Figure 5.2** presents a schematic of a conventional actuator design, *Hydraulic Servo Actuator* or HSA, that has been widely employed for more than 60 years (concept (a) of Table 5.3). The actuator is hydraulically supplied at constant pressure from a centralized power network. The control of power consists of wasting all the supply pressure in excess to that required to drive the load, by throttling at the metering valve (e.g., the power stage of the servovalve). This solution has low efficiency, especially at partial loads, but offers excellent dynamic response.

In concept (b), valve-controlled fixed-displacement hydraulic motors used for actuation of high-lift devices are replaced by yoke-controlled variable-displacement hydraulic motors. This design, introduced on the Airbus A380 *Power Control Units* (PCUs), provides a wide range of operation points (high speed at low force as well as high force at low speed) at high efficiency. However, the actuators remain hydraulically supplied at a constant pressure. The concept (c) applies to actuators involving hydraulic cylinders, for example, for primary flight controls. The hydraulic power is produced at constant pressure from the electrical network, as closely as possible to the actuation need, also called "zonal hydraulics". Direct drive of the load, low reflected inertia, and ease of power management functions can therefore be conserved. Although this solution was tested in the late 1970s as a backup power supply to increase the survivability of combat aircraft, it has entered into service with the Airbus A380 as *Local Electrohydraulic Generation System* (LEGHS). However, the actuators themselves remain hydraulically supplied and valve-controlled.

A significant step forward was achieved with the introduction of *Electro Hydrostatic Actuators* (EHAs), depicted in the middle of Figure 5.2, and called concept (d) in Table 5.3. In this design, the *Motor Control Electronics* (MCE, also called MPD for *Motor Power Drive* or *Power Drive Electronics* (PDE)) meters the power transferred from the aircraft electric network to the electric motor. The motor drives a positive displacement pump

TABLE 5.2 Natural advantages and drawbacks of hydraulic and electric technologies for actuation.

Technology for Power	Hydraulic	Electric
Principle of energy transfer	Power by pipe (requires mass transfer)	Power by wire (no mass transfer)
Principle of control of power	Throttling → speed on demand and power consumed for full force	Power on demand
Power density (actuator level)	Excellent, up to 10 kW/kg	Low, due to the presence of reducer, mechanical power management devices, and power electronics
Power transformation	Cylinders → Low speed, high force	Motors → High speed, low torque
Power transmission to load	Direct (lever arm effect)	Through mechanical speed reducer
Power density (power network level)	Low, specific network for actuation, mass transfer requires large pipes	High, networks can be shared
Integration of power network	Difficult: cabin forbidden, bend radius lower bounded for pipes, etc.	Medium, care to isolation, electromagnetic immunity, and partial discharges for high voltage
Reconfiguration/power management	Poor, strong segregation imposed for reliability due to mass transfer	Potentially high by switches and power drive electronics
Energy losses vs. load demand	Permanent leakage (servovalves) + mainly speed dependent (throttling)	Force dependent mainly
Sources of energy losses	Internal leakage, pressure losses, friction	Switching and commutation in power electronics Copper and iron losses in motors; friction in mechanical reducers, joints and bearings
Heat evacuation	Easy, by the fluid returning to reservoir	Hard, convective/radiative with ambient
Stiffness (open loop motion under external load)	Dominated by fluid compressibility	Dominated by backlash and mechanical compliance
Sizing	Easy, at rated points + fatigue	Difficult, from mission profile + fatigue
Damping at end-stop	Easy through snubbing	By control or specific mech device
Passivation in case of jamming	Easy through jack chambers bypass	Hard at electric level due to friction in backdriving mode and reflected motor inertia
Protection against overload	Easy, through pressure relief valves	Difficult, limited performance if performed at current level, may require torque limiter
Actuator inertia reflected at load level	Excellent, some % or 10% of load inertia (low mass of piston rod—direct drive)	Bad, some 100% to 1000% of load inertia due to high reduction ratio
Control design	By specialists, non-linear behavior and temperature sensitive	Easy, well-known, linear behavior, mainly intrinsic inner control loops (current, speed)
Cost	High, high accuracy machining but well-established processes, potential interest of additive manufacturing	High but decreasing with production rate augmentation
Maintenance	High, due to mass transfer (reserve, purging, filtering, bleeding, charging, temperature management)	Low, easy testability through built-in tests
Potential of improvement	Moderate	High, deployment just started
Environment friendliness	Aggressive, due to hydraulic fluid	Respectful
Maturity	High (more than 60 years development)	Increasing, through service hours accumulation

Unbalanced area cylinders or hydraulic motors not considered.

that is connected to the cylinder to make a hydrostatic loop. Although this design involves many power transformations, it takes the best of HSA in terms of transmission to load and ease of power management for secondary functions. The total removal of hydraulic fluid for power transmission requires the *Electro-Mechanical Actuators* (EMA), shown at the bottom of Figure 5.2 and as concept (e) in Table 5.3. In this design, the motor shaft

TABLE 5.3 Evolution towards hydraulic-less PbW actuation

	Concept	Power-by-Wire	Power on Demand	Hydraulic-Less
a)	Hydraulic servo actuator, constant-pressure hydraulic supply			
b)	Displacement control, constant-pressure hydraulic supply		X	
c)	Local electrohydraulic power generation, AC or DC electric supply	X		
d)	Electro Hydrostatic Actuator, AC or DC electric supply	X	X	
e)	Electro Mechanical Actuator, AC or DC electric supply	X	X	X

FIGURE 5.2 Main types of linear actuators for actuation in aerospace [5.2].

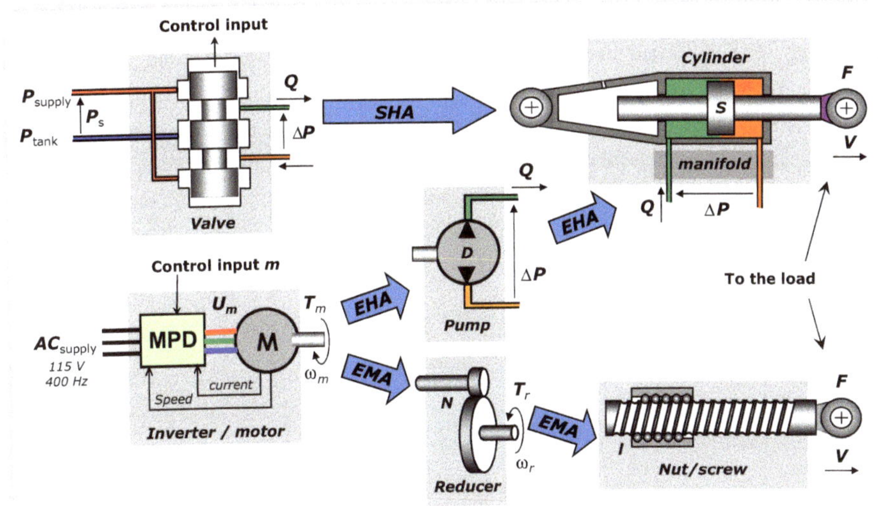

drives the load directly, or through a mechanical transmission that, for example, involves a gear reducer and a power screw. All drawbacks of hydraulic technology are removed, but also all its specific advantages regarding power management, reflected inertia, and cooling.

5.2.2 PbW Actuator Interface to Electric Power Networks

Most of the motor and control electronics technologies used for closed-loop controlled actuation with high power-to-mass ratio operate from a DC power source. As electric power is generally distributed in the aircraft through a three-phase AC network, 115 VAC, and more recently 230 VAC, PbW actuators require additional power interfaces to connect to the aircraft power sources (Figure 5.3). High power and constant voltage DC source is created locally after passive or active (typically called PFC for *Power Factor Correction*) rectification, making the DC link. As the rectification function is irreversible, the energy regenerated when the actuated load becomes aiding must be dissipated locally. This is achieved by connecting a brake resistance to a "chopper" power electronics circuit. The chopper is controlled to limit the increase of the DC link voltage that results from

FIGURE 5.3 Example of architecture of a PbW actuator regarding electric power [5.3].

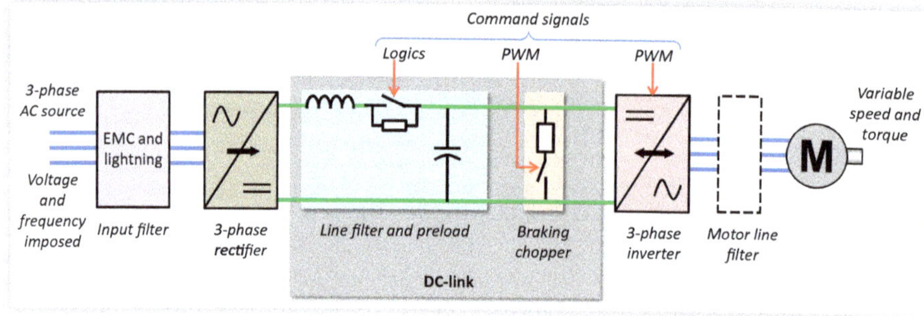

the charge of the DC link capacitor by the regenerative current. A preload resistance is also required to limit the inrush current charging the line capacitor when the actuator supply is enabled. Filters are installed to ensure electromagnetic compatibility, and to improve the power factor, the power quality, and network stability. In practice, all these electric and electronic elements required between electric supply and motor still result in a high penalty on actuator weight (typically 1-2 kg/kW for power electronics) and geometrical envelope, in comparison with conventional actuators.

5.2.3 Power Control in PbW Actuators

In brushed motors, the active coils of the rotor windings are switched mechanically by the commutator versus rotor/stator angular position. This introduces severe drawbacks that make brushed motors quasiabandoned:

- Sparking between brushes and commutator generates voltage drop and wear, and can result in ignition of combustible materials or explosions.
- Rotor inertia is high as it consists of copper windings and laminations.
- The motor housing acts as a barrier for heat rejection of electric and magnetic losses between the rotor and ambient air.

Conversely, the windings of permanent magnet, brushless motors are located on the motor housing while magnets are held by the rotor. Rotor inertia is reduced and heat rejection to ambient air is much more efficient. However, these motors require electronic switching of the active coils versus rotor/housing angular position. This is easily achieved by the power metering function performed by the power transistors, which can be driven according to this additional input. Most of in-service applications use 3-phase brushless motors with star connection of the phases.

Compared with conventional solutions, PbW takes benefit of the absence of mass transfer for power transmission. Power transistors (e.g., MOSFET or IGBT) are used as solid-state switches that can operate at high frequency (typically in the 10-20 kHz range with IGBT technology for aerospace actuation). Power transfer between the DC source and the motor windings is metered by controlling the transistors conduction and blocking times, generally through *Pulse Width Modulation* (PWM). It consists of acting on the duty cycle, the ratio between the conduction time and the total switching period, according to the actuator and motor control strategies (**Figure 5.4**). In PbW actuators

FIGURE 5.4 Example of architecture of a PbW actuator regarding control and power metering (saturations not displayed) [5.3].

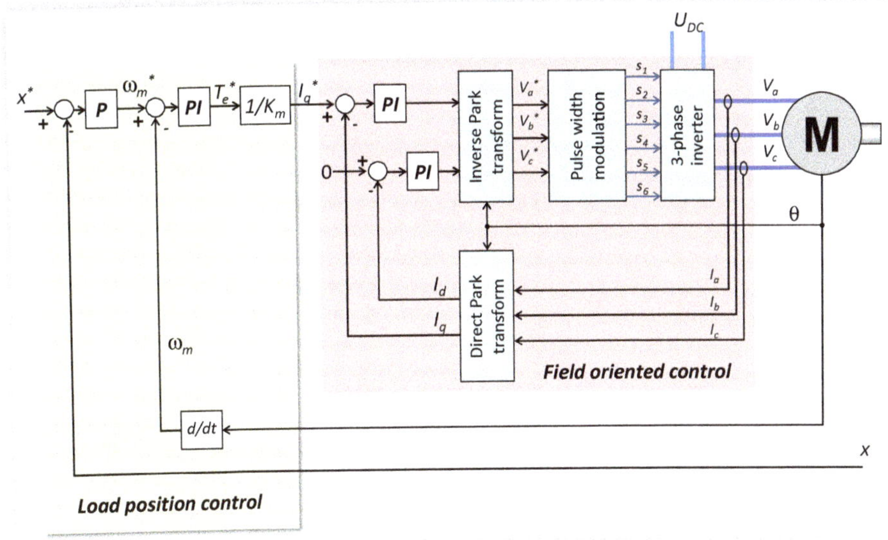

power is drawn "*on demand*" from the DC source: there is no functional power loss for power metering. Thus, the power electronics can be seen as a modulated power converter between the constant DC supply and the motor windings which is controlled by the duty cycle m. There are only very little energy losses due to conduction and switching at power electronics level. This is a real advantage in terms of efficiency over the conventional solution where power is drawn for "max force."

The power transistors operate in pairs to make a commutation leg that is dedicated to each phase of the motor. In a given leg, one and only one transistor is conducting at a given time. The motor phase current settles as a function of the average value of the pulse modulated voltage, thanks to the low-pass filtering effect produced by the electric time constant of the motor windings. From a functional point of view, the three legs of the MCE are controlled in order to generate a three-phase sine voltage (inverter function) at motor wires, in synchronization with the rotor/housing angular position. This position can be measured by discrete Hall effect sensors (*BrushLess DC* or BLDC motor having a trapezoidal back electromotive force, sine approached by a six-step command), or by a resolver or linear hall effect sensors (*Permanent Magnet Synchronous Machines* or PMSM with sine back electromotive force). Sensorless control is also possible for speed control, when the motor it not required to operate at low rpm or with high accuracy. The frequency of the three motor line voltages generated from the DC supply corresponds to the electrical frequency of the motor running: the motor speed times the number of pole pairs of the rotor, for example, 800 Hz at 12,000 rpm for a motor having 4 pole pairs. Controlling the phase lag between the sine voltages and the rotor/stator angular position offers another degree of freedom. The so-called *Field Oriented Control* (FOC), by opposition to scalar control, can, for example, maximize the transformation of current into torque, or even extend the speed range (at the expense of torque capability) through flux weakening.

The BLDC or PMSM motor associated with electronic commutation behaves globally as a DC motor. They can be modeled at functional level in a simplified way by the two following equations:

$$\text{Motor}\begin{cases} U_m = K_m \omega_m \\ T_m = K_m I \end{cases} \tag{1}$$

$$\text{Power electronics}\begin{cases} U_m = mU_{DC} \\ mI_m = I_{DC} \end{cases} \text{with } m \in [-1;+1] \tag{2}$$

with time variables: I_m current supplied to the DC motor, m modulation factor, T_m torque at motor shaft to load, U_m voltage applied to the DC motor, ω_m rotor/stator relative angular velocity and parameters: K_m equivalent DC motor constant

The motor acts as a power transformer of ratio K_m between the electric and mechanical domains. When only the functional effects are considered, the overall power transformation of the electric drive is therefore expressed by:

$$\begin{cases} \omega_m = m\, U_{DC} / K_m \\ T_m = \dfrac{1}{m} I_{DC} \end{cases} \tag{3}$$

A windings resistance and inductance, rotor inertia, and friction act as parasitic effects. The resistance produces the copper losses at motor windings while iron losses (hysteresis and eddy currents) are expressed as their equivalent friction torque. Power electronics also show imperfections not only in each leg (switching and conduction losses of diodes and transistors) but also at the level of the rectifiers and filters (e.g., leakage current and resistance of capacitors).

The position control of electric drives is well established. As shown in Figure 5.4, it typically involves a controller that comprises three nested control loops:

- The outer loop performs the position control with a typical bandwidth of several Hertz. For accuracy, it generally uses a position sensor that measures the actuator rod to body relative position x.
- The middle loop performs the speed control, for example, with a 100 Hz bandwidth that provides active damping to the position control. In brushless motors, a dedicated sensor may not be required due to the presence of the rotor position sensor that is used for electronic switching. The frequency of the motor sensor signal is directly proportional to the rotor/stator angular velocity.
- The inner loop is in charge of controlling the current (that fixes the electromagnetic torque) with a typical bandwidth of 1 kHz. At the timescale of the positioning function, it makes the motor behave as a piloted source of electromagnetic torque. The torque loop often implements the field-oriented control to maximize the transformation of current into electromagnetic torque. This usually involves two current loops for direct (I_d) and quadrature (I_q) currents that can be seen in Figure 5.4. The actual currents are calculated from the three motor phase currents and the angular position of the motor rotor relative its housing.

The required level of performance is generally obtained using conventional proportional-integral (PI) controllers in the different loops. Position control of EHAs may be more challenging as they face the combination of the motor and the hydrostatic transmission dynamics. Speed and current limitation of setpoints (ω_m^* and I_q^* respectively), implement in a simple manner overspeed and overcurrent protections (this equates to limiting the electromagnetic torque).

5.2.4 Reliability

The risk (or criticality) associated with the loss of a function is quantified by the combination of the consequence of the failure (severity), the frequency of occurrence and the inability to detect the failure. Therefore, the probability of loss of a safety critical function has to be extremely low (e.g., 1 event for every 1 billion flight hours for a given technical root cause). The first solution to reach the expected reliability level consists of using large safety margins to avoid failure during the operational life (*safe life*). However, this solution is often not sufficient and/or does not meet the weight and envelope targets. For this reason, the most common solution consists of calling for redundancy by installing multiple channels to perform the same function, for example, two channels simultaneously active (*active/active*) or 2 channels with only one channel active at a time (*active/standby*). In redundant designs, particular attention has to be paid to the response to failure of a given channel that has to be made *fail-safe* regarding the remaining healthy channel(s). Depending on the redundancy concept, the failed or inactive channel of an actuation system has to be either:

- *Fail passive* when the force or torque developed has to be annulled. This typically applies to force or torque summing redundant arrangements, for example, when two pinions engage on a single gear wheel.
- *Fail-freeze* when the channel output motion has to be forced to null. This typically applies to speed summing redundant arrangements, for example, implemented with an epicyclical gear set.
- *Fail damped* when a force opposing to motion has to be developed. For landing gears, this, for example, applies to shimmy (steering function) or free fall (extension).
- *Fail neutral* or fail locked when a given predefined position has to be reached, for example, for fuel metering or inlet guide vanes of engine controls.

Major failure conditions that may impact the power parts of a PbW actuator are: electrical component open-circuit, short circuits at motor windings, breakage or jamming in the mechanical transmission, or a non-response to the command of power semiconductors. The failure rate of PbW actuator parts lies in the range of 10^{-4} to 10^{-6} per flight hour. This makes simplex architectures, comprising a single power channel, though inconsistent with the reliability requirements of safety-critical applications. Although redundant power electronics and motors have been deeply investigated, the level of reliability is generally achieved by installing at least two independent power channels in a duplex architecture. As illustrated in **Figure 5.5**, there are several candidate topologies to implement this redundancy.

FIGURE 5.5 Main generic topologies with two redundant actuation channels [5.3].

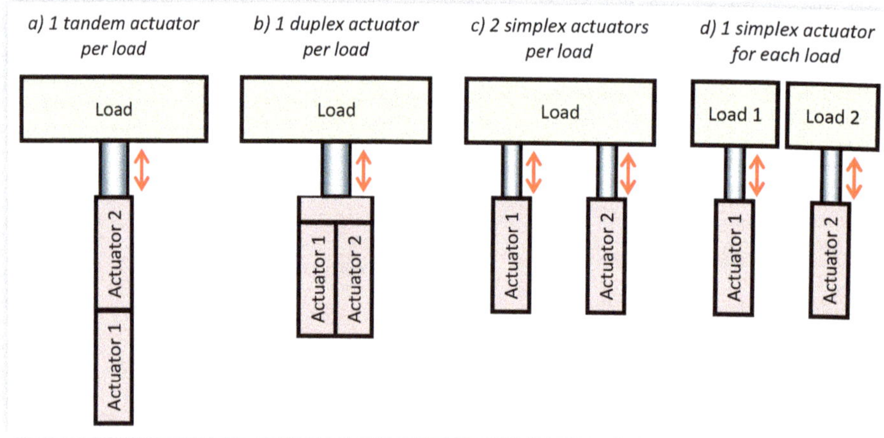

PbW offers an attractive complement or alternative to redundancy of electric elements. This comes from the calculation capacity, data storage, communication, and sensors that are already installed for the control of power. They provide an opportunity to implement *Health and Usage Monitoring* (HUM) or *Prognostic and Health Monitoring* (PHM). Usage monitoring consists of recording data that are representative on the effective use and the cumulated damage of the actuator and its elements at current time of its life. Health monitoring consists of checking the correct realization of actuators functions, detecting unpermitted variations (fault detection), determining the fault type and location (fault diagnosis or isolation), and predicting the evolution of the fault (prognostic) to estimate the *Remaining Useful Life* (RUL). A well-established approach consists of implementing a diagnostic function through Built-In Tests (BIT) that can be run either at power up or continuously to detect and isolate faults.

5.2.5 Integration and Mutualization in PbW Actuation

Many considerations drive the design choices related to the integration of the actuator: environment (e.g., thermal, electromagnetic, and vibratory), space envelope, maintenance, etc. The power control electronics can be separated from the motor and located in a less exposed area, as done for electric braking on the Boeing 787. This requires long power cables for connection to the motor(s) that introduce parasitic effects. Line inductance and capacitance, and electromagnetic emission due to high frequency chopping of high currents, therefore, require additional filters and shielding. Locating the power control electronics in the vicinity or even within the actuator removes these drawbacks but makes the environment constraints more severe. This is well illustrated by the case of aileron or spoiler actuators that are particularly exposed during landing mode to *High Intensity Radiated Fields* (HIRF) produced by airport radars and traffic. This type of integration may also require dedicated air intakes like scoops to improve the actuator cooling through

FIGURE 5.6 Integration options for redundant PbW actuators [5.3].

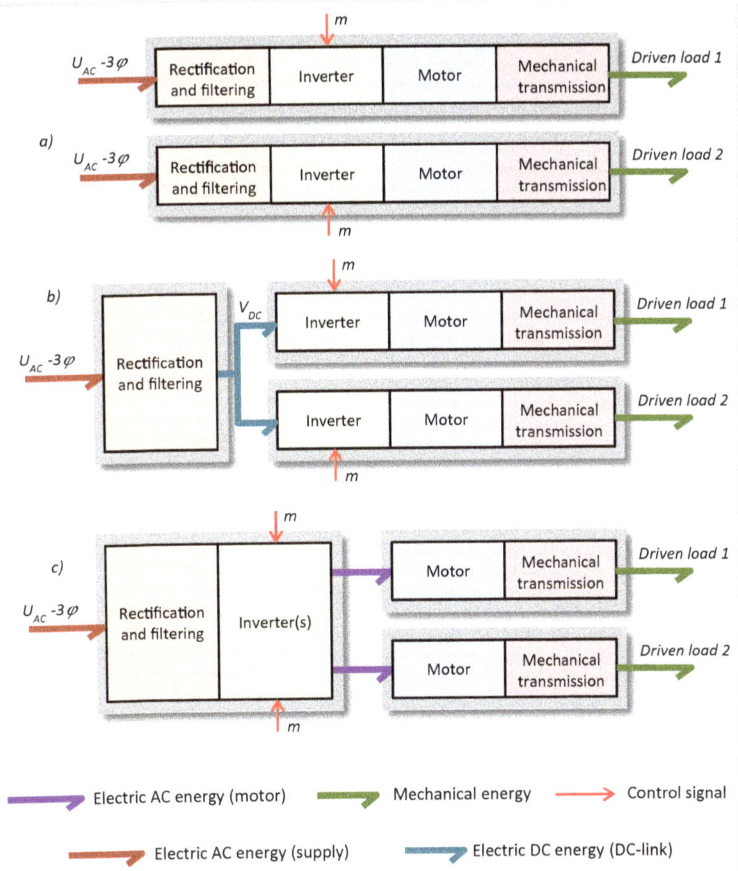

forced convection, as can been seen on the lower wing of an Airbus A350 at the aileron level. Separating the power control electronics from the motor can be driven by geometric constraints, for example, for recent aircraft having thin wing trailing edges.

Figure 5.6 summarizes the candidate architectures for actuator integration. When possible, mutualizing the electronics can bring a significant benefit in terms of weight and volume. In the first solution (a), each actuator is provided with its own rectification and filtering, power control electronics, motor, and power transformation to load. Design (b) mutualizes the elaboration of the DC link from the power source between actuators. It potentially enables regeneration under aiding load for downsizing the power source, as well as rectification and filtering components. Further integration (c) also merges the inverters in the same housing with the rectification and filtering functions. EMI protections and cooling can therefore be mutualized too, even some precautions at the inverter level should be carefully mitigated (e.g., specific current limitation). Power control electronics can potentially be shared between motors if they are used in sequence, for example, for actuating doors, locks, and leg in *Landing Gear Extension and Retraction Systems* (LGERS).

Significant experience has been acquired to make the integration of electronics robust and reliable. Packaging must ensure mechanical, electrical, electromagnetic, and thermal segregation between signal boards, power boards, and external environment. Temperature of power electronics should be monitored continuously in addition to that of motor windings. Cables carrying high power have to be replaced by bus bars, the power path has to be made a short as possible, and for high voltage lines, it shall be mitigated by design (e.g., sufficient distance between conductors) to avoid partial discharges.

5.3 Local Generation of Hydraulic Power for Actuation

The A380 is equipped with three LEHGS (**Figure 5.7**) that provide locally a backup source of hydraulic power from the electric power network, in order to supply the landing gear steering and braking functions. It combines:

- A hydraulic reservoir (9 L) equipped with a level indicator (LVDT type).
- An electric motor pump (9 kW at electrical input) that combines a BLDC motor and a fixed displacement axial piston pump (1 cm^3/rev) driven at variable speed (up to 15,000 rev/min).
- A manifold equipped to perform the fluid conditioning and power management functions.
- An *Electronic Control Unit* (ECU), not presented in the figure, that is supplied by the 3-phase 115 VAC network. It controls the motor to keep the high-pressure accumulator of the concerned centralized network charged. It is also in charge of monitoring and data exchange with the braking and steering computers.

FIGURE 5.7 One of the three LEHGS on the Airbus A380, adapted from an original image of [5.4].

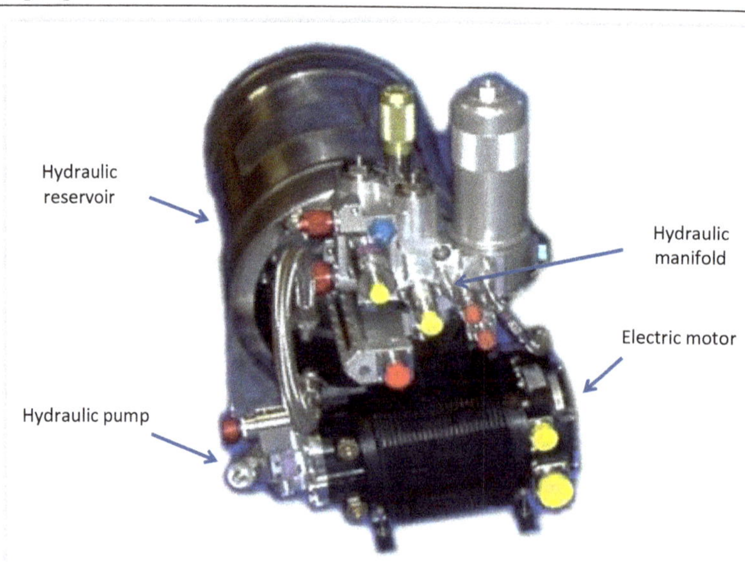

5.4 Electrohydrostatic Actuators

5.4.1 Functional and Architectural View

As displayed in **Figure 5.8**, electrohydrostatic actuators transmit power from the motor shaft to the driven load through a hydrostatic loop and a hydraulic cylinder (or motor, sometimes). A hydraulic manifold is inserted over the loop to integrate the power management functions, taking benefit of their easy realization in the hydraulic domain. The hydrostatic power transmission from motor shaft to the driven load acts functionally as a perfect power transformation. In practice, the power transformations (rotational mechanic to hydraulic, then hydraulic to translational mechanic) are affected by imperfections causing energy losses (pressure losses, internal leakage, and friction) and dynamic effects (hydraulic capacitance of the fluid volumes, inertia of the pump shaft, and mass of the cylinder rod).

In aerospace, the first EHAs were put in service for gun turrets in the late 1930s, and then applied to flight controls in the 1950s. As reliable and light power electronics were not available, the *Variable Displacement* (VD) concept was employed. In this design, an AC motor is directly connected to the aircraft electrical network. The pump is driven at constant speed and the power delivered to the load is controlled by action on the pump displacement.

Although this concept had been evaluated again in the 1980s and 1990s, it was put aside in favor of the *Fixed Displacement* (FD) concept that showed better efficiency and reliability. An EHA-FD uses a fixed-displacement pump that is driven at variable speed by a high-performance electric drive. The power transformation between the DC voltage supply and the driven load is metered by action on the modulation ratio m, as mentioned in Section 5.4.2.3. Figure 5.8 illustrates the power and signal architectures of the EHA-FD installed on the Airbus A380. The motor command electronics implements the speed and current loops to drive the motor power electronics (MPE). The actuator position

FIGURE 5.8 Power and signal architectures of a fixed-displacement EHA [5.3].

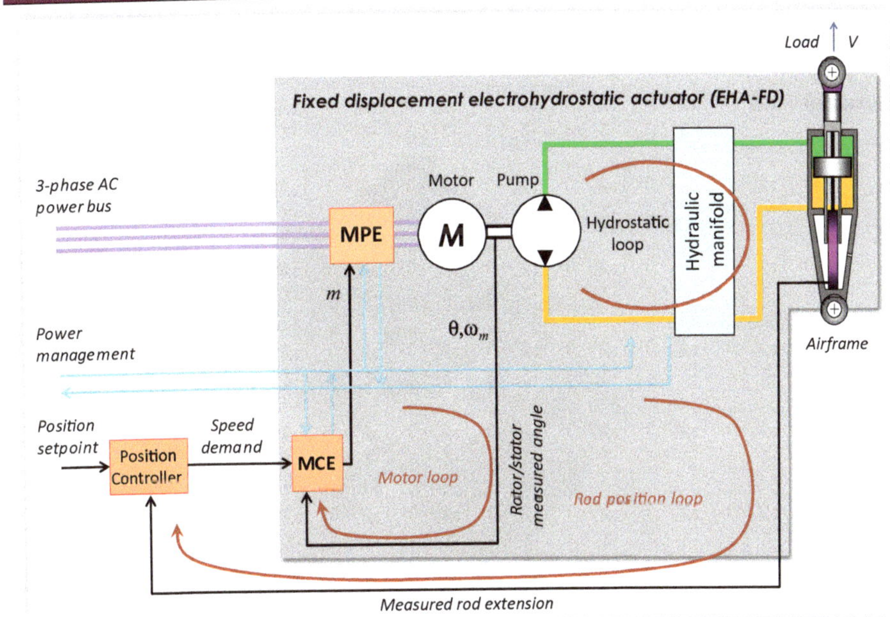

loop is performed by the *Flight Control Computer* (FCC). They use the rod extension position signal to calculate the speed command to the motor command electronics. Power is managed at both *Motor Control Electronics* (MCE), *Motor Power Electronics* (MPE), and hydraulic manifold levels. The MPE is also in charge of producing locally the DC link from the AC power source (see Figure 5.3).

5.4.2 In-Service EHAs

After 20 years of maturation, EHA-FD have been put into commercial service. This was initiated by the Airbus A380 double deck commercial aircraft, *Entry Into Service* (EIS) 2007 that used the EHA-FD concept for backup actuation. In combination with LEHGS for landing gear and *Electrical Thrust Reverser Actuation System* (ETRAS) and EHA/*Electrical-Backup Hydraulic Actuator* (EBHA), the Airbus A380 made a world's first by enabling flight with all electrically signaled and powered actuation. This required a step change in secondary power networks which involved two hydraulic and two electric channels (2H-2E architecture) instead of the well-established three hydraulic channels. Among its 43 flight control actuators, the Airbus A380 uses 8 simplex EHAs and 8 EBHAs that operate at a maximum pressure of 350 bar. The MCEs are commanded by the FCC like other HSA actuators: the command signal is representative of the speed demand while the position control loop is implemented in the FCCs. EBHAs offer hybrid redundancy by integrating in a single physical unit the elements to operate either as an HSA in normal mode or as an EHA in backup mode (Figure 5.2). A mode selector valve is inserted upstream the cylinder to select whether it is supplied by the HSA servovalve control ports or by the EHA pump ports. EHAs and EBHA integrate the MCE in the actuator physical unit. Similar architectures and designs were later employed by the European manufacturer on its A400 military transporter (Figure 5.9) (EIS 2013, EHA/EBHA) and its A350 extra-wide-body commercial aircraft (EIS 2015). The confidence acquired with electronics embedded in the EHA/EBHA actuators on the A380 enabled the designers to also implement the position control loop at the actuator level.

FIGURE 5.9 Airbus A400M EHA [5.3].

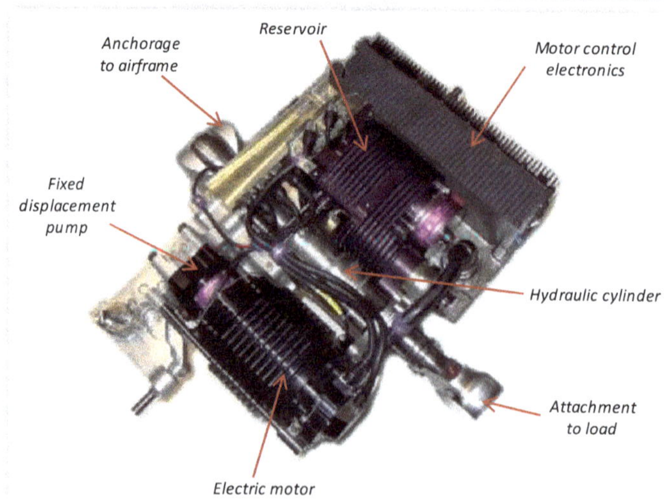

Since 2007, EHAs and EBHAs have been employed by other aircraft manufacturers. Lockheed Martin has totally removed centralized hydraulic power networks for primary flight controls off its F-35 Joint Strike Fighter. The *Electro Hydrostatic Actuation System* involves 6 EHAs that operate at a maximum pressure of 350 bar: a simplex EHA per rudder and a dual tandem redundant EHA per flaperon or elevator (Figure 5.10). The latter actuators integrate two power channels to convert electric power into hydraulic power that supply a single cylinder in a flow-summing configuration. The Gulfstream G650 private jet (EIS 2012) involves 7 EBHA operating at 210 bar maximum for driving elevators, rudder, ailerons, and outboard spoilers. The upcoming Embraer KC390 will use 6 EHAs for primary flight controls operating at a maximum pressure of 280 bar, 4 of them being of dual tandem type.

FIGURE 5.10 Dual tandem EHA for the flaperon of the Lockheed F-35 Joint Strike Fighter [5.3].

An example of the hydraulic circuit of an EHA is given in Figure 5.11. The main power transmission and control function is ensured by elements ❶ to ❹. The motor drive electronics (❶) meter the power delivered to the brushless motor (❷) from the aircraft

FIGURE 5.11 Functions performed hydraulically in an EHA [5.3].

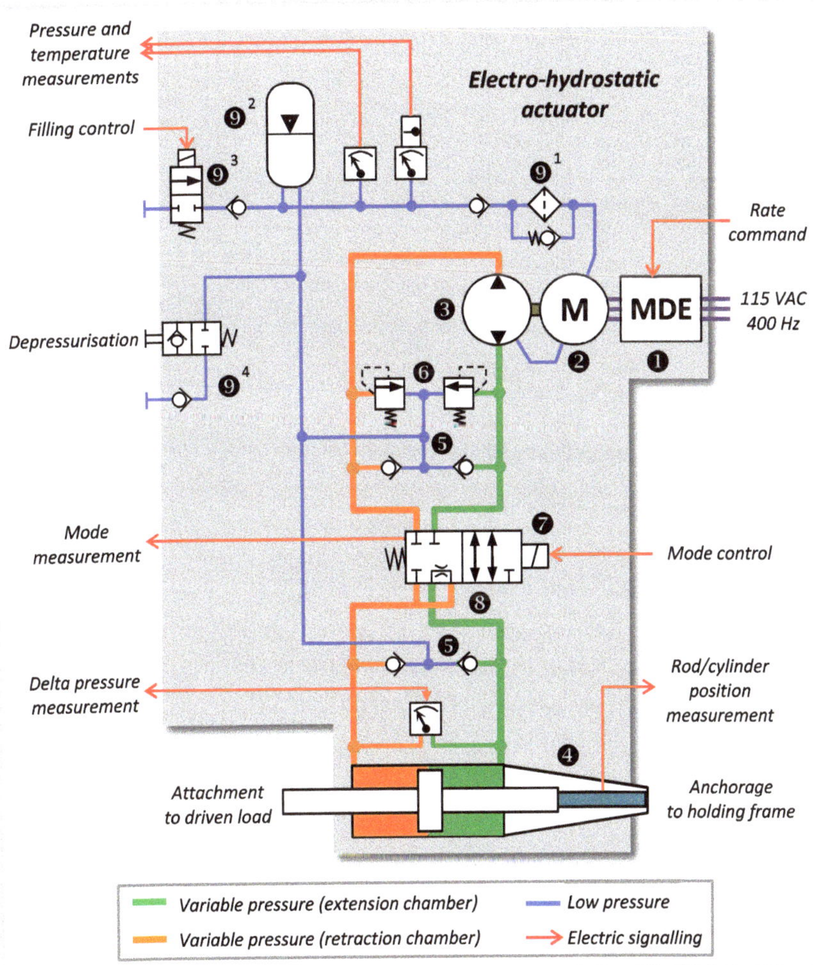

centralized AC electrical network. The motor drives the fixed-displacement pump (❸) that is connected to the cylinder (❹) to make the hydrostatic loop. The numerous other elements serve the power management and the fluid conditioning functions. Re-feeding valves (❺) protect against outgassing or cavitation. Pressure relief valves (❻) protect against excessive pressure (and excessive actuation forces due to the presence of the charge circuit). The operating mode, active or damping, is controlled electrically by the solenoid mode selector valve (❼). Damping in passive mode, for example, to avoid flutter, is achieved by inserting a flow restrictor (❽) in the bypass connection of the two cylinder chambers. Elements (❾) are in charge of fluid conditioning. The cleanliness of the hydraulic fluid is ensured by the filter (❾^1) that is installed on the pump drain line collecting its internal leakage. Fluid reserve and pressure charging are performed by the hydropneumatic accumulator (❾^2). The solenoid valve (❾^3) enables the actuator to be refilled on demand from the central hydraulic power network. The manually operated valve (❾^4) is used to depressurize the actuation prior to maintenance. Sensors are installed for closed-loop control, power management, and monitoring. The pump internal leakage is used to establish a permanent flow that facilitates the heat rejection to environment. It typically collects the heat produced at pump and motor levels then returns to the accumulator. The hydrostatic loop is kept charged by the corresponding flow taken from the reservoir that feeds its low-pressure line through the refeeding valve. This cooling circuit homogenizes the actuator temperature by taking advantage of the actuator surfaces in contact with ambient air. Fins and black color are used to increase the convective and radiative heat fluxes.

5.4.3 Main Issues for Extensive Use of EHAs

EHA-FD have been developed through numerous research projects since the mid-1980s with different objectives: in Europe, to remove heavy centralized hydraulic power networks in long-range commercial aircraft; in the United States, for decreasing vulnerability while facilitating maintenance in the theater of operations for combat aircrafts. A lot of lessons were learned from these programs:

- Thermal equilibrium is a major issue because the heat generated by actuator energy losses has to be dissipated locally. This highlighted the importance of the actuator integration, in particular when it is confined between wing spars and ribs. The impact of the mission profile was also pointed out, as conduction losses are directly proportional to the *Root Mean Square* (RMS) force delivered to the load.
- The hydro-mechanical stiffness of the hydrostatic loop has to be kept sufficiently high to reject load disturbances with sufficient damping. Therefore, the hydraulic loop has to be constantly charged at some 10 bar to avoid outgassing or cavitation.
- External leakage at cylinder rod seal cannot be strictly avoided for long life applications like commercial aircraft. Compensating this leakage for the whole aircraft service life by adequate sizing of the charge accumulator increases drastically its size and weight.
- EHAs and HSAs have different static and dynamic characteristics that are not totally wiped off by the control loops, in particular for rejection of load disturbances. When an EHA and a HSA simultaneously drive the same load (active/active configuration), they fight against one another to impose their own position. As this generates

significant internal loads and power consumption, the implementation in controllers of force equalization strategies significantly enhances the intrinsic performance without degradation of channel segregation and independence.

In practice, an EHA pump has to operate in very demanding conditions that reduce its service life. Firstly, it has to work in the 4 power quadrants, for example, between ±350 bar delta pressure and ±10,000 rpm. Secondly, it often has to hold high forces at almost null speed for long periods, for example, for aileron actuation where the mean force required is 20%-30% of the stall force (see Figure 5.1). In these conditions, the flow Q delivered by the pump to the cylinder is negligible while the pressure difference ΔP to be generated is high. Consequently the pump rotates under high pressure at extremely low speed (e.g., 1 revolution per second) to compensate its internal leakage at its ports. Quality of lubrication is then degraded. Thirdly, speed and pressure can change rapidly, for example, varying from 100 to 10,000 and again 100 rpm in 0.3 s during a step response. This induces strong inertial effects, and partly brakes lubricating films. This is one main reason why EHAs cannot yet meet the service life requirements for use in front line (normal mode) of primary flight controls of commercial aircraft.

On many aircraft, the EHAs are used as backup and act in force-summing configuration with a conventional HSA. Although EHAs operate most of the time in standby mode, their design raises particular issues. Firstly, the EHA temperature can fall to extremely low values in standby mode. In these conditions, the viscosity of the hydraulic fluid increases drastically, generating huge shear forces in the rotor/stator gap at motor level and increasing the cavitation risk at pump level. These effects strongly impact the ability of the EHA to rapidly take over the load positioning function when it is switched from passive to active mode in case of runaway of the normal HSA actuator. Secondly, the dynamic seals at rod level are continuously working as the EHA is backdriven by the normal HSA. Thirdly, the embedded MCE undergoes the pressure and temperature cycles at each flight whenever the EHA is active or not.

5.5 Electromechanical Actuators

5.5.1 Functional and Architectural Aspects

In electromechanical actuators, the power exchange between the electric drive and the load is achieved mechanically, through various means, as illustrated in Figure 5.12:

- Direct connection of the motor shaft to the load. As already mentioned, loads generally require high forces and low speed while the weight of electric drives increases drastically vs. their rated output torque. For this reason, an actuator driving the load directly from the motor shaft is far out of weight and envelope allocations.
- Use of rotary to linear transformation as a speed reducer. As far as linear motion is required for actuation, nut-screw systems offer interesting solutions to combine rotational/translational transformation and speed reduction (e.g., 1570 rpm for 0.25 m/s with 1 mm lead). Although they use a reducer effect, linear EMAs involving only a nut-screw system are called *direct drive (linear) EMAs*.

FIGURE 5.12 Generic types of EMAs [5.3].

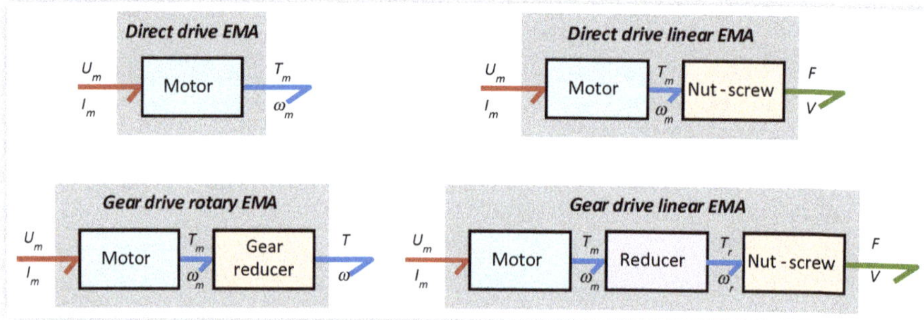

- Use of intermediate speed reducer for linear EMAs. Unfortunately, the speed reduction offered by nut-screws is generally not sufficient to meet the actuator's mass target. An additional rotary to rotary reducer is then inserted between the motor shaft and the rotation element of the nut-screw system, leading to the so-called *geared (linear) EMA*.
- Use of speed reducer for rotary EMA. Although most of the loads to be actuated have a rotational motion, they have been driven by linear actuators for more than 70 years to take advantage of hydraulic cylinders (see Table 5.2). Except the interest of nut-screws acting as speed reducers, there are less and less reasons to keep linear PbW actuators, as electrical drives produce naturally rotational motion. Switching to *geared rotary EMAs* however requires new approaches for design and integration of the actuator in the aircraft, in particular considering geometrical envelope, load path, and structural compliance.

As already mentioned, nut-screw systems are widely used to make linear EMAs. Indeed, they permit EMAs to be integrated between the airframe and the driven load as HSAs do. ACME nut-screws suffer from poor efficiency and wear because they involve sliding contacts between nut and screw threads. Conversely, ball-screws, or roller screws, insert rolling elements between nut and screw, which is of huge benefit to efficiency and service life. Roller screws are less established than ball-screws. They are more difficult to manufacture but they have the benefits of an increased number of rolling contacts providing greater load capacity. There are less manufacturing issues to offer lower leads than for ball-screws (**Figure 5.13**).

FIGURE 5.13 Ball-screw (left, Rollvis make) and satellite roller-screw (right, Umbra make).

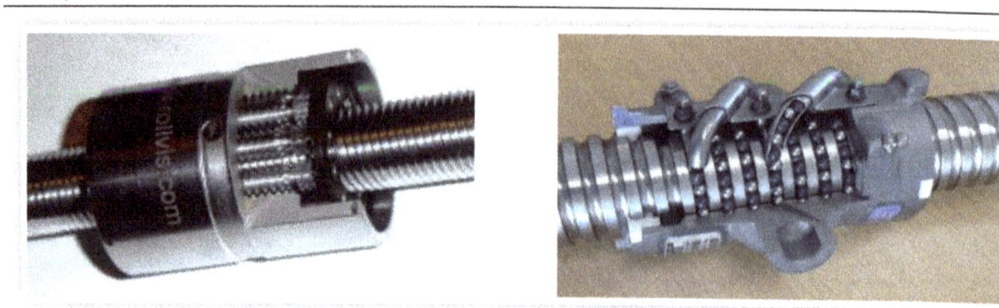

Courtesy of J.C. Mare

FIGURE 5.14 Top: Integration of a nut-screw in an EMA. Bottom: Integration of motor and mechanical transmission [5.3].

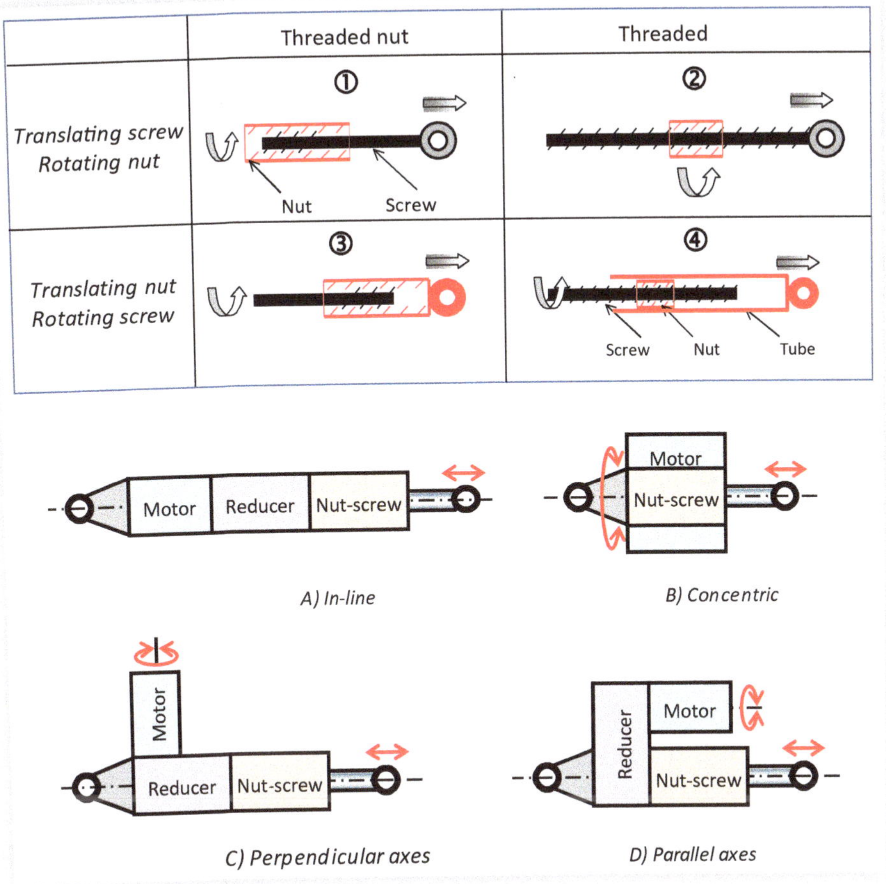

As shown at the top of Figure 5.14, four generic variants can be identified for integrating a nut-screw in a linear EMA. Designs 1, 3, and 4 facilitate the sealing between the actuator housing and the rod, but Designs 1 and 3 involve inverted screws that are more difficult to manufacture. Design 4 makes the actuator longer. Several options also exist for the integration of the electric motor and the mechanical power transmission elements, seen at the bottom of Figure 5.14. The concentric design is particularly interesting for direct-drive EMAs. The in-line design is well suited for the use of epicyclical reducers while they make the actuator long and slender. The parallel axis design is the most common solution for geared EMAs. They can use spur gears that are easy to integrate. They require less balancing forces than right-angle reducers, for example, bevel or worm gear sets that are involved in the perpendicular axes design. The option to be selected for geared linear EMAs depends highly on the allocated space envelope.

5.5.2 In-Service EMAs

Electromechanical drives have been put into service first when the jamming of the mechanical transmission was not critical. This typically concerned the secondary flight controls, whose mechanical power transmission is similar to that experienced for years with hydraulic motors: trimmable horizontal stabilizer actuators of the Airbus A350 or Boeing 787, one channel of the high lift power control units on Airbus A380 or Boeing 787. EMAs have been also introduced progressively in pure replacement of hydraulically powered jacks that were directly attached to the driven load. Four spoiler EMAs (**Figure 5.15**), over a total of 14 actuators, have been installed on the Boeing 787. The linear EMA is of parallel axis type with a remote ECU being in charge of 2 EMAs. Electrically powered brakes are already in service on the Boeing 787 and Bombardier C-Series (**Figure 5.16**). On the 787, each braked wheel is equipped with four linear *Electric Brake Actuators* (EBAs) of parallel axis type. The 4 motors' control electronics are merged in a single *Electrical Brake Actuation Controller* (EBAC) unit. Electromechanical actuation has also been employed with success on the Airbus A380 engines in the form of an *ETRAS* (**Figure 5.17**). Each engine reverser involves a single electric drive comprising an *Electrical Thrust Reverser Actuator Controller*

FIGURE 5.15 EMAs in service for primary flight controls [5.3].

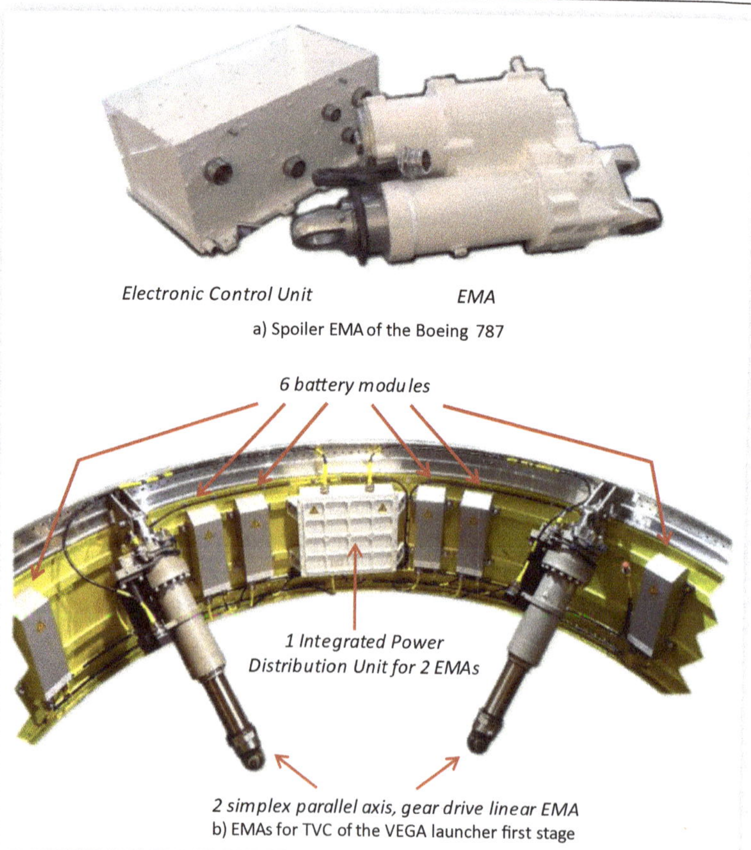

FIGURE 5.16 Example of in-service electric brakes.

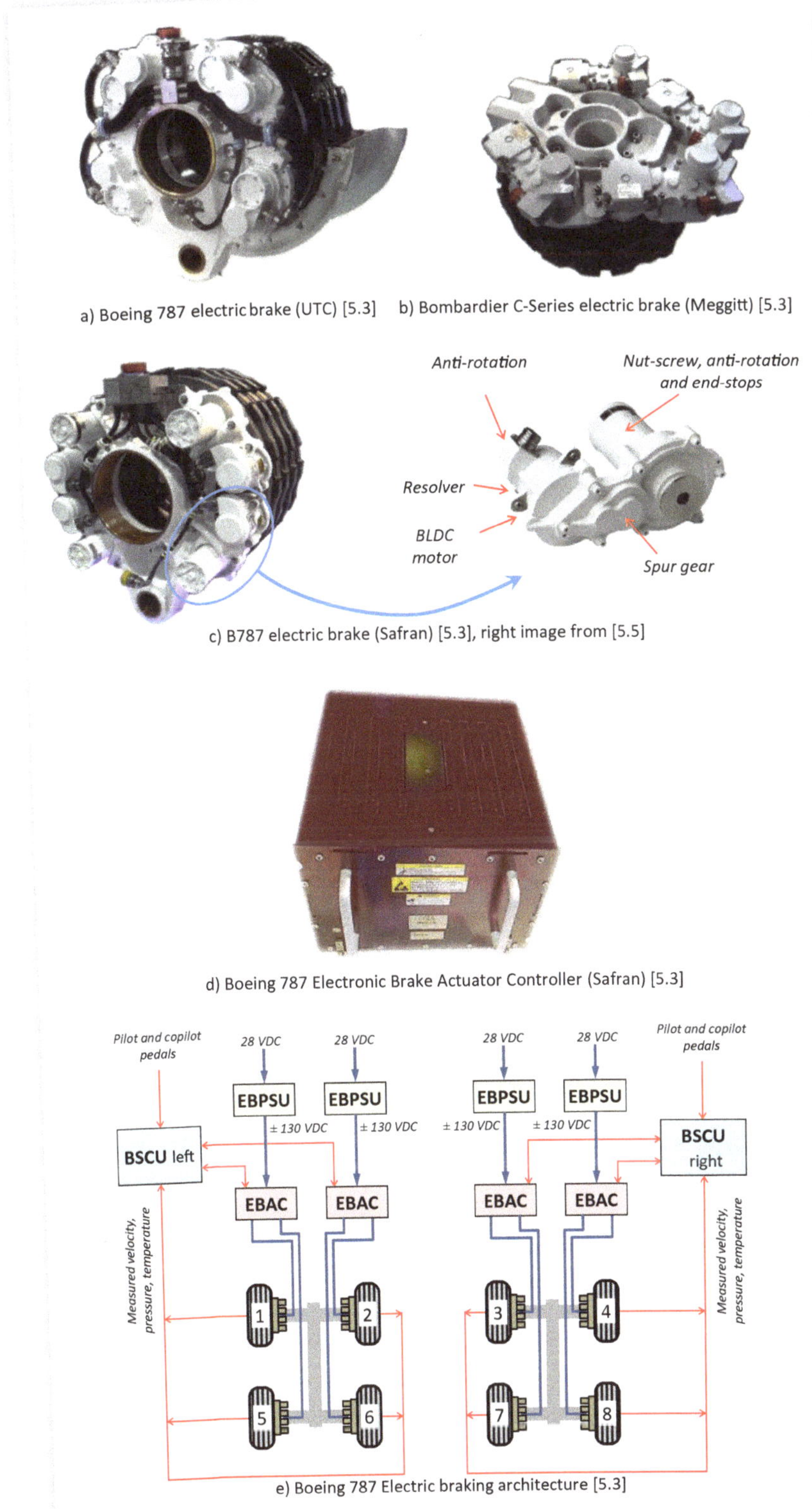

a) Boeing 787 electric brake (UTC) [5.3]

b) Bombardier C-Series electric brake (Meggitt) [5.3]

c) B787 electric brake (Safran) [5.3], right image from [5.5]

d) Boeing 787 Electronic Brake Actuator Controller (Safran) [5.3]

e) Boeing 787 Electric braking architecture [5.3]

FIGURE 5.17 ETRAS on Airbus A380 (per equipped engine) [5.3].

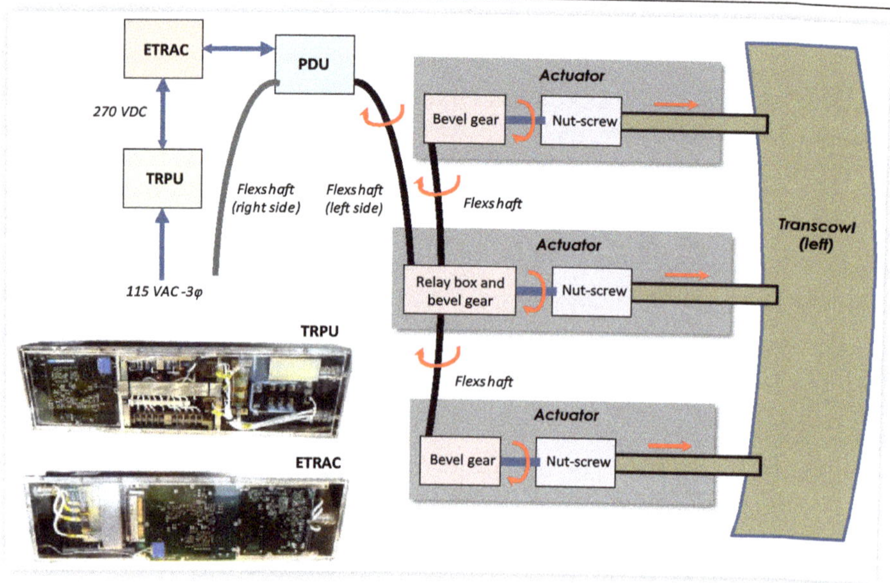

(ETRAC) for control loops, a *Thrust Reverser Power Unit* (TRPU) for control of electric power, and a *Power Drive Unit* (PDU) for motoring. The mechanical power is then transmitted to linear ball-screw jacks that actuate the two transcowls through flex-shafts and bevel gears in a torque sharing arrangement. With the LEAP engines of the COMAC, ETRAS will be used for the first time on a short-range aircraft. The ETRAS will drive in aft direction an innovative one-piece O-duct cascade reverser (Figure 5.18). EMAs have also been put into service for *Thrust Vector Control* (TVC) of space launchers, for example, M51, Atlas Centaur, Delta IV, and VEGA (Figure 5.15 bottom), which operate for a single very short mission (a matter of minutes). Many past and ongoing research programs focus on the use of EMAs for primary flight controls and landing gears (extension/retraction or steering). This kind of applications is still challenging due to the high criticality of a jamming event in the mechanical power transmission elements.

FIGURE 5.18 ETRAS with one-piece O-duct on COMAC 919.

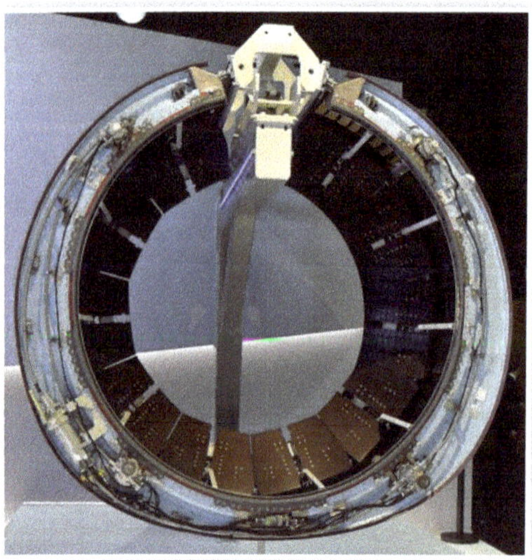

Source: Mediatheque SAFRAN

5.5.3 Imperfections of Technological Realization

Several parasitic effects alter the transformation and transmission of mechanical power between the electric drive and the load. Friction, compliance, backlash, and inertia may significantly impact performance, for example, in terms of force capability, energy consumption, stability, and accuracy.

Friction comes not only from power transformers (gears, nut-screws) but also from load balancing at bearings and joints, and from seals. For example, it is not uncommon that the axial thrust bearing holding in translation the rotating part of a nut-screw contributes to several tens of percent of the EMA overall friction losses. Friction is a complex phenomenon that is sensitive to magnitude and direction of relative speed between facing solids, normal force at contact, quadrant of operation (opposite or aiding load), and even rest time. Temperature plays also a significant role.

Compliance comes from the deformation of solids under loads. Mechanical compliance of EMAs is generally much less than hydro-mechanical compliance of HSAs and EHAs, except in the vicinity of null transmitted force due to backlash. Backlash helps to reduce friction losses and avoids effects of differential dilation but therefore decreases the effective mechanical stiffness of EMAs. It affects accuracy (no effect of drive on load) and stability (e.g., micro-vibrations for motor speed loop, flutter, or shimmy for load position loop). Preloading can be used to compensate the negative effects of backlash at the expense of additional friction. Accumulated service may increase backlash, reduce preload, or even turn it into backlash.

Inertia of mechanical power transmission elements play a much lessor role than the motor rotor inertia itself. It is however worth checking its effective contribution to the overall inertial effect of the EMA.

5.5.4 Maturation of EMAs

As already mentioned, an actuator has to perform many "other" functions than the pure power metering, transmission, and transformation between the electric power supply and the driven load. This firstly concerns the protection functions, like soft end-stop or limitation of overload. It also deals with power management, depending on the operating mode and redundancy concept that may, for instance, require braking, declutching, or damping.

In HSAs or EHAs, these functions are realized in the hydraulic domain in a simple, light, and compact way. One attractive option for EMAs consists of performing them in the electronic, electrical, or electromagnetic domains: for example, damping through iron losses at motor, protecting against overload by limiting the motor current at MCE, soft end-stop by commanding a speed reduction when the end of stroke is approaching, etc. Unfortunately, several parasitic effects often make these options less attractive in practice in meeting the expected level of performance:

- Friction in mechanical transmission, bearings, and joints, in particular due to the reduced mechanical efficiency in backdriving mode
- Inertial effect of moving parts, in particular the rotor inertia
- Compliance or backlash in the mechanical transmission, bearings, and joints

A first approach may consist of measuring the force transmitted to the load (or even the reaction force to hold the actuator housing) for signaling the MCE. It is therefore possible to modify the control loops to actively produce the damping effect or the protection against overloads. Unfortunately, this option is often not satisfactory in terms of reliability or performance. In these cases, the functions have to be implemented mechanically, and located optimally, to get the best compromise between weight and performance. The shock on stops is absorbed internally within the EMA without loading the rolling elements, for example, using dog stops or friction springs. Protection against overloads is achieved by a torque limiter. Declutching is performed upstream in the mechanical power path, as far as backdriving is possible. **Figure 5.19** illustrates the complexity introduced by the integration of these mechanical functions (in blocks with bolded border) in the power path. The half arrows represent power flows, while the full arrows are associated with signal flows. Dashed arrows indicate the reaction forces or torques that balance the forces or torques flowing in the functional power path.

Reaction loads have to be considered with care when using linear EMAs. There is no torque generated in linear HSAs involving hydraulic cylinders because they act as a cylinder pair joint with variable axial motion. Rotation being free between the rod and the housing, it introduces a degree of freedom that helps keeping the kinematics isostatic, even in the presence of structural deformations under transmitted load. Conversely, linear EMAs require several reaction torques to be balanced, for example, at housing for reducers, at motor stator, and at the translating part of the nut-screw. These reaction torques are balanced by the kinematic loop externally to the actuator (actuator rod ↔ driven load ↔ holding structure ↔ actuator body), and the axial degree of freedom is removed. Rod anti-rotation can be however performed internally to the actuator, for example, as shown in the AR block of Figure 5.19.

When it is not acceptable to block the load in case of actuation fault (e.g., landing gears or primary flight controls), jamming of the mechanical transmission still raises strong design issues. Despite the numerous past or ongoing research programs dedicated to actuators' resistance or tolerance to jamming, the inability to declutch the actuator (or the actuation channel) under permanent load with a testable design is still a major obstacle to the replacement of HSAs by EMAs.

FIGURE 5.19 Example of mechanical power architecture to perform power management functions [5.3].

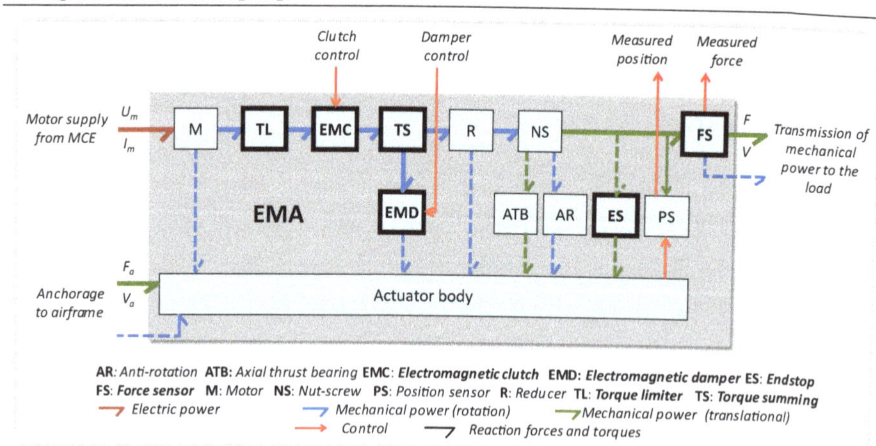

5.6 Challenges with Generalization of PbW Actuation

5.6.1 Important Considerations for Use of PbW Actuation

Some major considerations should not be forgotten when dealing with the introduction of electrically powered actuation:

a. Transmission of energy
In conventional systems, the hydraulic fluid conveys energy in hydrostatic form: the fluid is pressurized by pumps and transported by pipes to users. Therefore, energy transfer requires mass transfer, which is absent when energy is transported by electric wires. This has several major consequences. Beneficial to the conventional solution, the fluid can also be used as a heat conveyor to take away the heat produced by energy losses at the actuator level. It acts also as a lubricant, for example, for hydraulic motors or gear reducers used in secondary flight controls. The main advantage of PbW actuation comes from the power distribution and management. There is no need for the hydraulic fluid to be in sufficient quantity and condition that would require fluid reserve, filtering, purging, bleeding, charging, temperature management, maintenance procedures, etc. The power distribution is lighter than in conventional solutions where it is penalized by the weight of the hydraulic fluid (typically 500 kg on a Boeing 747) and the high-pressure hydraulic pipes (e.g., 14-25 g/m/Kw, 3000 psi network, with an energy loss of 30-90 W/m at 38°C). Moreover, electric power networks can be mutualized between actuation and other users. Yet, the hydraulic power networks quasi[1] exclusively serve the actuation need and cannot be mutualized. The absence of mass transfer in electric actuation offers a lot of possibilities for power management at the aircraft level when segregation and independence of power networks are considered. Important to note, hydraulic power networks have to be highly segregated to avoid external leakage or fluid contamination to be propagated. Last but not least, transmitting energy by moving electrons is much more environment-friendly than using aerospace hydraulic fluids that can be very chemically aggressive.

b. Force (or torque) generation
For the power need to be met, electric actuators have a much lower power density than conventional ones when power exceeds a few kW. Moreover, in electric motors, the torque comes from current, turns, and flux density. Current and turns define the quantity of winding wires (generally copper) while flux density is produced by the combination of magnets and iron laminates, all of them being

[1] On some airplanes, they also supply alternate means to produce electric power, for example, Constant Speed Motor Generator (CSMG) or Backup Power Supply (BPS).

weighty materials. This is why weight saving leads electric motors to be designed for developing low torque at high speed, and to be associated to a high-ratio mechanical reducer. As a direct consequence, this strongly increases the inertia or mass reflected by the motor rotor at load level (the rotor inertia times the squared reduction ratio). It is common that this reflected inertia is 15-50 times the load inertia itself, and 3000-10,000 times the reflected inertia that would be produced by a conventional actuator.

c. Power management at the actuator level
As mentioned earlier, several power management functions have to be performed at the actuator level. Hydraulic cylinders have an excellent mechanical efficiency and a low moving mass and can be directly connected to the driven load. Consequently, cylinder pressure is a good representation of the force that is transmitted to the driven load. This enables the designers to perform these functions in the hydraulic domain, as closely to the load as possible, with extremely low weight and volume. On their side, electromechanical actuators generally involve a high-ratio mechanical reducer, in which energy loss mainly depends on load, velocity, power quadrant, and temperature. The high inertia reflected by the motor rotor at load level, combined with the variable mechanical efficiency of the mechanical transmission, makes the motor current an inaccurate representative of the instantaneous force delivered to the load. As mentioned in Section 5.5.4, this generally requires using mechanical devices to perform passively the power management functions and/or a sensor to measure the force transmitted to the load to perform them actively, that is, through the motor control loops.

d. Service life
Hydraulic actuators involving cylinders generally do not use bearings with rolling elements (only plain or hydraulic bearings). Therefore, their service life is mainly determined by wear and leakage of dynamic seals and fatigue due to repeatedly applied loads. Electromechanical actuators employ many rolling contacts (e.g., ball or roller bearings, anti-rotation linear spline, and axial thrust bearing, ball or roller screws, gears). Their service life is highly impacted by the rolling contact fatigue. The sizing of mechanical elements therefore requires the effective actuator mission to be known in terms of combination of force and velocity. The effective mission plays also a major role for sizing the motor and the MCE with regard to thermal balance and thermal fatigue. Finally, this generates a new need at specification level because defining a generic mission for sizing is a challenging task.

e. Natural dynamics and closed-loop control
The natural dynamics of HSAs, EHAs, and EMAs are far different. In hydraulics, the fluid compliance and the inertia of the driven load combine to produce an underdamped second order having a natural frequency of several tens of Hz. In HSAs, the servovalve dynamics acts upstream the hydraulic transmission as a well-damped second order in the 50-100 Hz range. In PbW actuation, electric

motors show two first orders in series. They are characterized by the electrical time constant $\tau_e = L/R$ coming from the combination of the windings resistance R and inductance L, and by the mechanical time constant $\tau_m \cong JR / K_m^2$ that comes mainly from the rotor inertia J in combination with the windings resistance and the motor electromagnetic constant K_m. The bandwidth of the current (or torque) control loop performed at MCE level is typically around 1 kHz. Compliance and friction in the mechanical transmission of EMAs impact the actuator transfer function, for example, by introducing transmission zeros. The position control of EMAs is generally easy with the 3 current/velocity/position loops implemented in the MCE (Figure 5.4). In very specific cases where structural compliances have a significant effect, dynamic force feedback may improve the rejection of load disturbances.

5.6.2 Evolution Towards All-PbW Actuation

Assessing the interest in PbW actuation is a difficult task. It requires consideration of the global impact on end users, operators, and the environment, and needs to address the total acquisition and operation costs. Besides the easier reconfiguration and monitoring offered by electrical technology, the deletion of centralized hydraulic power networks and the easier maintenance compensate the increased weight and cost of PbW actuators.[2] Electric drives for high power and safety-critical applications are very promising because there is a great potential of performance improvement and reduction of costs and weight in several directions:

- At the system level, through advanced mutualization, modularization, and standardization
- At the motor control electronics level, by improving cooling (e.g., through heat pipes and cool plates) and power electronics (e.g., with extensive use of SiC semiconductors), and by maturation of matrix converters
- At the motor level by optimized combination of the design options regarding the type of motor, the number of poles and slots, the shaping of the stator slots, the magnets arrangement at rotor, and the type of winding
- At the mechanical transmission level, by maturing jam tolerant/resistant designs and magnetic gears, and progressing in tribology (e.g., solid or dry lubrication)
- At the monitoring level, by developing and improving robustness of HUM Systems

The rapid growth of the UAV market and the increasing production rate of EHAs and EMAs will undoubtedly accelerate the evolution towards all-PbW actuation.

[2] It is commonly said that EMAs are twice heavier and more expensive than HSAs, but have twice less components than EHAs.

References

[5.1] Radio Technical Commission for Aeronautics, Environmental Conditions and Test Procedures for Airborne Equipment, RTCA DO-160G, August 12, 2010.

[5.2] Maré, J.-C. and Budinger, M., "Comparative Analysis of Energy Losses in Servo-Hydraulic, Electro-Hydrostatic and Electro-Mechanical Actuators," *The 11th Scandinavian International Conference on Fluid Power, SICFP'09*, Linköping, Sweden, June 2-4, 2009.

[5.3] Maré, J.-C., *Aerospace Actuators*, Volumes 1 to 3, ISTE/WILEY Editors, 2016-2018.

[5.4] Dellac, S. and Ternisien, D., "Airbus 380 "Electro-Hydraulic Back-Up Architecture" for Braking and Steering Systems", *Proceedings of the 2nd International Conference on Recent Advances in Aerospace Actuation Systems and Components*, Toulouse, France, November 24-26, 2004, 103-108.

[5.5] Chico, P., "Electric Brake," *Proceedings of the 6th International Conference on Recent Advances in Aerospace Actuation Systems and Components*, Toulouse, France, April 2-3, 2014, 25-28.

[5.6] Biedermann, O., "Development, Qualification and Verification of the A380 Spoiler EBHA," SAE A-6 Committee, San Diego, CA, USA, October 19, 2005.

SAE- and ISO-Related Documents

Raymond, E.T. and Chenoweth, C.C., *Aircraft Flight Control Actuation System Design*, (SAE Press, 1993).

ARP5724 - Testing of Electromechanical Actuators, General Guidelines For.

ARP5754 - Electromechanical Actuators Specification Guide.

ARP5879 - Test Methodology for Electrohydrostatic Actuators.

ARP 6025 - Duty Cycle Considerations for Electrohydrostatic Actuators.

ARP6154 - Aerospace Fluid Power Electrohydrostatic Module, Design, Performance and Test Recommendations.

ISO 22072:2011, Aerospace—Electrohydrostatic Actuator (EHA)—Characteristics to be Defined in Procurement Specifications Projects.

ARP7490 - General Guidelines for Motor Control Electronics for Electrically Powered Actuation.

Acronyms

BIT - Built-In Test
BLDC - BrushLess Direct Current
BPS - Backup Power Supply
CSMG - Constant Speed Motor Generator
FCC - Flight Control Computer
EBA - Electric Brake Actuator
EBAC - Electric Brake Actuator Controller
EBHA - Electrical-Backup Hydraulic Actuator
ECU - Electronic Control Unit

EHA - Electro Hydrostatic Actuator
EMI - Electro Magnetic Interferences
EIS - Entry Into Service
EMA - Electro Mechanical Actuator
ETRAS - Electrical Thrust Reverser Actuation System
FbW - Fly-by-Wire
FC - Flight Control
FD - Fixed Displacement
FOC - Field Oriented Control
HIRF - High Intensity Radiated Fields
HSA - Hydraulic Servo Actuator
HUM - Health and Usage Monitoring
LEHGS - Local Electro Hydraulic Generation System
LG - Landing Gear
LGERS - Landing Gear Extension and Retraction System
LVDT - Linear Variable Differential Transformer
MCE - Motor Control Electronics
MPE - Motor Power Electronics
PbW - Power-by-Wire
PCU - Power Control Unit
PDU - Power Drive Unit
PFC - Primary Flight Control
PFC - Power Factor Correction
PHM - Prognostic and Health Monitoring
PWM - Pulse Width Modulation
RCCB - Remote Control Circuit Breaker
RMS - Root Mean Square
SFC - Secondary Flight Control
SiC - Silicon Carbide
THS - Trimmable Horizontal Stabilizer
THSA - Trim Horizontal Stabilizer
TRPU - Thrust Reverser Power Unit
TVC - Thrust Vector Control
VD - Variable Displacement

CHAPTER 6

Propulsion Options for the Electric Aircraft

Pascal Thalin
Chair and Member - SAE Electric Aircraft Steering Group

This chapter provides insight into various propulsion technologies and architectures foreseen for the electric aircraft. To begin with, we will see how gas turbines have been at the center stage of propulsion ever since the early days of aviation. Technology advances have pushed modern-day engine performance to extremes. Depending on the type of aircraft considered, research into the future of air transport propulsion is focusing on technologies and architectures wherein gas turbines may partially or totally give way to electrical solutions. Nevertheless, gas turbines may still find a sweet spot in this gradual paradigm shift towards electric propulsion (EP). A completely rethought design of the aircraft and its propulsion system involving their synergistic integration is a key element in stepping up overall energy efficiency while shrinking the footprints of greenhouse gas emissions and noise.

6.1 Conventional Engines

Depending on aircraft size, range, and payload, various types and sizes of engines are required for their propulsion. Starting off from the general aviation segment up to long-range commercial aircraft, engines come with different technologies, but all use kerosene-based fossil fuel.

FIGURE 6.1 Different types of engines for different aircraft segments.

Piston engines, also called reciprocating engines, often found in general aviation, extract propulsion power from Avgas aviation fuel, whereas larger aircraft use gas turbine-based turboprop or turbofan engines running on Jet-A fuel (Figure 6.1).

As regards the business aviation sector, 2012 U.S. deliveries show that the majority of aircraft are equipped with turbofan engines (>90%) while a small proportion are powered by either turboprop (~7%) or piston engines (~2%). Regional aircraft (20-99 passengers) use both turboprop and turbofan engines. Turboprops accounted for roughly half of the regional aircraft delivered in 2013. Short- and long-range large commercial aircraft (>100 passengers) exclusively use turbofan engines.

6.1.1 Gas Turbine

Turboprop and turbofan engines use gas turbines. The gas turbine provides the necessary thrust to move the aircraft through the air. It converts fuel stored in tanks into propulsive power on the one hand and, on the other, non-propulsive power which is supplied to aircraft systems.

The basic principle of a gas turbine consists in extracting energy from fuel by means of combustion of kerosene, converted from liquid to gaseous phase and mixed with air. The mixture gets heated up in the process due to high-pressure conditions. In order to do this, there is a compressor and a combustion chamber followed by a turbine (Figure 6.2). The turbine is actually driven by this outflow from the combustor and powers the compressor via temperature reduction of the same exhaust gas. Therefore energy is extracted from fuel in the form of shaft power, compressed air, and thrust.

Logically, the more the temperature is increased during combustion the more efficient the gas turbine gets, provided components and materials have the higher temperature and pressure capabilities.

FIGURE 6.2 Gas turbine operating principle [6.7].

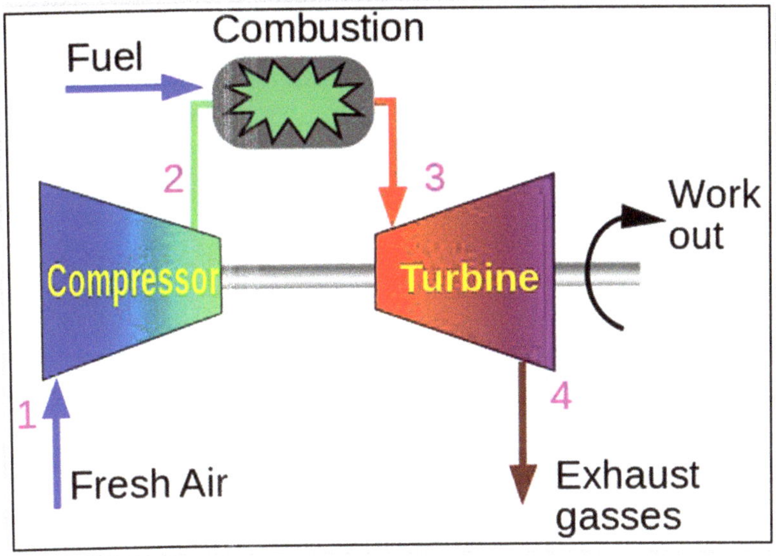

6.1.2 Turboprop Engine

The arrival of the turboprop engine in the early 1940s brought a step-change in power, reliability, and efficiency over the piston engines then being used on regional aircraft. It is actually a gas turbine driving a propeller and is best suited to aircraft cruising at less than about 800 kph, compared to a turbofan-powered jet aircraft which flies at around 800 kph. Turboprop technology has potential economic and environmental performance benefits, especially among regional aircraft manufacturers.

Although only a single-shaft configuration of turboprop engines is shown in Figure 6.3, they can also feature a two-shaft configuration, wherein the high-pressure (HP) shaft drives the compressor up to speeds of 30,000 rpm and the low-pressure (LP) shaft drives the propeller via a gearbox at constant speed, usually in the range of 900-2500 rpm depending on the aircraft.

A modern turboprop can consume 25%-40% less fuel than an equivalent turbofan engine on short-haul routes. But they cannot be scaled up to the higher thrust needs of faster, larger capacity, and longer-range aircraft.

Turboshaft engines widely used on helicopters are akin to turboprops except that the shaft power is used to drive the rotor instead of a propeller.

6.1.3 Turbofan Engine

While piston and turboprop engines rely on a propeller, the turbofan engine uses a fan in a different operating principle, allowing for higher speed and thrust ranges that turboprop technology cannot deliver. Moreover, the introduction of turbofan engines offered a less noisy and more efficient alternative to the turbojet engines originally used on faster and longer-range commercial jets, as well as military fighter aircraft.

FIGURE 6.3 Operating principle of a turboprop engine (single shaft) [6.8].

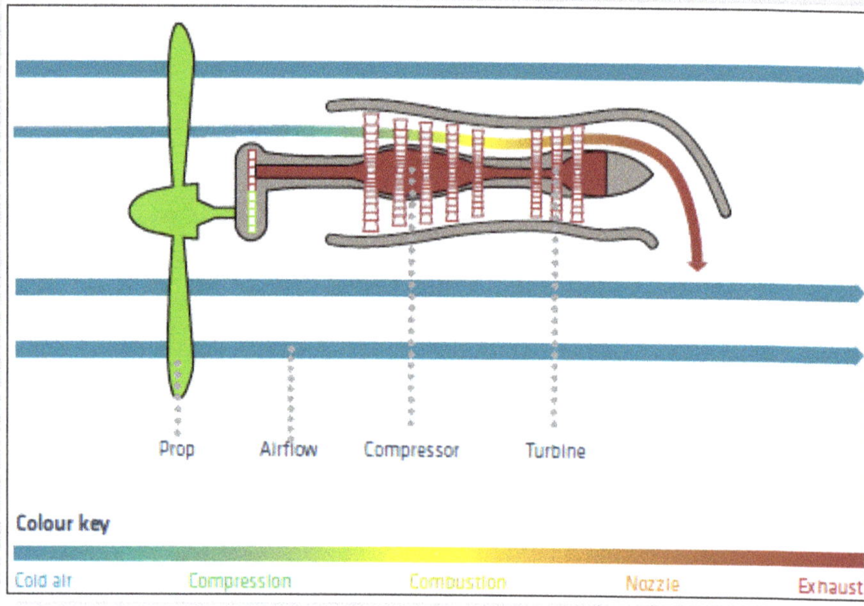

Both turbojet and turbofan engines use gas turbine exhaust to produce thrust. In a turbojet engine all incoming air is directed to the compressor and ignited with fuel to create all the thrust and drive a turbine. This differs from the more efficient operating principle of turbofan engines (**Figure 6.4**), which use a ducted fan, coupled to the gas turbine entry stage, to drive a proportion of the air more slowly around the core of the engine, called bypass air flow. In fact, the fan pulls in a huge amount of air into the engine intake, some of which is directed into the hot core of the engine—where it is compressed and then ignited—but most of which bypasses the core where it creates a majority of the engine's thrust. If there is twice as much cold air bypassing the core as the hot air going through it, the bypass ratio is 2:1.

Therefore, the total turbofan engine thrust is generated by the combination of gas turbine exhaust and bypass flow. The higher the BPR, that is, the ratio between the mass flow rate of the bypass stream to the mass flow rate entering the core, the better the engine fuel efficiency, as more thrust is being generated without burning more fuel. Moreover, turbofans are quieter than turbojets in part because the flow of cold air surrounding the exhaust from the engine core reduces the noise produced by the exhaust gases.

Figure 6.4 shows the architecture of a turbofan engine with two independent shafts. The LP shaft connects the inlet fan, the LP compressor, and the LP turbine. Externally to the LP shaft, there is the HP shaft, which connects the HP compressor to the HP turbine. The speeds of the two shafts are independent and variable, depending on operating conditions of the aircraft. Typically, for a two-shaft turbofan engine, the LP shaft speed can approximately change in the 600-6000 rpm range and in the 7000-20,000 rpm range for the HP shaft.

FIGURE 6.4 Operating principle of a turbofan engine (two shafts/conventional architecture) [6.8].

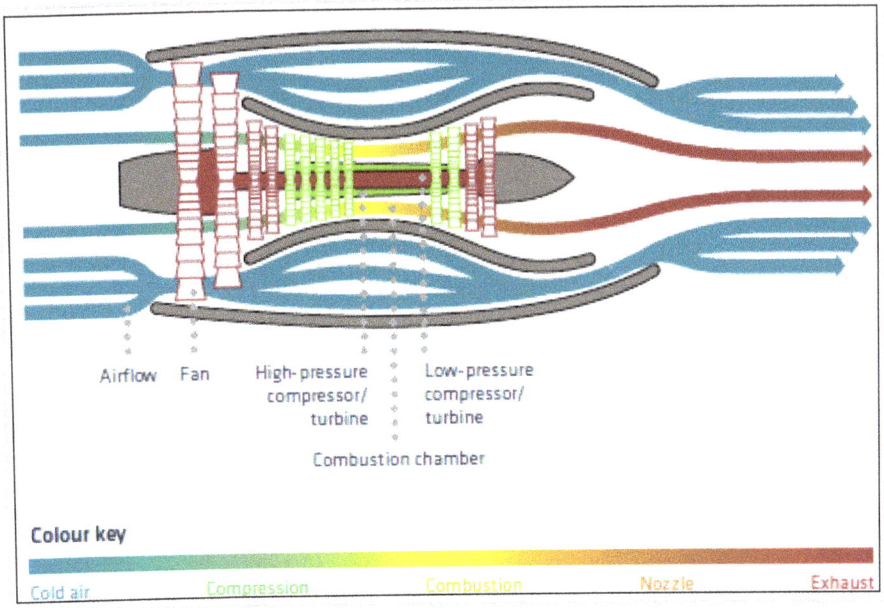

It is to be noted that the majority of turbofan engines come in a two-shaft configuration. Exceptions are Rolls-Royce long-range engines which all have three shafts because of the intermediate pressure (IP) section interposed between the LP and HP sections.

The appearance of the high BPR turbofan engine in the late 1960s revolutionized air transport by offering a much quieter and efficient alternative to the turbojet engine.

Gas turbine and fan technology advancements as well as novel engine architectures have also pushed new boundaries. One such architecture called geared turbofan (GTF) has entered the market recently. Unlike all other turbofans which come in a "conventional" architecture where the fan and the low pressure turbine (LPT) rotate at the same speed, a reduction gearbox is introduced in the GTF in order to decouple fan and LPT speeds (ratio of 1:3). This configuration allows for a larger fan to rotate more slowly than the rest of the engine to push larger volumes of air around the jet core, thereby improving the BPR.

As Figure 6.5 illustrates, turbofan BPRs have seen a significant boost in recent years and top out at above 12:1, more than double the level of the very first turbofan to enter into service.

Most recently, the short-range aircraft segment has seen the entry into service of high-bypass turbofan engines in both their conventional and GTF architectures. High-bypass LEAP engines are conventional just like the original lower bypass CFM56 engines, both engine generations being produced by CFM International. On the other hand, high-bypass PurePower engines from Pratt and Whitney are of the GTF type. It is interesting to see how these new engines stack up against their predecessors (CFM56 and V2500) when considering key engine parameters like thrust, BPR, and fan diameter.

FIGURE 6.5 Evolution of turbofan engine BPR [6.9].

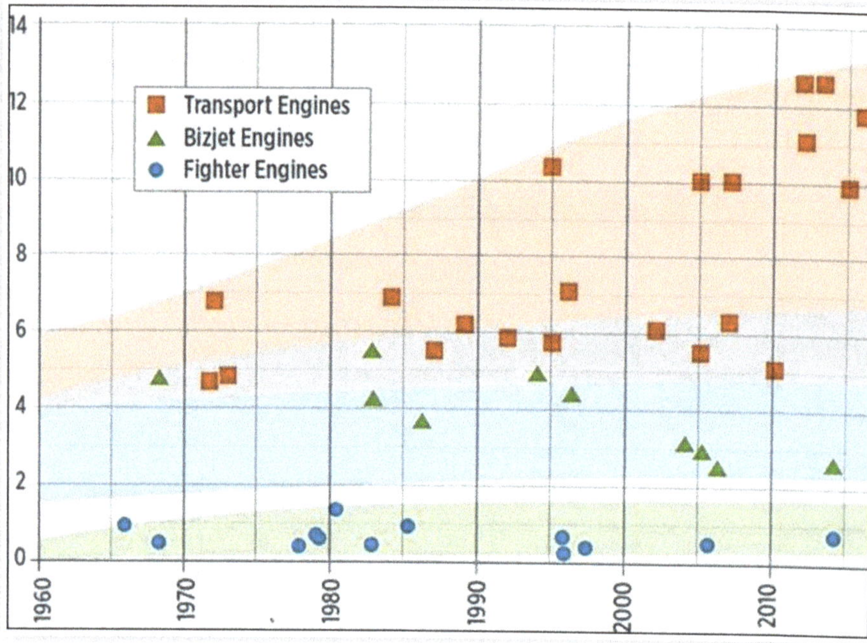

The original CFM56 engine has a thrust rating of up to 33,000 lbf, a BPR of up to 6, and a fan diameter of 68.3″. Figures 6.6 and 6.7 provide for each type of replacement engine a reminder of the operating principle and pictures of the engine before its podding, and after completion of podding and aircraft integration. Data shown are maximum values for the engine family considered and as sized for its target short-range aircraft.

FIGURE 6.6 High-bypass conventional turbofan: LEAP engine (35,000 lbf thrust, 11 BPR, 78″ fan) [6.10, 6.11, 6.12, 6.13].

FIGURE 6.7 High-bypass GTF: PurePower engine (33,000 lbf thrust, 12.5 BPR, 81″ fan) [6.10, 6.14, 6.15].

6.1.4 Efficiency

Apart from their weight, additional drag, and reliability, gas turbine engines are characterized by the overall efficiency with which they convert fuel energy into kinetic energy allowing aircraft displacement. This metric is calculated by just multiplying two other engine parameters, namely, the "motor thermodynamic efficiency" (conversion of fuel flow power to shaft power) and the propulsive efficiency (conversion of shaft power to propulsive power).

In Figure 6.8 motor thermodynamic efficiencies of piston, turboprop, and turbofan engines are plotted against their takeoff power. Also shown is how, depending on the type of engine used, propulsive efficiency varies with airspeed.

Propulsive efficiency increases as aircraft airspeed goes up, except for the turboprop engine whose efficiency starts plummeting after reaching a tipping point close to airspeeds around 350 mph suitable for regional operations. Also clear is that the propulsive efficiency of a turbofan engine can be enhanced by increasing its BPR. For turbofan engines, propulsive efficiency is in the range of 70%-80%, whereas turboprops are about 10% more efficient at their cruise airspeeds.

The high power/thrust ratings of turbofan engines have enabled the design of large aircraft. Their market penetration spans regional, short- and long-range segments. Given their overall airspeed capability and thrust/power scalability, turbofan engines have currently become mainstays of aircraft propulsion. With takeoff thrusts of 20,000 lbf and above, the most recent commercial aircraft gas turbine engines are of the high bypass turbofan type. They have the best overall efficiency of over 40% at cruise where they operate with propulsive efficiencies of well over 70% and motor thermodynamic efficiencies of up to 55%.

Thanks to a steady investment in advanced gas turbine technology and variant architectures (GTF), the turbofan engine has come a long way in improving overall efficiency. One key enabler has been the evolution of the BPR.

Improvements in overall engine efficiency allow cutbacks on fuel burn which help in reducing both greenhouse gas emissions and operating costs.

FIGURE 6.8 Efficiency, airspeed, and power comparisons for various engine types [6.16, 6.17].

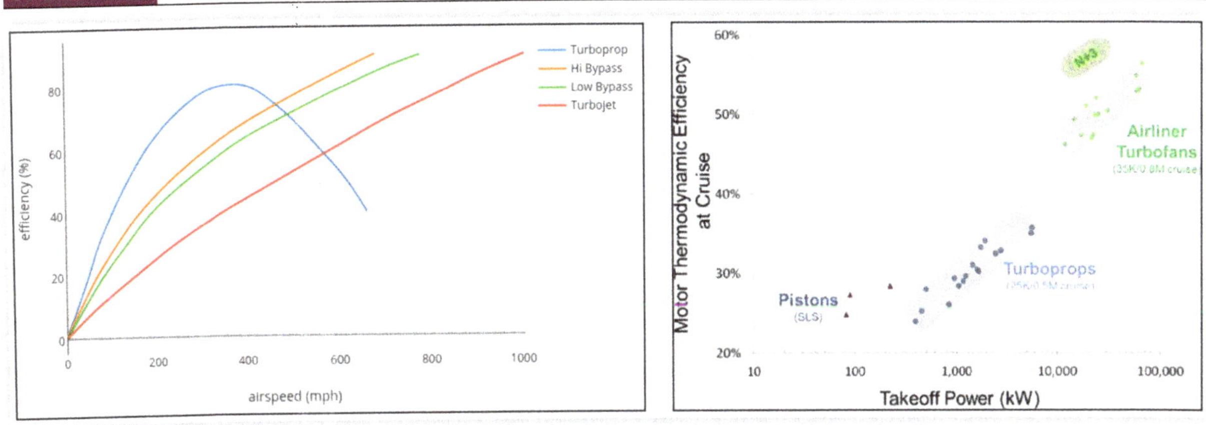

6.1.5 Noise

Over the past 50 years, substantial improvements in engine technology have significantly reduced aircraft community noise levels (see **Figure 6.9**), particularly with the introduction of second-generation turbofan engines with advanced nacelles in the 1970s.

Apart from fuel burn and emissions, noise levels have also been slashed over time, thanks to a drastic increase in BPRs allowing for larger fans to rotate at lower speeds and hence generate lower noise. Nevertheless, a larger engine design should consider constraints from higher fan tip speeds and aircraft integration issues due to ground clearance limitations on conventional "tube and wing" aircraft.

The latter issues are very likely to be encountered with incremental design approaches such as aircraft re-engining, which consists in shifting the aircraft propulsion baseline from one engine generation to the next while limiting aircraft changes to the strict minimum. While re-engining developments may deliver the best bang for the buck and a shorter time to market, they may run into design constraints, worse may even face showstoppers. Energy efficiency, emissions, and noise targets for the future are so steep that the only way out may be to adopt a totally different aircraft design approach. When all major components of the aircraft such as structures, engines, and systems are to independently contribute to overall performance, there are always missed opportunities and ones that may go untapped. Synergistic optimizations achieved in an aircraft development seamlessly integrating airframe, propulsion, and system designs are definitely more likely to yield the best benefits. This approach will help push the envelope harder and leverage aircraft efficiency and noise performance. This ultimately will put the industry on the path of a departure from the old order of wing-and-tube aircraft powered exclusively by kerosene-fueled engines. The transition to alternative aircraft configurations will open up new avenues and opportunities even if that means dealing with totally different aircraft

FIGURE 6.9 Evolution of noise performance [6.18].

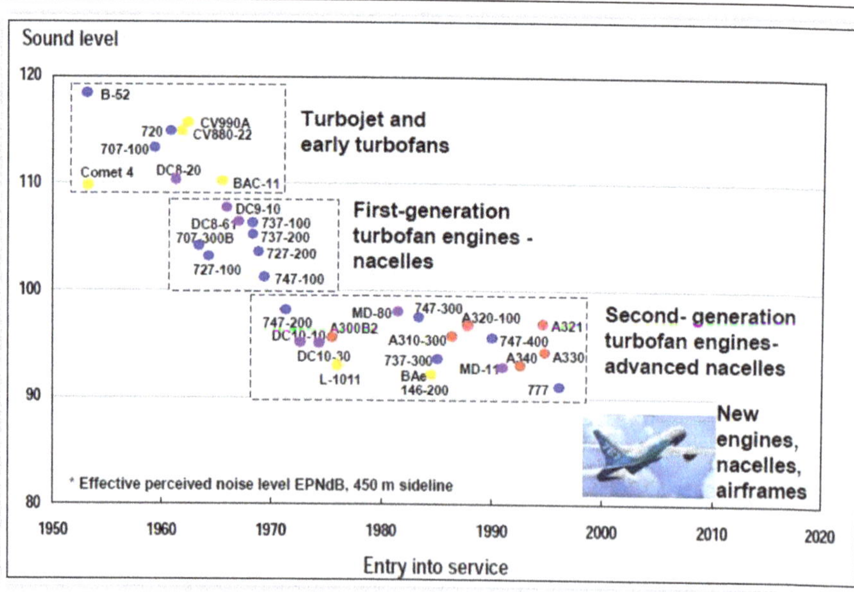

shapes and aerodynamics so long as they can help unlock the potential of breakthrough propulsion concepts and technologies currently under intensive research worldwide. The electric aircraft is one such configuration, and the following sections will get into further detail on how propulsion systems may deliver solutions catering to its needs.

6.2 Bleedless Engines for the More Electric Aircraft

Parallel to the advent of the latest technology high-bypass turbofan engines, gradual electrification of aircraft systems has also taken place. Acting as an airborne energy utility, via electrical power generators coupled to them, engines have always been the means of supplying non-propulsive energy to operate aircraft systems. Therefore, electrical energy produced by the engines from fuel has seen an upward trend especially in the more-electric aircraft, a stepping stone towards the all-electric aircraft. This has all but impacted engine design and integration, not from a propulsion standpoint but from a non-propulsive energy production one. Engine-produced electrical energy reached its apex recently with the more-electric Boeing 787. This long-range aircraft's high-bypass turbofan engines (78,000 lbf thrust, 11 BPR, 112″ fan) produce a staggering 1 MW of electrical power.

Depending on the degree of electrification adopted for aircraft systems and propulsion, propulsion system developments fall broadly into three main categories. First comes the bleedless engine devoted to the More Electric Aircraft (MEA) with a high level of systems electrification. Next in line are the electric propulsion (EP) and hybrid-electric propulsion (HEP) systems for the electric aircraft wherein all systems are assumed to be totally electrified. The following sections will go into further detail on these engine categories. If on an MEA intensive electrification is applied to systems in a bleedless architecture, like in the case of the Boeing 787, then more efficient "bleedless" engines can be used, but they still derive propulsive power from the operation of their gas turbines. The Boeing 787 has embraced electric power for the Environmental Control System (ECS) and Wing Ice Protection System (WIPS), replacing their pneumatically powered predecessors. By moving to electrification of these functions, the Boeing 787 has eliminated the requirement for bleed air to be extracted from the engine, hence engine starting is also facilitated by using the same electric machine as the one which generates aircraft electrical power, once the start sequence is completed. Nevertheless, the Boeing 787 has provided the biggest challenge for the engine since replacement of the bleed air-driven functions with electric alternatives has a dual impact on the engine:

- Removal of the bleed air, currently used for some engine functions (cooling, sealing, etc.)
- Increase in the electrical power required from the engine powerplant, requiring a significant increase in the mechanical torque offtake from the engine via the accessory gearbox

Figure 6.10 shows one of the Boeing 787 bleedless turbofan engine options. Moreover, with different colors for the compressor and turbine stages, the typical Rolls-Royce three-shaft engine topology is also highlighted.

FIGURE 6.10 Rolls-Royce Trent 1000 engine powering the Boeing 787 [6.19, 6.20].

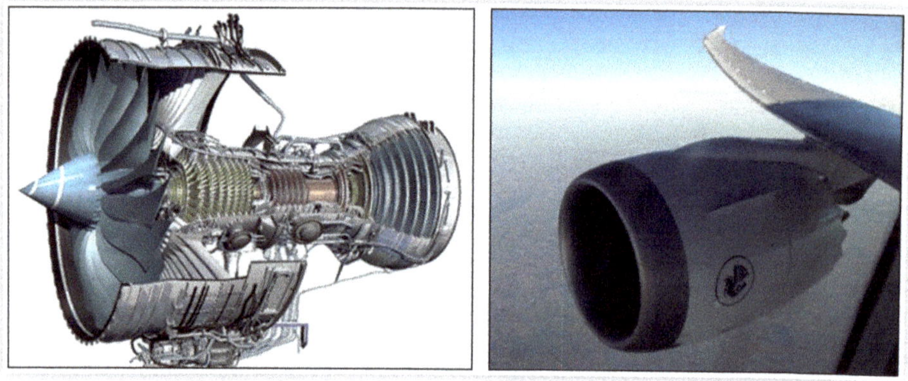

FIGURE 6.11 Elimination of Boeing 787 engine bleed components [6.21].

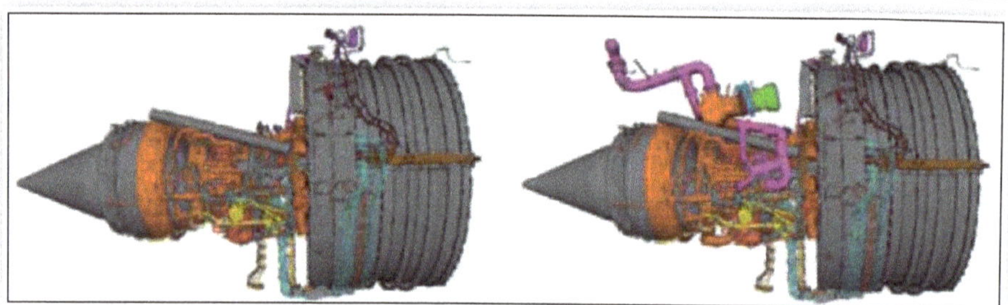

The elimination of pneumatic (bleed) portions on the Boeing 787 engines (bleedless architecture) is illustrated in **Figure 6.11**.

6.3 Propulsion Systems for the Electric Aircraft

The previous section shows that the extent to which aircraft systems are electrified defines the changes impacted on the powerplant, but the basic assumption is to rely only on gas turbines for extracting propulsion power, even though they are "bleedless," at least when the aircraft considered is large enough in size. This is partly explained by the fact that power capability, weight, volume, efficiency, and maturity aspects of EP alternatives have been a far cry from gas turbine engine performance which has continually benefitted from technology advances (architectures, materials, and thermal capabilities). However, this is set to change in the future.

Compared to the needs of aircraft systems, power requirements for EP are much higher, as illustrated in **Figure 6.12** comparing the conventional, more-electric and hybrid-electric/all-electric aircraft.

FIGURE 6.12 Evolution of aircraft-level electrical power requirements (propulsion and systems) [6.22].

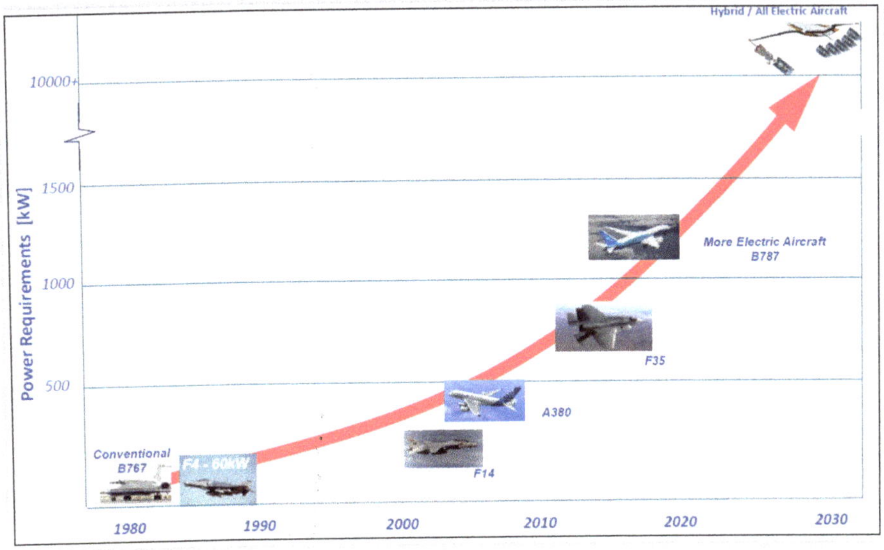

Up to the more-electric Boeing 787, electric power generated by turbine engines is solely used to power aircraft systems having undergone electrification on top of usual electrical systems. Hybrid-electric/all-electric aircraft requirements are in the major part linked to propulsion but also include the needs of aircraft systems.

With EP, because gas turbines are removed, electrical energy for both propulsion and systems has to be sourced differently. Energy storage systems (e.g., batteries) are therefore swapped with aviation fuel and become the principal source of electrical energy. Alternatives (e.g., hydrogen) to kerosene or Avgas may be used if fuel cells are chosen to be the baseline. One situation though, where aviation fuel may still be required on board, is when using fuel cells relying on "reformers" to produce hydrogen from aviation fuel.

In HEP, an electrical system dedicated to propulsion is introduced as a complement to the gas turbine in such a way that propulsive power is drawn from either or both of the two. Aviation fuel and energy storage systems will have to cohabitate on board as energy reservoirs of two different types.

In fact, a paradigm shift is occurring thanks to EP and HEP systems under study targeting the electric aircraft. To complete this picture the following section explains how in the process of its electrification, aircraft propulsion is enabled by novel architectures and technologies.

6.3.1 Enabling Architectures

From the drawing board to research demonstrators, there is an array of possibilities with EP. This section highlights the operating principles, benefits, and challenges.

6.3.1.1 Electric Propulsion: As explained in Section 6.1, the conventional engine can be schematized by the block diagram in Figure 6.13. It is to be noted that the thrust

FIGURE 6.13 Conventional propulsion (adapted from [6.23, 6.22]).

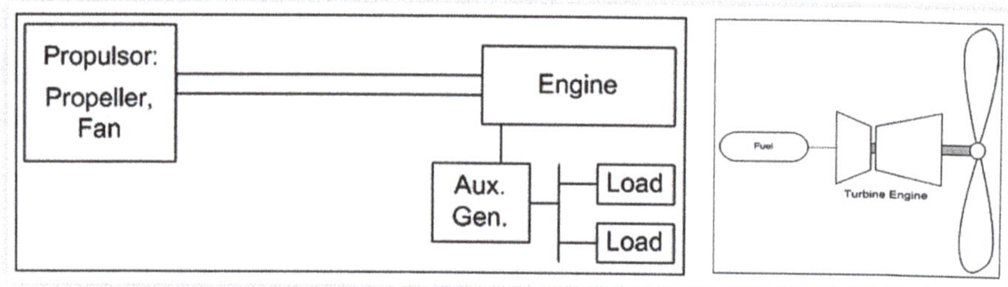

provided via the propulsor, which actually is a propeller on a turboprop engine or a fan in the case of a turbofan, comes exclusively from the gas turbine which also supplies power to non-propulsive system loads.

With EP, the gas turbine is deleted and the propulsor is driven by an electric motor using energy extracted from energy storage systems such as batteries or fuel cells (**Figure 6.14**).

It is worthwhile to see how efficiencies of conventional and EP systems stack up against each other. Orders of magnitude of efficiencies are shown in **Figure 6.15**.

FIGURE 6.14 Electric propulsion (adapted from [6.23, 6.17]).

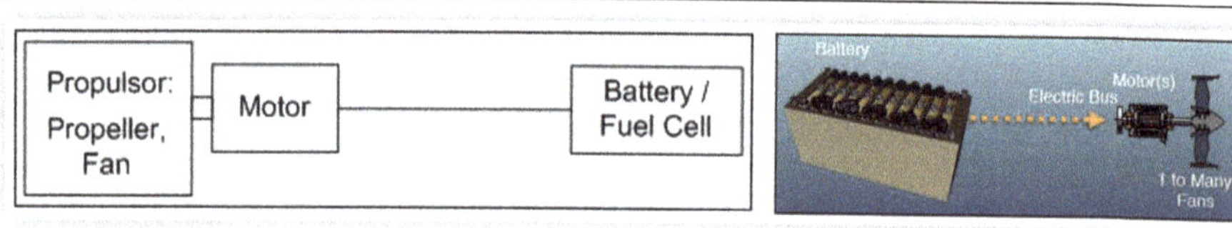

FIGURE 6.15 Overall propulsion efficiency (old-generation turbine engines vs. electric) [6.24].

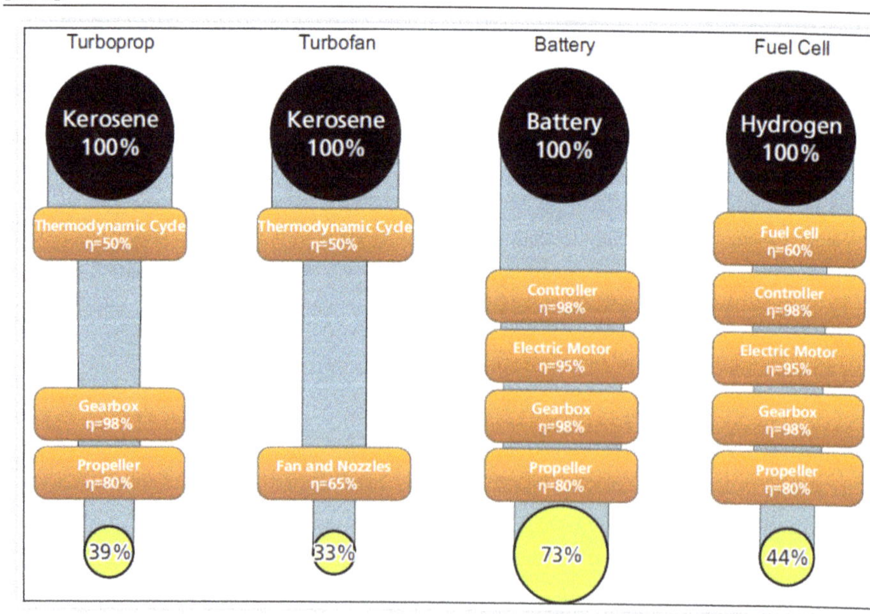

From an overall efficiency perspective, it is clear that battery-powered EP has an edge over conventional propulsion. It avoids the conversion of fuel to electricity in the airplane and delivers efficiencies upward of 70%, significantly above the 40% hardly reached with old-generation turboprop engines. As seen in Section 6.1.4, the latest turbofan engines achieve better thermal efficiencies which at almost 60% raise overall efficiency from 33% shown in Figure 6.15 to 39% [6.24].

6.3.1.2 **Hybrid-Electric Propulsion:** Let us see how HEP may constitute a compelling stepping stone in achieving the performance goals. As opposed to EP, where not even a single drop of aviation fuel may be required in most instances, the HEP energy is drawn from both aviation fuel and other energy sources by a shrewd association of gas turbines, electric systems, and energy storage in either a "parallel" or "series" configuration.

In parallel HEP, power flow down to the propulsor originates in part from a gas turbine, similar to the situation in a conventional engine, and is complemented with power provided by either a motor/generator drawing energy from another energy source or just a motor/battery setup. This is called parallel hybrid operation, for both gas turbine and motor/generator (or motor/battery) can invariably deliver power to the propulsor at any given time of aircraft operation. The two independent power flow pathways for a turbofan with motor/battery setup are shown in **Figure 6.16**. It is to be noted that reversible operation is possible with the motor/generator setup enabling power supply to other loads. Batteries can stay afar from the powerplant with the motor/battery setup.

Let us now consider a parallel-hybrid turbofan for a short-range aircraft [6.26]. Within the powerplant, a conventional electric motor and a power conditioner work hand in hand with an advanced two-spool direct-drive turbofan engine with a fan pressure ratio of 1.45. Electrical energy is drawn from the battery and, via a power conditioning unit and an electric motor, delivered as mechanical energy to the gas turbine LP spool (**Figure 6.17**). Motor rating depends on the hybrid-electric operational mode chosen for the design of the powerplant. A "balanced" hybrid operation of motor and gas turbine throughout the mission would call for a 1.3 MW motor. On the other hand, a "core shutdown" hybrid mode, wherein the gas turbine is switched off halfway through cruise, would require a higher motor rating of 5.3 MW.

FIGURE 6.16 Parallel HEP (adapted from [6.23, 6.25]).

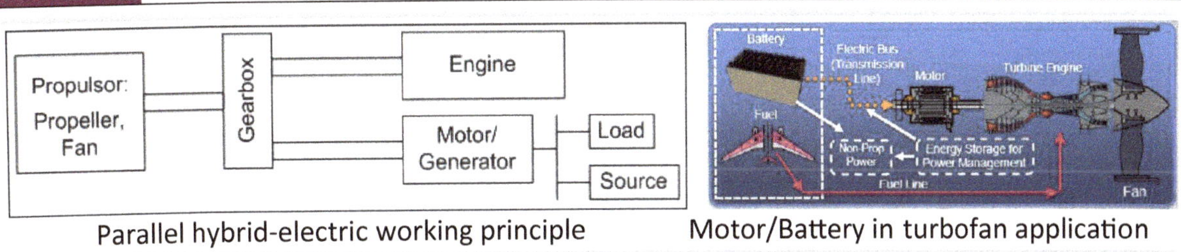

FIGURE 6.17 GE Parallel hybrid-electric turbofan (balanced and core shutdown modes)—short-range aircraft [6.26, 6.22].

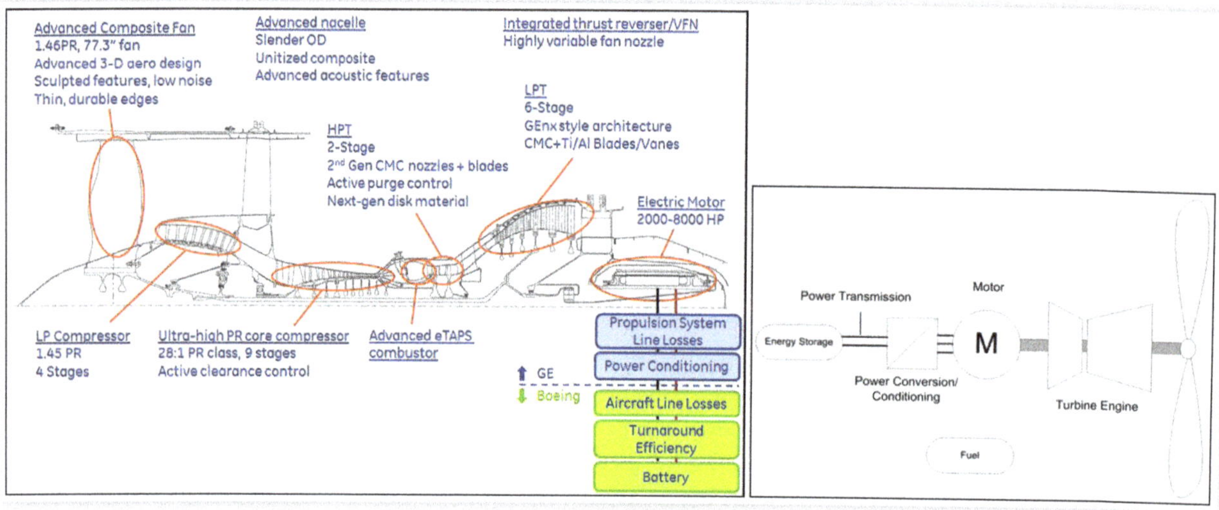

Actually, at the HEP system level, the utilization of electric power will decrease fuel consumption at a specified thrust level. Electric fan power augmentation causes the fuel flow to cut back, resulting in substantial improvements in the SFC of the propulsion system, as shown in **Figure 6.18** [6.26]. Under cruise conditions, a 6000 hp electric motor allows for a sizeable SFC reduction in the order of 75%.

The hybrid-electric system is designed in such a way that battery and electrical energy usage are optimized to minimize fuel burn and manage battery energy. The goal for fuel burn reduction against the conventional CFM56 turbofan is set at 60% on a 900 nm mission.

The HEP detailed here utilizes a conventional electrical system. This includes, for either side of the aircraft, a power conditioner, solid-state circuit breaker, heat exchanger,

FIGURE 6.18 Specific fuel consumption (SFC) vs. electric power augmentation at cruise [6.26].

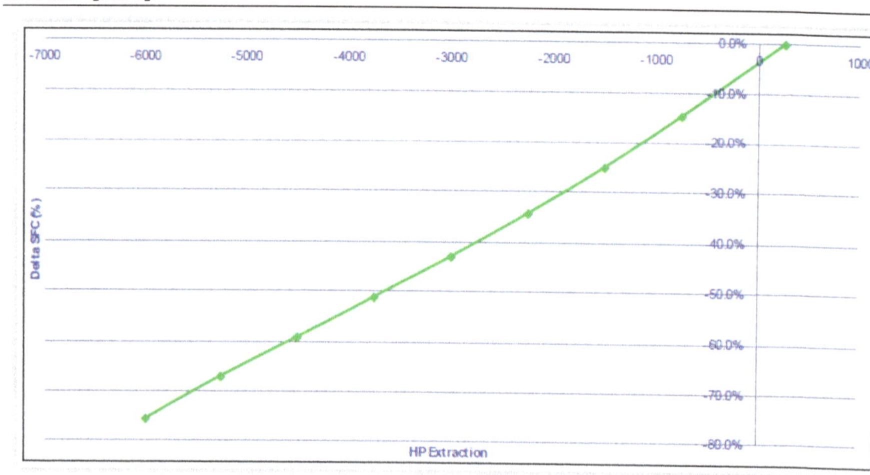

and conventional electric motor. The motor is a normally conducting high power density Switched Reluctance Motor (SRM) with liquid cooling for continuous peak electromagnetic power operation. Advanced composite materials are used for the shaft and motor case. The power conditioning unit utilizes high voltage SiC devices, advanced high-temperature plastic film capacitors, and an advanced composite liquid-cooled heat sink. Additionally, the power conditioner has low loss inductors and an all composite enclosure. The resulting conventional electric system has a specific power in the range of 2-3 hp/lbm and an overall efficiency of ~93%.

The implementation of a superconducting electrical system in lieu of the conventional system may unlock additional capabilities, notwithstanding its more stringent requirements. Unlike the conventional system, the electric motor, wires, and power conditioner use superconducting technology requiring a coolant. The proposed coolant is liquid natural gas (LNG). The motor includes conductors that are held at LNG temperatures and are superconducting with alternating current. The motor requires high-strength insulating structural materials and an evacuated ironless design. The heat sink of the power conditioning unit is cooled by LNG. The resulting superconducting electric system has a specific power in the class of 5-6 hp/lbm and an overall efficiency of 99%.

From an aircraft standpoint, this type of HEP system can come in monolithic and stand-alone configurations similar to conventional turbine engines.

Departing from the engine design studied in [6.26] and described above, the parallel hybrid-electric principle considered may also be applied to the high pressure spool instead of the lower pressure spool of the gas turbine. Moreover, it may also be applied with different operating modes to GTFs (**Figure 6.19** [6.27]) and turbofans with alternative motor locations, or for that matter to turboprop engines.

In total turboelectric propulsion, all thrust power is extracted from fuel by a gas turbine and transmitted through a generator to the motor driving the propulsor(s). Integration may happen within the conventional powerplant envelope or the motor/propulsor setup can be installed remotely from the powerplant. In the latter configuration

FIGURE 6.19 Parallel hybrid-electric turbofan variants and operating modes—short-range aircraft [6.27].

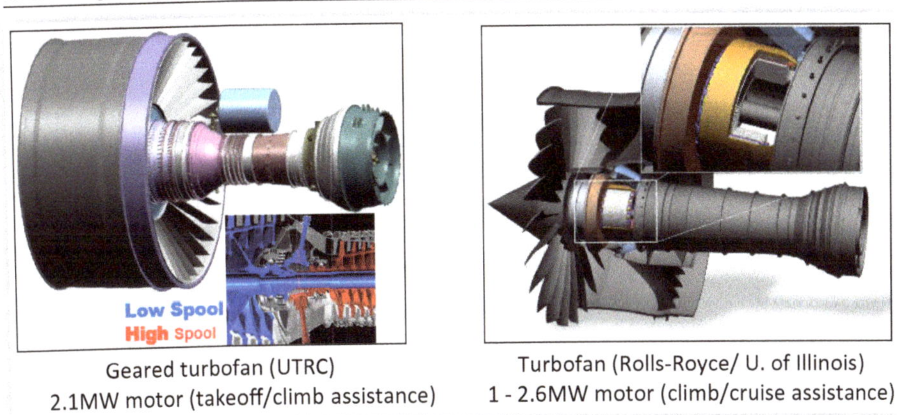

FIGURE 6.20 Turboelectric (total) and series hybrid propulsion (adapted from [6.23, 6.25, 6.17]).

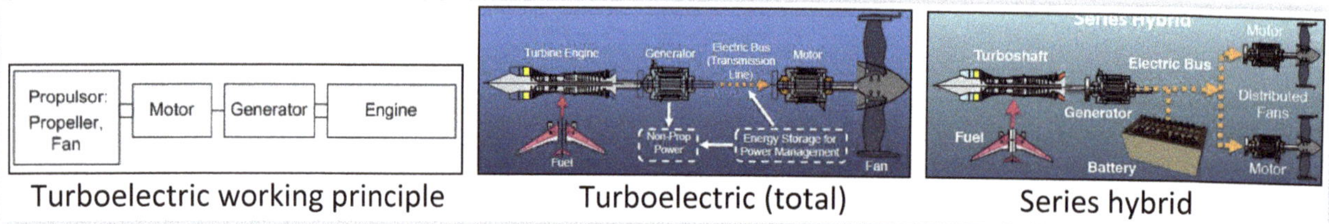

fan airstream and the core airstream are physically separated. When batteries are also used, the configuration is called series hybrid propulsion (**Figure 6.20**).

Other power sources together with the turbine engine can also be used to define a redundant architecture like the one shown in **Figure 6.21**, where either gas power units (gas turbine + generator + rectifier) or batteries can be used to power the propulsor motors.

Architecture tradeoffs at aircraft level have to be performed in order to down select the best configuration.

The optimization of hybrid-electric systems requires increased efficiency and specific power in electric drive systems, thermal management systems, power extraction, and/or energy storage. Let us now see how hybrid-electric architectures may enable the following configurations:

- **Boundary Layer Ingestion (BLI)**: Allows propulsion systems to energize boundary layers without distorted flow entering turbine core
- **Distributed Propulsion (DP)**: Allows effective increase in fan BPR through distributed propulsors
- **Lower Carbon Designs**: Reduces combustion-based propulsive power (and emissions) using electric motors and/or onboard "clean" energy storage

BLI is the ability of an engine to have the boundary layer coming off another surface such as the aircraft fuselage to contact the fan and not cause disruptive interference. If this can be accomplished, then the propulsion units can be moved closer to the wings or fuselage which would decrease structural weight and drag associated with large pylons.

FIGURE 6.21 Redundant HEP architecture [6.28].

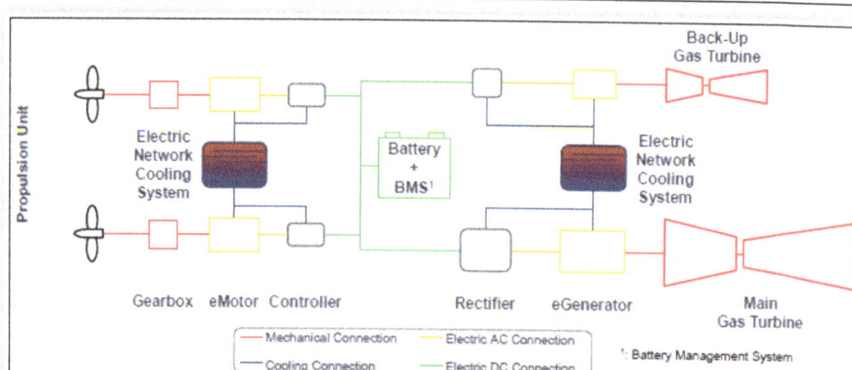

Most of the viscous drag is contained in the boundary layer. If small fan(s) close to the fuselage are used, this region can be ingested and accelerated, therefore removing some of the drag. Hence, the overall vehicle propulsive efficiency can be improved by reenergizing low-energy low-momentum wake flow (**Figure 6.22**).

The benefits of BLI may be leveraged using partial turboelectric systems. This is achieved by using a conventional or advanced turbine engine for producing thrust and driving a gearbox-mounted electrical generator powering one or more remote BLI fan motors. The BLI motor sized for the application shown in Figure 6.22 is in the range of 2-3 MW.

Instead of tapping energy exclusively from the turbine engine, another energy source (battery or fuel cell) may also be utilized (**Figure 6.23**).

However, the implementation of a boundary-layer ingesting system means that the airflow into the fans is not uniform. For the potential benefits to materialize, the turbomachinery—and in particular, the fan blades—must be able to withstand the associated unsteady conditions due to the distorted intake flow. A specifically optimized fan design is necessary in order to deliver the best performance in the distorted flow conditions that are experienced in a BLI configuration.

DP maximizes the opportunity for BLI and offers the following features:

- Low specific thrust propulsion with installation facility
- Structural efficiency/optimized propulsion system weight
- Minimized asymmetric thrust, reduced vertical fin area
- Reduced jet velocity and jet noise

FIGURE 6.22 Boundary layer ingestion (BLI) [6.22, 6.29].

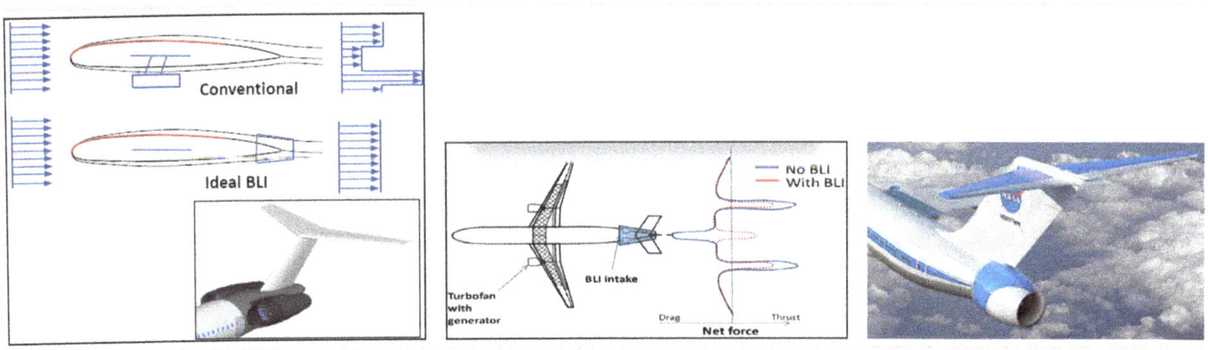

FIGURE 6.23 Partial turboelectric: turbofan for thrust and motor(s)/fan(s) for BLI [6.27].

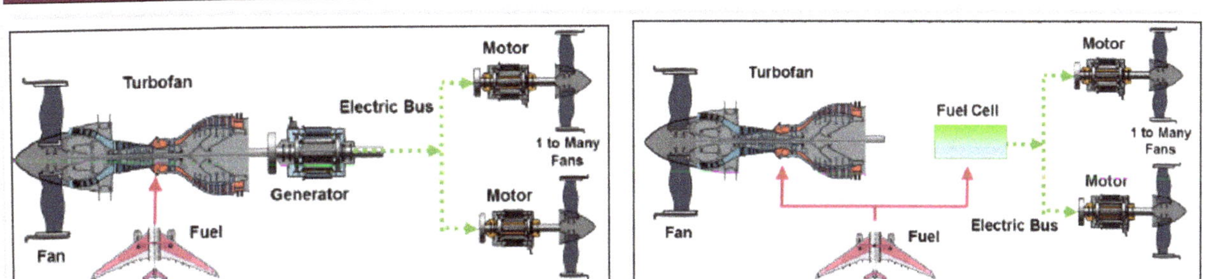

Aircraft design optimizations with reduced cruise thrust requirements are possible with the following configurations of DP with BLI:

- Blown wing (Deflected Slipstream)
  A large number of small propulsors accelerate air evenly over the wing to provide additional lift at low aircraft speeds allowing for lower drag wings.
- Airframe design
  Without the constraints of two large high BPR engines, new and more aerodynamic designs are possible such as Blended Wing Body (BWB).

Let us see how the flexibility of HEP during the flight cycle can offer design and operational benefits. In other applications, such as automotive and marine, hybrid systems deliver benefits around the vehicle's operating cycle. In fact, the conventional system loses efficiency away from its operating point, and its hybrid-electric variant can correct this. Nevertheless, these benefits are most significant for types of transport with complex operating cycles including stop-starts and variable loads. Hence, hybrid ships bringing very significant benefits now dominate the marine sector.

However, aircraft do not have frequent stop-starts like cars and ships and deal with flight sequences such as taxi, takeoff and cruise, descent, and landing. A typical flight cycle can be characterized by its very asymmetric form with ups and downs in power demand (**Figure 6.24** [6.30]).

This makes it difficult to optimize the conventional propulsion system for these different requirements. This is where a hybrid-electric system could come into play and deliver improvements. Where the conventional gas turbine engine has to be sized for the maximum power demand (takeoff), optimal sizing of a hybrid-electric design is possible by leveraging the varying requirements.

In fact, the electrical architecture of HEP offers different choices of energy sources and their mix to power a fan (or a propeller). This flexibility allows for the smoothing of the energy peaks of the flight cycle and a boost capability in demanding phases.

Let us focus now on how BPRs may get a boost when dealing with HEP systems. Looking back, high oil prices have constrained aircraft manufacturers in recent years to resort to re-engining solutions in order to alleviate their negative impact on aircraft economics and move towards greener solutions in the same process. Engine manufacturers have been able to push boundaries in gas turbine, propeller, and fan technologies

FIGURE 6.24 Typical flight cycle [6.30].

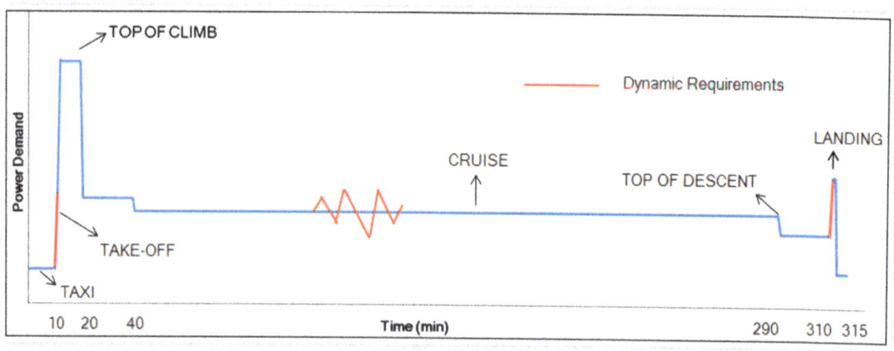

in order to hit the incremental performance targets of the re-engining approach. One of the main areas of improvement in turbofan engines is the BPR where values of 12 have been reached. We know from Figure 6.8 that increasing the BPR improves overall efficiency. But this adds weight and drag and limits installation possibilities. The large fan diameter and weight of conventional turbofan engines limit where they can be located on an airframe—usually under the wing. Their location does not enable advanced aerodynamic efficiency techniques to be used.

On way to step up the effective BPR without excessively large diameter fans is to use distributed electric propulsion (DEP).

6.3.1.3 Distributed Electric Propulsion (DEP): A fan (or propeller) DP system may be an attractive alternative to conventionally podded turbine engines. A DP configuration literally separates overall propulsion into multiple smaller sub-systems spread across the aircraft. DEP and distributed hybrid-electric propulsion (DHEP) comprise a number of electrically driven fans (or propellers) that are integrated into the airframe, thereby improving weight and aerodynamics.

In DHEP, total turboelectric distributed propulsion (TeDP) (Figure 6.25) has several upsides, especially in the form of decoupled energy management (DEM). TeDP might be more accurately considered DEM as the overarching concept that enables HEP. The major benefit that DEM brings is an abundance of options pertaining to the aircraft's configuration and operation.

Decoupling the gas turbine from the thrust producer (distributed) allows for several unique advantages that could very well improve aircraft performance and efficiency. Turbofan engines are designed with a mechanical linkage between fan and gas turbine that forces all components—fan, low pressure and high pressure compressor, and turbine—to compromise on their individual peak operating points. Removing the mechanical linkage, or shaft, and replacing it with an electrical system allows components to be optimized with less concern for each other especially with regard to rotational speed.

Not only does this yield the possibility for reduced component weight and fuel burn, but other advantages can be garnered, such as:

- Use of supplemental power (batteries, super capacitors, fuel cells) for failure modes or downsizing of gas generators
- Effective BPR increases without excessively large diameter fans
- Finer control over power distribution during any flight segment

FIGURE 6.25 DHEP (total turboelectric) [6.27, 6.31].

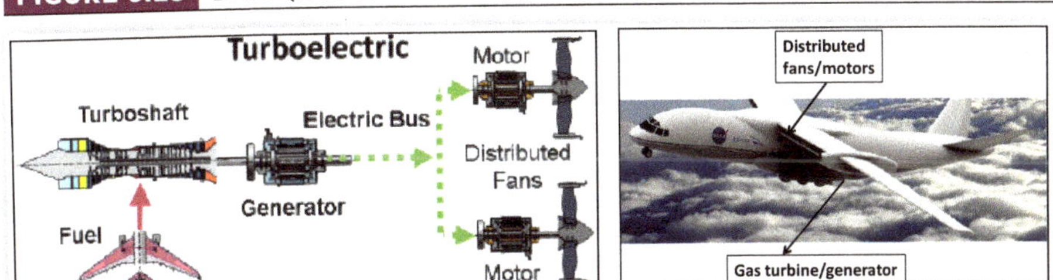

Introducing this method of DEM does bring with it high transmission losses and an overall increase in propulsion group weight. From an efficiency standpoint, propulsive efficiency can be increased. But until technology of conventional machines is improved, the increase cannot offset the weight penalty. Savings in fuel burn must be much higher than the increase in propulsion system weight for a fixed takeoff gross weight if TeDP is to be economical. The generator sized for the application shown in Figure 6.25 is in the range of 7.5-12 MW while the four fan/motor assemblies per wing have a unit rating of 1.1-1.5 MW.

In addition to BPR increase, BLI is one other key vector for the improvement of overall efficiency of propulsion systems. Disruptive architectures such as DP, by enabling BLI and step increases of BPR, may unlock further aircraft potentials.

With DHEP, the effective system BPR can be pushed up to 20, far beyond the values of the most recent turbofans.

6.3.2 Enabling Technologies

This section provides an overview of technology developments enabling EP and HEP, their challenges, and foreseen timelines.

6.3.2.1 Motors: In EP for small general aviation, the electric motors replacing the engines may obtain power-to-weight ratios of up to six times that of piston engines. From an efficiency standpoint, fourfold improvement is achievable and noise impacts are alleviated when using electric motors. Motors with power densities of up to 5-6.5/kW/kg and 97% efficiency have been developed and are on their way into small general aviation (GA) aircraft. Figure 6.26 shows a few examples of motor prototypes devoted to big and small propulsion demonstrators.

The power density of a 350 hp gasoline piston engine may not exceed 2 kW/kg. In comparison, the feasibility of a replacement 260 kW electric motor, with more than double that density at 5 kW/kg, has been demonstrated. Furthermore, the 2500 rpm motor speed rotation allows for propeller direct drive, therefore helping eliminate the cumbersome and inefficient gearbox usually found between the engine and the propeller.

FIGURE 6.26 High density and efficiency motor prototypes.

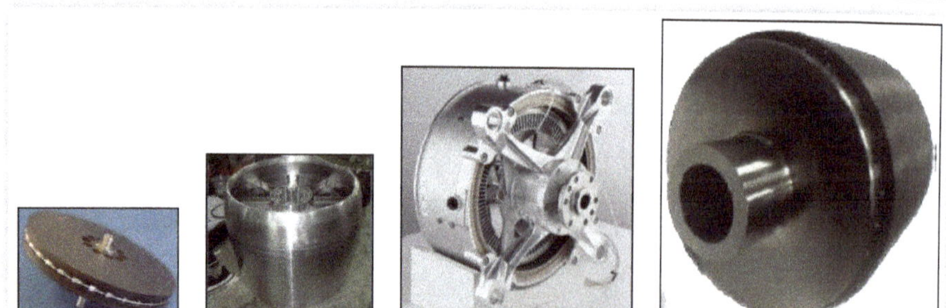

VTOL UAV (HEP) 6 kW generator (Launchpoint) [6.32]

GA (DEP) 60 kW motor (Joby aviation) [6.33]

GA (EP) 260 kW motor (Siemens)

Short range (HEP) 1 MW motor (U. of Illinois/Rolls-Royce) [6.27]

FIGURE 6.27 EP and motor ratings for various aircraft segments [6.25].

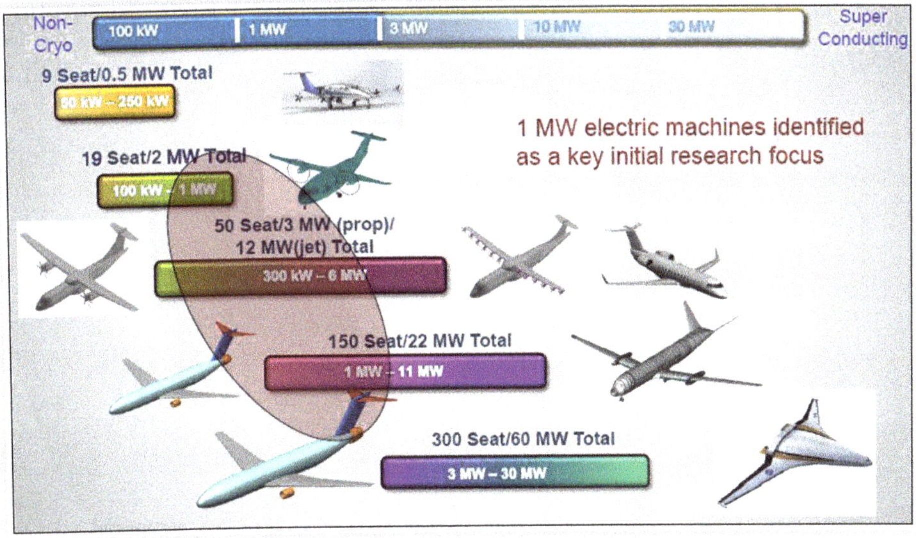

Figure 6.27 [6.25] shows motor sizing requirements captured by NASA for a variety of aircraft segments. The ~1 MW electric motor appears to intersect with several aircraft segments, hence research is focused around it.

In order to cover a broad range of applications, motor design has to go through radically different design approaches in order to achieve long-term power density targets of 10-15 kW/kg at system level as shown in Figure 6.28 [6.28].

FIGURE 6.28 Motor power density (current vs. future needs)—system level [6.28].

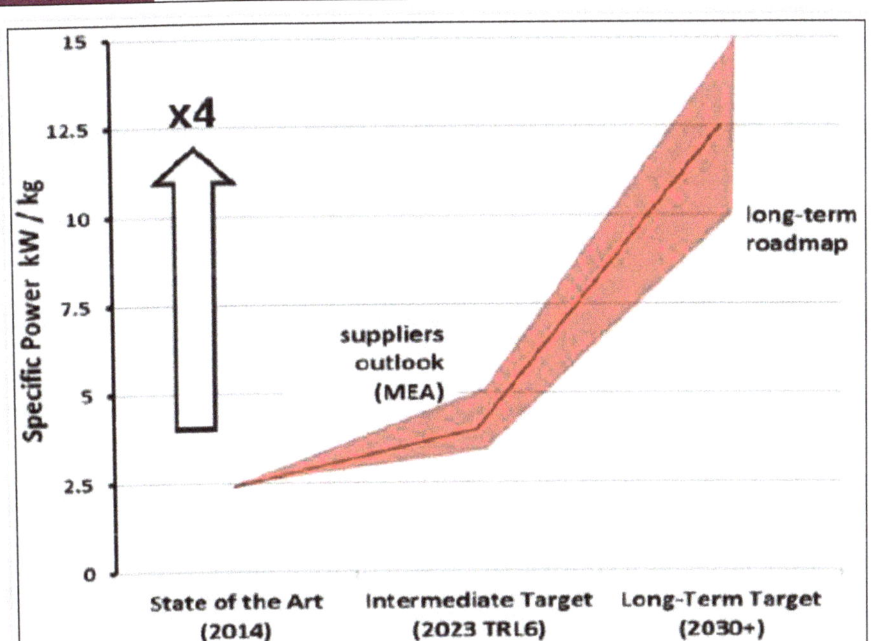

6.3.2.2 **Motor Controls:** Power electronics required for motor controllers need to reach densities in the range of 20-25 kW/kg. Liquid-cooled and cryogenically cooled inverters are under study. On top of silicon, wide bandgap semiconductors based on Silicon Carbide and Gallium Nitride are mainstays in the research of power switching devices to be implemented in multi-level motor controller (inverter) topologies. Very aggressive efficiency goals (>99%) are targeted for the converters which would have to work from a high-voltage DC bus (>1 kV).

6.3.2.3 **Motor and Motor Control Demonstrators:** EP is leveraging the headway being made in motor controller technology in other industries. For instance, building upon its expertise in automotive electric-vehicle motor controls, Siemens, in collaboration with Valeo, has designed liquid-cooled power-dense controllers, suitable for urban mobility and general aviation EP. This has been possible thanks to the introduction of Silicon Carbide (SiC) switching devices (MOSFETs), in lieu of Silicon (Si) IGBTs, within the power electronics (inverter) section of the controller. The following two aircraft applications help illustrate how this technology shift ushers in much improved motor and controller power-to-weight ratios:

Urban mobility—CityAirbus: The eight propellers of this four-seat multi-rotor eVTOL aerial vehicle are driven by eight 100 kW low-speed/high-torque motors. The motor, overrated at 200 kW for demonstration purposes, is shown in **Figure 6.29**. The SiC motor controller shown features a scaled-up power density of ~12 kW/kg for 100 kW.

General Aviation—Extra 330LE: World's first CS23-approved electrically propelled test version of the Extra 330LX, a two-seat aerobatic general aviation aircraft. This aircraft's power requirements are covered by the 260 kW motor shown in Figure 6.26, using two controllers like the one shown in Figure 6.29. The redundant motor's optimization ensures a power density of ~6 kW/kg (**Figure 6.30**), yet to be confirmed through testing. Moreover, the utilization of two motor controllers allows to operate the motor windings independently. An overall motor controller density of ~15 kW/kg is therefore achieved, partially closing the gap with long-awaited expectations of aircraft manufacturers.

FIGURE 6.29 Urban eVTOL, propulsion motor and controller.

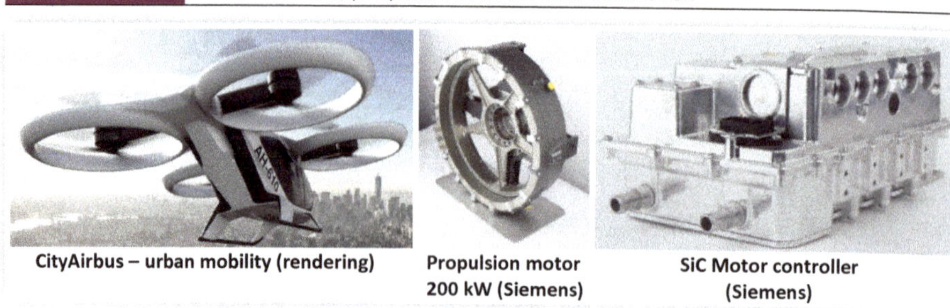

FIGURE 6.30 Electric GA test aircraft + redundant propulsion motor.

Extra 330 LE test aircraft - General aviation

260 kW redundant motor (Siemens)

6.3.2.4 **Materials:** A non-exhaustive list of key research areas in materials is shown below [6.27]:

- New soft magnetic materials for improved performance of converter filters and electric machines
- Electrical insulation with better thermal transfer for electric machine performance enhancement
- High-Conductivity Copper/Carbon Nanotube Conductor for cabling weight reduction
- AC superconducting windings/wiring for electric machines or distribution

6.3.2.5 **Superconducting Electrical Systems:** When aircraft size goes up, propulsion power levels in the megawatt range are required and electrical efficiency requirements get so stringent that new high-voltage superconducting electrical components have to be designed and validated.

For instance, at high power ratings even advanced conventional electrical machines show efficiency drawbacks due to sizable heat losses in the windings and the cabling caused by their electrical resistance. One way to root out this issue is to bring the electrical resistance to zero value thanks to superconductivity. Superconductivity is a quantum mechanical phenomenon of exactly zero electrical resistance, which occurs in certain materials when they are cooled below a critical temperature. It allows for electrical components to be much smaller, lighter, and more efficient compared to conventional copper or aluminum-based technology. For example, light-weight Magnesium Diboride ($MgB2$) superconducting wires are already used in medical applications. Therefore, superconductivity can be leveraged in the distribution of electrical power from energy sources (gas power unit and energy storage) to propulsors by cooling cables, generators, motors, and other components to extreme levels (–252°C).

In summary, specific energy and efficiency of EP for large-sized aircraft can be increased thanks to superconducting electrical systems, which have to be cooled to very low temperatures. This cryogenic cooling can be achieved through heat transfer to a cryogenic fluid (e.g., liquid hydrogen, liquid helium, or liquid nitrogen) used as a heat sink. This may open new avenues for synergy with "cryogenic" fuels that may be required

FIGURE 6.31 Cryocooler weight-to-power ratio forecast [6.22].

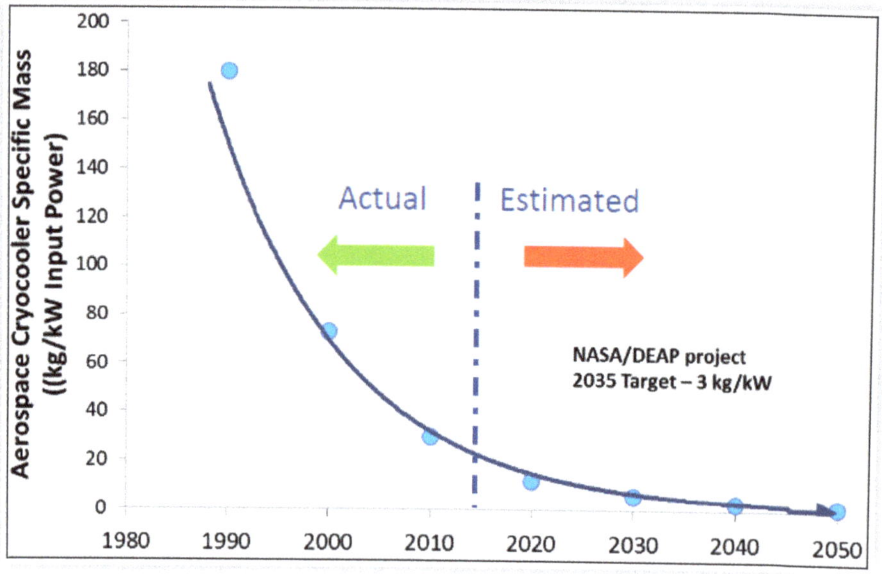

on board. Remember, fuel cells produce electrical energy from hydrogen which in its liquid form has to be kept cooled down to –252°C.

An alternative cooling solution consists in producing autonomously the necessary cold temperatures by way of an electrical cryocooler. Space and other applications already integrate cryocoolers, but aircraft applications require their weight-to-power ratio to be dramatically cut back, from ~20 kg/kW down to ~3 kg/kW (**Figure 6.31** [6.22]).

6.3.2.6 **Fuel Cells:** If hydrogen is used in lieu of aviation fuel, a fuel cell can be used as an electrical energy source for propulsion and systems. Fuel cells are also capable of producing energy from aviation fuel provided a reformer is used to convert fuel to hydrogen. There are two types of fuel cell technology, the proton exchange membrane (PEM) and the solid oxide fuel cell (SOFC). The latter consumes only 80% of the fuel processed and needs inlet air preheating of up to 1000°C, the reason why it is deemed more suitable for hybrid-electric rather than EP. Hence, PEM fuel cells working at lower temperatures (~100°C) and lower pressures may constitute EP replacement of gas turbine powerplants, so long as the required power densities, currently off the mark, are reached in the future.

SOFCs are generally thought to work best as part of a hybrid system with a gas turbine, where the resulting combined cycle can theoretically achieve thermal efficiencies in the 70% range. The SOFC/gas turbine combination offers several advantages over conventional turbofan or turboprop propulsion systems. These advantages include increased energy efficiency, reduced emissions, fuel savings, noise reduction, and efficient integration. Nevertheless, SOFC weight, volume, and cost have to be mitigated in the face of steep specific energy challenges in aviation propulsion.

Traditionally, SOFC development has focused on low-cost, high-efficiency stationary power, with less emphasis on weight and volume. Additionally, a key disadvantage of using SOFCs for aviation applications is that they have not specifically been designed for high specific power. Thanks to promising new materials and packaging techniques, power density and specific mass of SOFCs could be pushed higher, to more than double the values of the technology status considered in [6.34]. Thus, a twofold improvement in performance, coupled with a twofold decrease in mass yields a fourfold improvement in specific power.

Figure 6.32 [6.34] shows that PEM fuel cells operate most efficiently at lower powers and currents densities. For applications where cruise is ~1/3 of takeoff power, this results in a potential fuel cell stack efficiency of 71% in cruise and takeoff stack efficiency of ~59%. This makes it well suited for some applications as cruise fuel consumption is a predominant contributor to mission fuel burn.

Fuel cells are more efficient than gas turbines, therefore fuel load may be alleviated. Furthermore, multifunctional integration on an aircraft may allow for fuel cells to ensure vital processes like de-icing, cabin air conditioning, water supply, or fire suppression, thanks to green byproducts from fuel cell operation such as heat, water, or oxygen-depleted exhaust air.

Figure 6.33 provides a comparison of power and energy densities of current and future fuel cell EP for aviation on the one hand and, on the other, current turbofan engines [6.35]. Values shown are at system level taking into account fuel load and the tank required to carry it; therefore back-to-back comparison is made between fuel cells/motors/hydrogen fuel and tank and turbofans/kerosene and tank.

FIGURE 6.32 PEM fuel cell efficiency (current vs. future) [6.34].

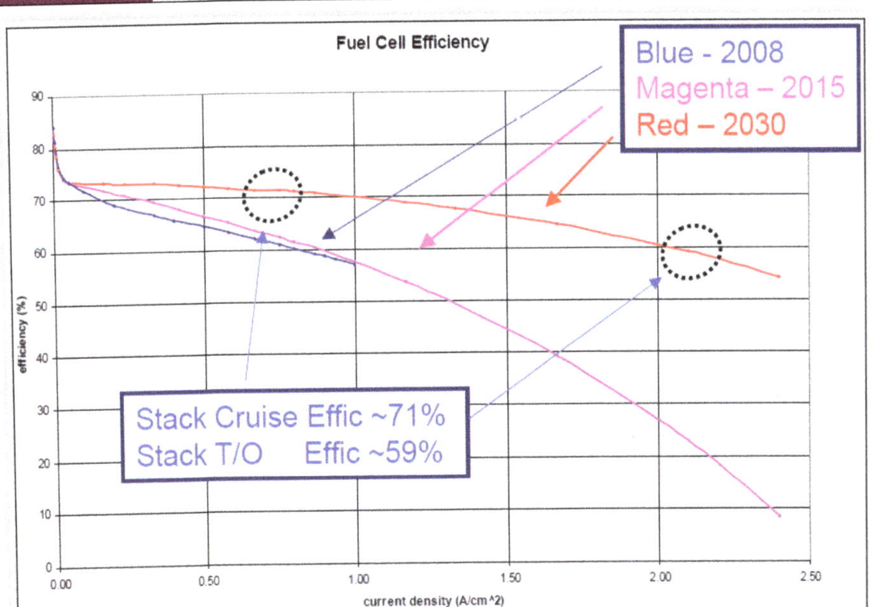

FIGURE 6.33 Fuel cell EP (current/future) and turbofan specific power and specific energy [6.35].

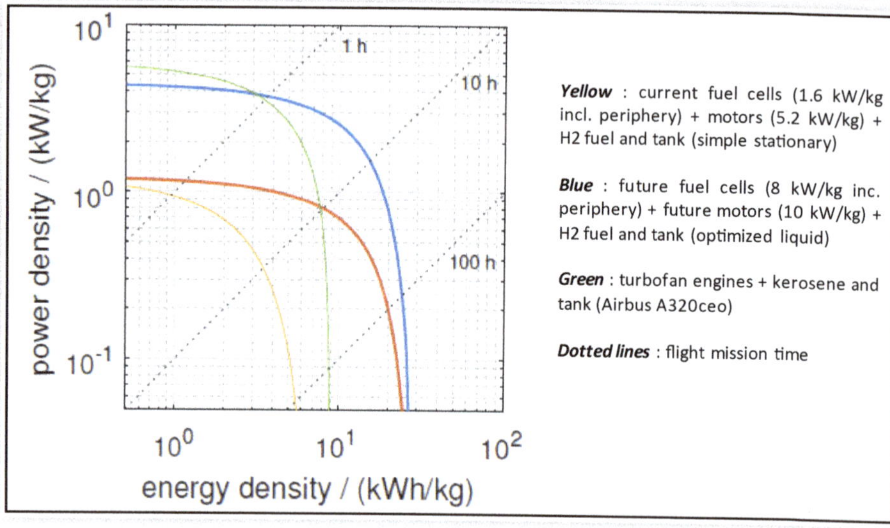

The analysis in [6.35] shows that future fuel cells, with much higher specific power and specific energy, can become compelling alternatives to turbofan engines. The longer the flight duration, the better they stack up against turbofan engines.

6.3.2.7 **Batteries:** Lithium-ion (Li-Ion) batteries, with much improved specific energy compared to predecessors, have been leveraged on recent aircraft developments. But they usually are backup energy storage systems. Newer battery technologies may make inroads as aircraft primary energy sources able to supply EP or HEP. Nevertheless, their implementation on large aircraft boils down to significantly high energy-to-weight ratios, even with just HEP. Additionally, battery cycle life can be a significant business driver.

While there is still quite a long way to go (**Figure 6.34** [6.28]), scientists expect the specific energy of new generations of energy storage systems to exceed 1000 Wh/kg within the next two decades. This is a sure giant leap from today's best performance, but still necessary to cover system-level long-term targets for large aircraft.

Lithium-air (Li-Air) batteries are poised to offer a promising solution for energy storage requirements. Their specific energy is better than that of lithium-ion technology thanks to a lighter cathode. Moreover they use oxygen to operate, freely available in the environment.

Developments in other transportation sectors (automotive) also benefit the aerospace industry. Because of power range and size similarities between automobiles and small general aviation aircraft, direct synergies become feasible.

FIGURE 6.34 Battery specific energy (current vs. future needs)—system level [6.28].

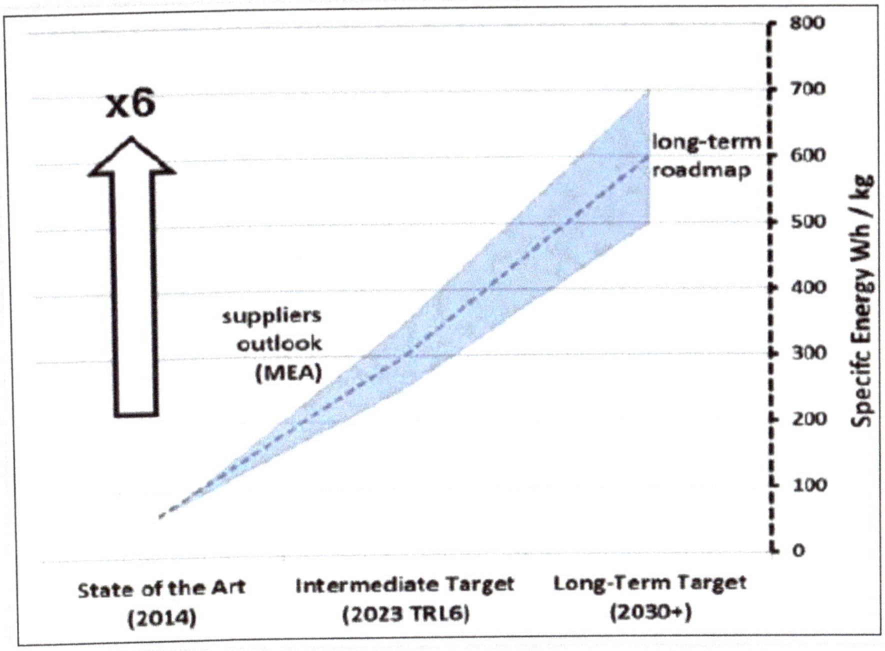

FIGURE 6.35 Battery and gasoline engine specific energy—powertrain level [6.36].

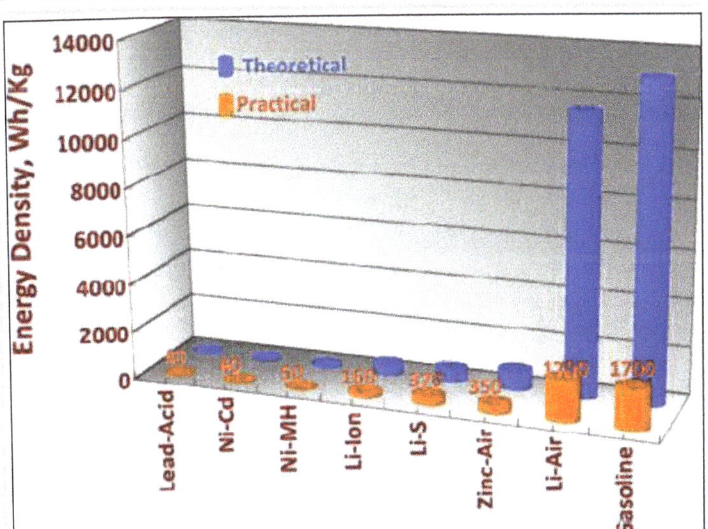

To illustrate this, Figure 6.35 [6.36] compares specific energy between various battery technologies and highlights the potential of Li-Air to reach 1700 Wh/kg, putting it on par with gasoline-powered automotive internal combustion engines, absent the gas emissions. "Practical" values correspond to a system perimeter including powertrain and energy source (battery or gasoline), whereas "Theoretical" refers to the specific energy of the raw material behind the energy source (lithium or gasoline).

6.4 Conclusion

In summary, before large aircraft manufacturers can opt in, at least two decades of research and development are foreseen on enabling EP technologies (motors, superconducting systems, fuel cells, and batteries). Figure 6.36 shows timelines foreseen for EP systems of various aircraft segments to achieve technology readiness level (TRL) 6 [6.25].

In fact, short- and long-range aircraft set the bar high for maturity and performance that technology roadmaps have to deliver prior to their inclusion in electric aircraft developments. Nevertheless, when addressing small general aviation aircraft, targets may be hit ahead of time because of lesser complexity and constraints. They may find a sweet spot and therefore be poised to become early adopters of EP. No superconducting technology is required for these aircraft, and they can make do with less stringent energy storage densities.

FIGURE 6.36 Electric propulsion: Timescale vs. aircraft segments/power (adapted from [6.25]).

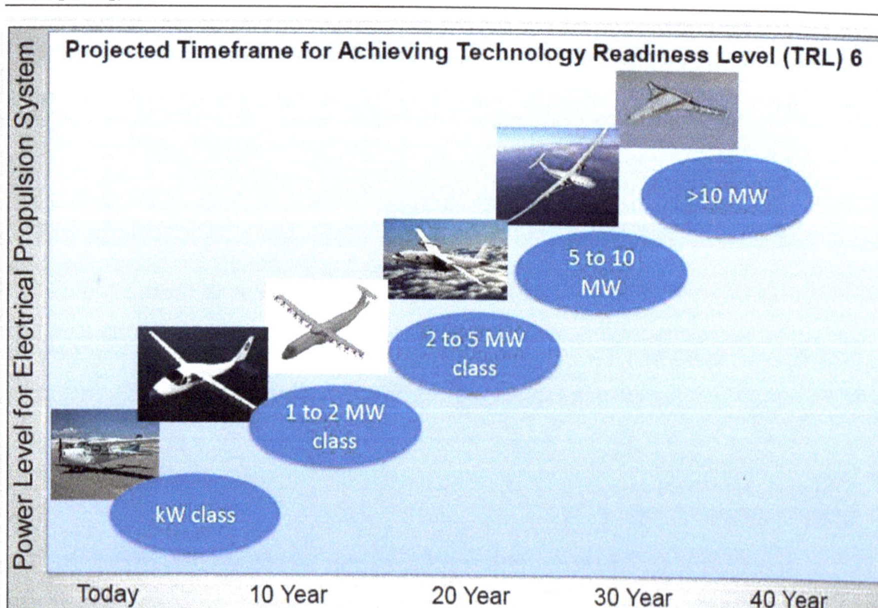

On a global scale, various studies and technology development programs focusing on propulsion architectures discussed previously are in full swing. Future electric and hybrid-propulsion baselines, wherein dependence on conventional aviation fuel will retreat and the potential of alternative energy sources will be parlayed, are expected to drastically cut back on fuel consumption and greenhouse gas emissions.

References

[6.1] https://www.cgtrader.com/3d-models/aircraft/part/aircraft-engine-of-continental-io-550, accessed Feb. 12, 2018.

[6.2] http://www.algavia.top/cirrus-sr-22/, accessed May 1, 2018.

[6.3] https://www.aelisgroup.com/news/aelis-group-arranged-the-sale-of-3x-pw127, accessed May 1, 2018.

[6.4] http://saspozatlse.free.fr/ATR%2072.html, accessed February 2, 2018.

[6.5] http://www.aeroexpo.online/fr/prod/safran-aircraft-engines/product-170451-869.html, accessed February 12, 2018.

[6.6] http://theflight.info/airbus-a320-american-airlines-photos-and-description-of-the-plane/, accessed February 12, 2018.

[6.7] https://www.turbinesinfo.com/gas-turbines/, accessed October 18, 2018.

[6.8] Air Transport Action Group (ATAG), "Beginner's Guide to Aviation Efficiency," Air Transport Action Group (ATAG), November 2010, https://www.atag.org/component/attachments/attachments.html?id=615, accessed November 20, 2018.

[6.9] https://www.compositesworld.com/articles/composites-in-commercial-aircraft-engines-2014-2023, accessed Feb. 18, 2018.

[6.10] https://www.economist.com/news/science-and-technology/21684775-quieter-more-economical-jet-engine-fitted-gearbox-about, accessed February 14, 2018.

[6.11] https://www.cfmaeroengines.com/engines/, accessed February 16, 2018.

[6.12] Image source https://thepointsguy.com, https://www.google.com/search?q=737-MAX-engine&client=firefox-b&source=lnms&tbm=isch&sa=X&ved=0ahUKEwiUoIb0peXaAhXFthQKHToiBuoQ_AUICigB&biw=1024&bih=694&dpr=1.25#imgrc=QGmwlihkqg1zlM, accessed February 16, 2018.

[6.13] http://www.mro-network.com/manufacturing-distribution/cfm-confirms-initial-leap-1a-and-leap-1b-assembly-allocation, accessed February 16, 2018.

[6.14] Image source https://aviator.aero/, https://www.google.com/search?q=pw1100g-jm-engine-4&client=firefox-b&source=lnms&tbm=isch&sa=X&ved=0ahUKEwj1hICIq-XaAhWLtRQKHexcBR0Q_AUICygC&biw=1024&bih=694&dpr=1.25#imgrc=yvpEL-IaDbpgdM, accessed February 16, 2018.

[6.15] http://www.air-cosmos.com/pratt-whitney-envisage-des-changements-sur-son-gtf-102223, accessed February 16, 2018.

[6.16] https://commons.wikimedia.org/wiki/File:Gas_turbine_efficiency.png, accessed February 14, 2018.

[6.17] Committee on Propulsion and Energy Systems to Reduce Commercial Aviation Carbon Emissions; Aeronautics and Space Engineering Board; Division on Engineering and Physical Sciences; National Academies of Sciences, Engineering, and

Medicine, *Commercial Aircraft Propulsion and Energy Systems Research: Reducing Global Carbon Emissions*, (The National Academies Press, 2016), ISBN 978-0-309-44096-7, DOI 10.17226/23490, http://nap.edu/23490, accessed February 19, 2018.

[6.18] http://www.cimne.com/vpage/2/2189/Objectives, accessed February 17, 2018.

[6.19] http://www.foundrymag.com/moldscores/alcoa-s-100-million-investment-casting-expansion-completed, accessed May 1, 2018.

[6.20] https://www.knaviation.net/air-france-boeing-787-inaugural-part-2, accessed May 1, 2018.

[6.21] Sinnett, M., "787 No-Bleed Systems—Saving Fuel and Enhancing Operational Efficiencies," *AERO Magazine* Q4, 2007, published by Boeing.

[6.22] Armstrong, M., "Superconducting Turboelectric Distributed Aircraft Propulsion," Rolls-Royce, *Cryogenic Engineering Conference/International Cryogenic Materials Conference*, Tucson, AZ, USA, June 28-July 2, 2015.

[6.23] Garrigan, N., Aviation Electrification, GE Aviation, Power Systems Track Panel—Electric Flight, *SAE 2016 Aerospace Systems and Technology Conference (ASTC)*, Hartford, CT, USA, September 27-29, 2016.

[6.24] Hepperle, M., "Electric Flight—Potential and Limitations," prepared by German Aerospace Center (DLR) for *NATO STO Workshop "Energy Efficient Technologies and Concepts of Operation"*, 2012, doi:10.14339/STO-MP-AVT-209.

[6.25] Heidmann, J., "NASA Investments in Hybrid-Electric Technologies for Large Commercial Aircraft," NASA Glenn Research Center, *Electric & Hybrid Aerospace Technology Symposium 2015*, Bremen, Germany, November 17-18, 2015.

[6.26] Bradley, M.K. and Droney, C.K., "Subsonic Ultra Green Aircraft Research: Phase II—Volume II—Hybrid Electric Design Exploration," prepared by the Boeing Company for NASA Langley Research Center, Contract NNL08AA16B—Task Order NNL11AA00T, NASA/CR–2015-218704/Volume II, April 2015.

[6.27] Jansen, R.H., Bowman, C., Jankovsky, A., Dyson, R. et al., "Overview of NASA Electrified Aircraft Propulsion Research for Large Subsonic Transports," NASA Glenn Research Center, *EnergyTech 2017*, Cleveland, OH, USA, October 31-November 2, 2017.

[6.28] Rostek, P., "Hybrid Electric Propulsion—A European Initiative for Technology Development," Airbus, *Electric & Hybrid Aerospace Technology Symposium 2015*, Bremen, Germany, November 17-18, 2015.

[6.29]. http://www.aircosmosinternational.com/aurora-explores-turboelectric-aircraft-concept-102265, accessed March 6, 2018.

[6.30] Malkin, P., "Hybrid Electric Distributed Propulsion Aircraft—A Hybrid Like No Other?", Newcastle University, *Electric & Hybrid Aerospace Technology Symposium 2015*, Bremen, Germany, November 17-18, 2015.

[6.31] Based on image from http://aviationweek.com/awin/turbo-electric-concept-forms-latest-focus-esaero accessed March 6, 2018.

[6.32] Ricci, M., "Gen-Sets for Hybrid-Powered Transformative Flight," LaunchPoint Technologies, *Transformative Vertical Flight Concepts: Second Annual Joint Workshop on Enabling New Flight Concepts through Novel Propulsion and Energy Architectures*, AHS-AIAA-NASA, NASA Ames Research Center, CA, USA, August 3-5, 2015.

[6.33] Clarke, S., Papathakis, K., Samuel, A., Lin, Y., and Ginn, S., "NASA SCEPTOR Electric Concept Aircraft Power System: X-Plane Electric Propulsion System Design and Qualification for Crewed Flight Testing," NASA Armstrong Flight Research Center, *IEEE Transportation Electrification Conference and Expo (ITEC)*, June 27-29, 2016, Dearborn, MI, USA, doi:10.1109/ITEC.2016.7520287.

[6.34] D'Angelo, M.M., Gallman, J., Johnson, V., Garcia, E. et al., "N+3 Small Commercial Efficient and Quiet Transportation for Year 2030-2035," prepared by GE Aviation, Cessna Aircraft and Georgia Institute of Technology for NASA Langley Research Center, Contract NNC08CA85C, NASA/CR-2010-216691, May 2010.

[6.35] Kadyk, T., Krewer, U., Winnefeld, C., and Hanke-Rauschenbach, R., "Analysis and Design of Fuel Cell Systems for Aviation," *Energies* 11 (February 6, 2018): 375, doi:10.3390/en11020375, published by MDPI.

[6.36] Nikolić, Z. and Živanović, Z., "The Contribution and Prospects of the Technical Development on Implementation of Electric and Hybrid Vehicles," http://dx.doi.org/10.5772/51771, published by Intech.

CHAPTER 7

Aircraft Applications—Part I

Electric Propulsion, Electric Taxiing

Pascal Thalin
Chair and Member - SAE Electric Aircraft Steering Group

Sven Taubert
Lufthansa Technik

Case studies, demonstrations, and developments of various electric and hybrid-electric (HE) propulsion concepts applied to different aircraft segments are presented in this chapter. They are expected to be ever more environment-friendly and in some instances may deliver drastic fuel burn reductions falling in line with air transport goals set by NASA in the United States and Advisory Council for Aeronautics Research in Europe (ACARE) (**Figure 7.1**). The European Commission has established a 2050 target goal in its highly ambitious Flightpath 2050 Vision for Aviation of reducing CO_2 by 75%, NOx by 90%, and noise by 65%.

FIGURE 7.1 ACARE (Flightpath 2050) and NASA (N+3) air transport goals [7.1].

GOALS	TECHNOLOGY BENEFITS RELATIVE TO A YEAR 2000 REFERENCE AIRCRAFT	
	VISION 2020	FLIGHTPATH 2050
CO_2 reduction per passenger km	-50%	-75%
NOx reduction	-80%	-90%
Noise reduction[2]	-50%	-65%
Taxiing		Emission-free
Manufacturing and design		All aircraft recyclable
Sustainable alternative fuels		Europe established as center of excellence
Atmospheric research		Europe at the forefront

GOALS	N+1 = 2015 Technology Benefits Relative to a Single Aisle Reference Configuration	N+2 = 2020 Technology Benefits Relative to a Large Twin Aisle Reference Configuration	N+3 = 2025 Technology Benefits
Noise (cumulative below stage 4)	-32 dB	-42 dB	-71 dB
LTO NOx Emissions (below CAEP 6)	-60%	-75%	better than -75 %
Performance: Aircraft Fuel Burn	-33%	-50%	better than -70 %
Performance: Field Length	-33%	-50%	exploit metroplex concepts

7.1 Battery Electric Propulsion—Small General Aviation

Since 2015 **Pipistrel** has been selling *Alpha Electro*, a commercial two-seat electric aircraft for pilot training aligned with the needs of flight schools. Key capabilities include 80-mile range, short takeoff distance, powerful 1000+ fpm climb rate, and endurance of 1 h plus 30 min of reserve. The *Alpha Electro* is optimized for traffic-pattern operations, where 13% of energy is recovered on every approach, increasing endurance and at the same time enabling short-field landings. In the face of rising fuel costs, technologies developed specially for this aircraft cut back pilot training costs by as much as 70%. Training operations with zero gas emissions and minimum noise can therefore be based in smaller airfields closer to towns. The *Alpha Electro* meets microlight and American Society of Testing and Materials (ASTM) light sport aircraft (LSA) criteria, as well as standards for electric propulsion.

The electric powertrain is operated with one simple lever, thereby easing pilot burden. Design features of this aircraft include an almost silent cockpit, a responsive powertrain, and the total absence of gas emissions. The engine is rated at 50+ kW with rotating speeds in the range of 2100-2400 rpm (Figure 7.2).

FIGURE 7.2 Battery electric propulsion—two-seat training aircraft (**Pipistrel** *Alpha Electro*) [7.2].

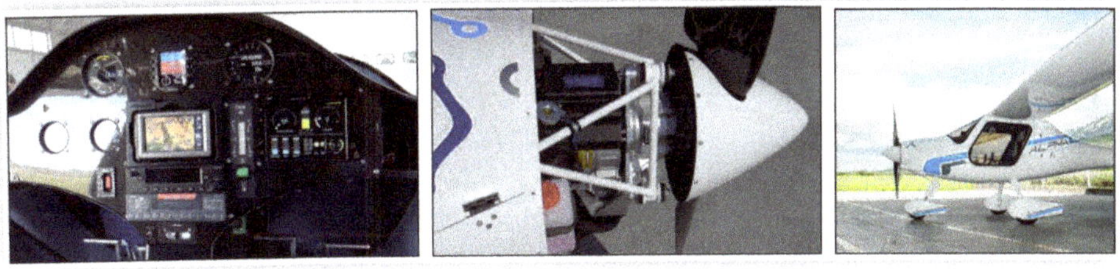

Bye Aerospace has also built a cost-effective two-seat electric airplane similar to the *Alpha Electro* for pilot training.

In 2016, *Solar Impulse*, a one-seat long-range experimental solar-powered aircraft, completed the first ever round-the-world series of flights (with recharging and repair stops) and set numerous records. It had no fuel on board but batteries working exclusively from solar energy.

Using renewable energy (solar) to power the batteries, **Aero Electric Aircraft Corporation (AEAC)** is focusing on electrically propelled two- and four-seat training aircraft called *Sun Flyer*. It is to be FAR 23-certified. Lithium-ion batteries run the electric propulsion system, which directly drives the composite propeller. Engine performance is controlled by an electronic control unit ensuring optimal use of battery energy. Compared to a conventionally powered aircraft, reduced cooling drag is achieved and the nose area is smaller thanks to reduced motor size and cooling intake requirements. On the other hand, propeller efficiency gets improved by the utilization of additional blade area. Emission-free and low-noise (55 dB) performance allows for environment-friendly operations at airports close to populated areas. Engine controls are simplified and piloting made easier by the single control lever to intuitively tune the electric motor's throttle with no need to adjust mixture richness or monitor cylinder head temperature as in aircraft with internal combustion engines. The operating costs of the *Sun Flyer* are five times lower than the costs associated with similar training airplanes equipped with fuel-burning internal combustion engines. Only about 3 USD of electricity is needed for each flying hour.

In 2014-2015 Airbus conducted flight demonstrations on a two-seat aircraft called the *E-Fan*, a 1100 lb (empty weight) training aircraft equipped with electric propulsion with an autonomy of 1 h. The propulsion system uses 250 V lithium-ion polymer batteries powering twin 30 kW electric fan motors producing a thrust of 340 lbf.

As can be seen, small general aviation (GA) aircraft with smaller thrust needs are perfect first candidates for electric propulsion. This is proven by commercially available two-seat trainer aircraft found in the LSA category. Nevertheless, depending on battery densities to be reached in the short term, larger capacity aircraft may also reap the benefits of electric propulsion. Research in this area, underpinned by feasibility studies and real-life validation testing, is enabling faster technology development.

FIGURE 7.3 Cirrus SR-22—GA aircraft [7.3].

By the horizon of 2025, motor solutions for GA discussed in Chapter 6 may find their way into either electric or HE propulsion of GA aircraft such as the four-seater Cirrus SR-22 with a maximum takeoff weight of up to 4000 lbs (**Figure 7.3**).

FIGURE 7.4 Conventional GA aircraft redesigned into an electric aircraft with DEP [7.4].

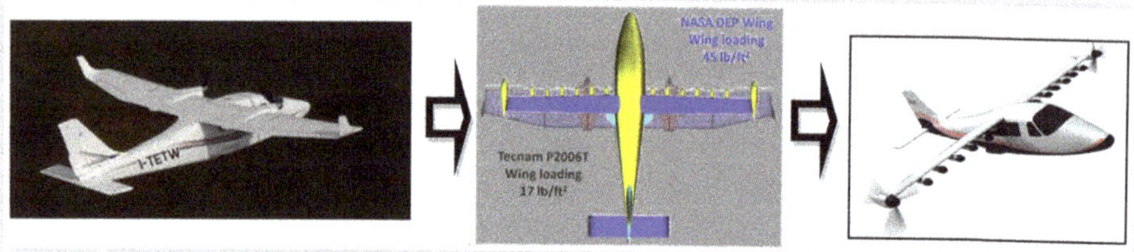

In the same timeframe, electric propulsion of such GA aircraft with battery densities of ~500 Wh/kg at cell level may allow for ranges of up to 500 nm. This would come with 90% reduction in greenhouse emissions at ~30% lower operating costs. An earlier milestone by 2020 timeframe may also be possible with ranges of above 300 nm with lower battery densities around 400 Wh/kg. Last but not the least, significant engine noise reduction is possible with electric propulsion.

Electric propulsion winds up offering dependable designs. Also the inherent integration benefits of electric propulsion allow compact aircraft sizing applicable to all aircraft segments including the small scales in play in GA. On top of this, when redundant Distributed Electric Propulsion (DEP) is implemented, the additional integration benefits enable closely coupled synergies across aerodynamics, propulsion, control, acoustics, and structures.

Conventional GA aircraft are only aerodynamically efficient at low speeds in cruise due to the wing oversizing necessary to meet constraints related to stall conditions and airfield lengths but which unfortunately compromises the lift-to-drag ratio. When converting to an electric aircraft, wing downsizing in conjunction with DEP, offers better wing loading, more resilient aerodynamics, lower drag, and higher lift, therefore allowing higher speeds during cruise (**Figure 7.4**).

Since 2014 NASA has been validating the DEP concept through the SCEPTOR project dedicated to the research of elements of electric propulsion. It included wind tunnel-like ground validation of the DEP high-lift system (**Figure 7.5**). Tests showed that the distribution of power throughout the leading edge motors generated more than double the lift at lower speeds than traditional propulsion systems.

The X-57 Maxwell, a modified 4-seat P2006T GA aircraft, will enable flight testing of DEP, with the goal to demonstrate a fivefold reduction in energy use at high-speed

FIGURE 7.5 NASA ground and flight DEP demonstrators.

Ground validation of DEP high-lift system [7.5]

X-57 Maxwell [7.6]
Battery DEP test aircraft (under construction)

cruise compared to traditional propulsion, zero in-flight carbon emissions, and flight that is much quieter for the community on the ground. The test aircraft comprises 14 electric motors and propellers (12 high-lift motors along the leading edge of the wing and two large wingtip cruise motors).

The high-aspect ratio experimental wing will feature a large reduction in area, with wing loading increasing from 17 to 45 lb/ft^2. The reduction in wing area will also contribute to more efficient cruise flight through decreasing friction drag. Testing has to ensure that the modified wing with the integrated high-lift motor system will allow the X-57 to take off and land at the same speed as the baseline P2006T, and the aircraft will also be less sensitive to gusts and turbulence, leading to a smoother flight.

One of the most critical, and noticeable, elements of boosting cruise efficiency through electric propulsion is the relocation of the large cruise motors to the tips of the wing. The 100 hp (75 kW) original Rotax 912S piston engines will be replaced by 60 kW motors developed by Joby Aviation, reducing the weight of each motor and propeller from approximately 125 lbs to about 57 lbs. The much lighter-weight electric motors allow for their relocation outboard. By moving the cruise motors from inboard position to the wingtips, the cruise motors will recover energy that would otherwise be lost in the wingtip vortices. The X-57 has the following characteristics:

- Aircraft weight: circa 3000 lbs
- Cruise altitude: 9000 ft
- Cruise speed: ~172 mph
- Batteries: Lithium Ion—69.1 kWh
- 2 Cruise out-runner motors and propellers: 60 kW from Joby. Air-cooled, direct drive, permanent magnet 6-phase
- 12 High-Lift in-runner motors and propellers: 5-blade, folding propeller, 10 kW, air-cooled

Leading-edge high-lift motors and propellers would be designed to operate, along with the wingtip cruise motors, to get the X-plane airborne. When the plane levels out for cruise mode, the high-lift motors would then deactivate. The five propeller blades for each motor would then stop rotating, and fold into the nacelles, so they don't create unwanted drag during cruise. The two wingtip cruise motors would maintain flight during this phase of the flight. When the time comes to land, the motors would then reactivate, and centrifugal force would cause the propeller blades to unfold and create the appropriate lift for approach and landing.

The first piloted DEP X-plane will also seek to reach the goal of zero carbon emissions in flight, which would surpass the NASA N+3 fuel burn reduction targets. For small aircraft, electric propulsion provides not only a five-to-ten times reduction in life-cycle greenhouse gas emissions with current electricity, essentially zero emissions with renewable-based electricity, but it also provides a technology path for small aircraft to eliminate 100 Low Lead Avgas, which is the foremost contributor to current lead environmental emissions [7.7].

Therefore, since the X-57 will be battery powered, it can run off renewable-based electricity, making clear the environmental and economic advantages.

DEP technology takes advantage of the scale-invariant nature of electric motors to achieve high specific power and efficiency at any size, and combined with electric motor compactness and high reliability, DEP permits thrust to be optimally positioned on the airframe where propulsion can enhance the aerodynamic and control characteristics. As such the distributed propulsion concept can be applied to larger aircraft as will be discussed further below.

7.2 Urban Air Transportation

Rethinking the future of urban transportation is at the heart of many current research studies. Deeper market penetration of greener and more silent electric ground vehicles will result in notable enhancements with regard to noise and pollution.

It is understood that disruptive road traffic management solutions with driverless electric cars will certainly help squeeze out better traffic fluidity from increasingly congested road networks and ensure better traffic flow for an entire region, or metroplex. However, in the future, another method to ease the saturated roads of megacities is to take to the air with green flying vehicles wiggling in and out of crammed urban environment, where not even helicopters would be up to the task, noise being one major barrier. What is needed is a silent environment-friendly VTOL vehicle concept that can square with the evolving on-demand needs of city dwellers of large metropolitan areas. Drawing from technologies developed for electric GA aircraft, novel concepts of small inner-city flying vehicles are gathering momentum.

In the future, people would be able to hail such flying cars via a smartphone app. A flight, five times faster than driving, with zero emissions, may be booked on such vehicles with one click. It will pick customers up from a nearby VTOL hub and fly them safely to their destination. On-demand aviation has the potential to radically improve urban mobility, giving people back time lost in their daily commutes by saving up to an hour a day in day-to-day travel.

As discussed in [7.8], just as skyscrapers allowed cities to use limited land more efficiently, urban air transportation will use three-dimensional airspace to alleviate transportation congestion on the ground. A network of small, electric VTOL aircraft, will enable rapid, reliable transportation between suburbs and cities and, ultimately, within cities. The development of infrastructure to support an urban VTOL network will likely have significant cost advantages over heavy-infrastructure approaches such as roads, rail, bridges, and tunnels. Repurposed tops of parking garages, existing helipads, and even unused land surrounding highway interchanges could form the basis of an extensive, distributed network of "vertiports" (VTOL hubs with multiple takeoff and landing pads, as well as charging infrastructure) or single-aircraft "vertistops" (a single VTOL pad with minimal infrastructure). As costs for traditional infrastructure options continue to increase, the lower cost and increased flexibility provided by these new approaches may provide compelling options for cities and states around the world.

Furthermore, VTOLs do not need to follow fixed routes. Trains, buses, and cars all funnel people from A to B along a limited number of dedicated routes, exposing travelers to serious delays in the event of a single interruption. VTOLs, by contrast, can travel towards their destination independently of any specific path, making route-based congestion less prevalent.

Recently, technology advances have made it practical to build this new class of VTOL aircraft.

Over a dozen companies, with as many different design approaches, are passionately working to make VTOLs a reality. The closest equivalent technology in use today is the helicopter, but helicopters are too noisy, inefficient, polluting, and expensive for mass-scale use. VTOL aircraft will make use of electric propulsion so they have zero operational emissions and will likely be quiet enough to operate in cities without disturbing the neighbors. At flying altitude, noise from advanced electric vehicles will be barely audible. Even during takeoff and landing, the noise will be comparable to existing background noise. These VTOL designs will also be markedly safer than today's helicopters because VTOLs will not need to be dependent on any single part to stay airborne and will ultimately use autonomy technology to significantly reduce operator error. Therefore, urban mobility concepts consider both piloted and autonomous versions of flying vehicles, similar to the trend seen in the automotive sector.

Technological advancements in electric propulsion have already achieved turbine-like specific power, but with more than triple the efficiency of small helicopter turboshaft engines. DEP enables a fivefold improvement in the vehicle efficiency for a GA Conventional Takeoff and Landing (CTOL) aircraft. Importantly, large efficiency improvements are possible because DEP enables fixed-wing VTOL aircraft that avoid the fundamental limitations of helicopter edgewise rotor flight, while wings provide lift with far greater efficiency than rotors. Because helicopters have such poor aerodynamic efficiency, far lower than CTOL aircraft, DEP VTOL aircraft will likely be able to achieve a tenfold efficiency improvement compared to existing helicopters.

In the recent years multiple air-taxi startups have come to life and are in the process of designing and testing VTOL aircraft prototypes. They have entered into partnerships with aircraft manufacturers, system suppliers, and blue chip/tech companies. Wealthy entrepreneurs are also financially backing some product development initiatives. Venture capital firms are eyeing investment opportunities to jump into. Airlines and ride-sharing firms have joined the fray as well.

Like the evolution taking place in the field of on-demand mobility solution offered by Uber or Lyft, future urban mobility is poised to expand airborne services. Today, on-demand mobility services offered by such firms consist in matching availability of fleets/individual vehicles with customer transportation needs. These needs, ranging from personal and professional mobility, access to services (health), or goods delivery (food, purchases, etc.), can be quickly satisfied thanks to airborne mobility.

Over a dozen startups around the world have projects at various stages of design as do Airbus and Boeing. Uber is working on an air-taxi service, tentatively called Uber Elevate, based on various concepts of VTOL aircraft and the vertiport/vertistop infrastructure they require (**Figure 7.6**) [7.8].

The right VTOL vehicle is the linchpin to opening up a new market for short-hop air transportation and must feature:

- Unparalleled safety through redundant systems
- Low noise
- Range and speed optimized for moving groups of people quickly to and from their destinations

The analysis in [7.8] compares the performance of a future urban mobility VTOL with an electric car for a 50-mile trip. The VTOL power requirement for takeoff is 500 kW and weighs 4000 lbs including batteries with specific energy of 400 Wh/kg at pack level.

FIGURE 7.6 Rendition of hub and stops for urban mobility VTOL aircraft [7.8].

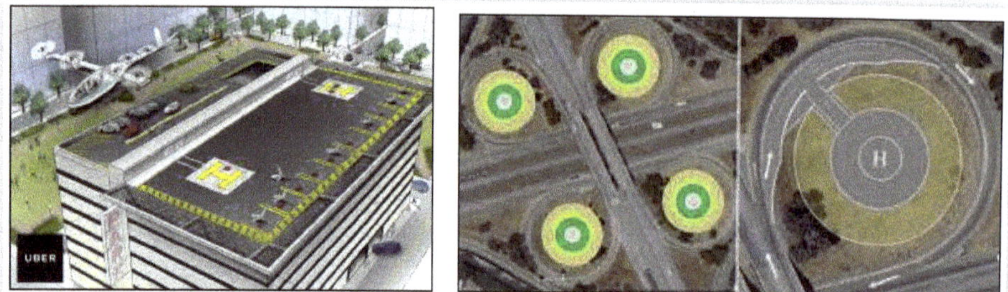

Rooftop "vertihub" "Vertistops" integrated to roads

The analysis considers vehicle "burdening" accounting for the vehicle inefficiencies associated with the extra energy required for a car to accelerate and brake multiple times relating to traffic lights/stop signs and, for the VTOL to take off, accelerate to cruise speed, and land vertically. Also considered is trip "burdening" which accounts for the trip inefficiencies of congestion on roads for the car and the more direct routing that a VTOL experiences on most trips with no congestion assumed.

Compared to an electric car, at shorter trip distances, the VTOL is less energy efficient per mile because it spends less time in the more efficient cruise mode while the power required for vertical takeoff and landing remains constant. As such, vehicle "burdened" VTOL motion efficiency improves at longer distance and decreases at shorter distances. Trip distances longer than 50 miles would experience high cruise efficiency for a larger portion of the trip, making the VTOL compare even more favorably to the car.

At higher car and VTOL speeds, the VTOL's motion efficiency gains an advantage over the electric car. Electric VTOLs present a transport mode that is as efficient or has higher efficiency than attempting to use cars at high ground speeds.

It is expected that daily long-distance commutes in heavily congested urban and suburban areas and routes underserved by existing infrastructure will be the first use cases for urban VTOLs. This is due to two factors. First, the amount of time and money saved increases with the trip length, so VTOLs will have greatest appeal for those traveling longer distances and durations. Second, even though building a high density of landing site infrastructure in urban cores (e.g., on rooftops and parking structures) will take some time, a small number of vertiports could absorb a large share of demand from long-distance commuters since the "last mile" ground transportation component will be small relative to the much longer commute distance.

Fuel costs and other operating expenses (like maintenance and depreciation) are a significant component of air travel costs. A small, efficient helicopter on a short, 25-mile trip can use more than 40 USD in fuel. A fully-electric vehicle can cover the same distance at twenty times lower energy cost with significant reductions in other operating expenses. These savings can be delivered directly to customers.

While electric VTOLs may be more expensive on a per-unit basis than traditional combustion planes or helicopters, the end cost to customers can be lower if the vehicle is both high capacity and highly utilized. This is how existing commercial airlines are able to drive down costs.

They keep their vehicles filled, moving a lot of people most of the time. As a result, they can spread the fixed costs of the vehicle and the variable costs of pilots and maintenance over a larger number of passenger trips at progressively lower ticket prices. In the long term, VTOLs may be an affordable form of daily transportation for the masses, even less expensive than owning a car. Normally, people think of flying as an expensive and infrequent form of travel, but that is largely due to the low production volume manufacturing of today's aircraft. Even though small aircraft and helicopters are of similar size, weight, and complexity to a car, they cost about 20 times more.

Ultimately, if VTOLs can serve the on-demand urban transit case well—quiet, fast, clean, efficient, and safe—there is a path to high production volume manufacturing (at least thousands of a specific model type built per year) which will enable VTOLs to achieve a dramatically lower per-vehicle cost. The economics of manufacturing VTOLs will become more akin to automobiles than aircraft.

Direct ownership of such flying cars may not be the panacea because it is thought that people will share flying cars, much like Uber and Lyft with ground vehicles. Initially, of course, VTOL vehicles are likely to be very expensive, but because the ridesharing model amortizes the vehicle cost efficiently over paid trips, the high cost should not end up being prohibitive to getting started. And once the ridesharing service commences, a positive feedback loop ultimately reduces costs and thus prices for all users, that is, as the total number of users increases, the utilization of the aircraft increases. Logically, this continues with the pooling of trips to achieve higher load factors, and the lower price feeds back to drive more demand. This increases the volume of aircraft required, which in turn drives manufacturing costs down.

Uber estimates the initial price of an urban mobility VTOL to be double that of a helicopter, but with high volumes (5000 units/year) could be cut to less than half the helicopter price. In the long term an autonomous VTOL will have a competitive direct cost of 35 cents per "ground mile" at the vehicle level.

Although electric propulsion has the potential to mitigate aircraft noise and feature zero operational emissions, in order for the on-demand VTOL urban mobility concept discussed above to reach the market, several feasibility barriers in the following areas remain to be transcended (success factors are mentioned within parentheses):

- Certification (timely new processes for new VTOL aircraft)
- Battery technology (high specific energy, fast charge rate, and low cost per kWh)
- Vehicle efficiency (commercially viable efficient vehicles)
- Vehicle performance and reliability (high cruise speed and low takeoff/landing times, all-weather operation)
- Air Traffic Control (ATC) (high frequency and airspace density over metropolitan areas)
- Cost and affordability (reduced operational and affordable acquisition costs)
- Safety (better safety than with cars)
- Life-cycle emissions (reduced hydrocarbon fuel-based electricity generation for battery recharge)
- Vertiport/vertistop infrastructure in cities (ideal location and sufficient number)
- Pilot training (reduced training times)
- Autonomy (vehicle management)

Delivering progressively better service to customers, with steadily lower fares, would shift an ever greater percentage of trips to zero-emission vehicles. Point-to-point transportation that is significantly better than driving and steadily more affordable over time is the foreseen evolution of urban mobility.

Building the necessary infrastructure with vertiports and vertistops will certainly boost job creation in the construction sector and allow for the increase of revenue streams for towns and cities. Let us now look at several VTOL aircraft of various sizes for urban mobility and the strategic partnerships behind to bring them to the market.

The *S2* from **Joby Aviation** features 16 propellers, 12 of which are tilting electric propellers that give the aircraft its VTOL abilities (**Figure 7.7**). Once it is in the air and up to speed, the plane's rotors fold away into aerodynamic bullet shapes, and then the aircraft can reach speeds of up to 200 mph, traveling up to 200 miles using four additional cruise-optimized props on the backs of the wings and tail fins (**Figure 7.8**).

The electric propellers are fixed pitch, giving them less moving parts than systems used by helicopters. VTOL cruise efficiency is estimated to be about three times better than a helicopter because of the design bias towards achieving high cruise efficiency through wing-borne flight instead of using rotors.

The *S2* will also cost considerably less. The initial cost to build one is estimated at 200,000 USD, and the electric motors mean the operating cost over time will be much lower than that of a helicopter.

The **Airbus A³** *Vahana* shown in Figure 7.7 features the following technical characteristics:

- 8 motors: 15 kW each, self-piloted, variable pitch
- Batteries: Lithium-ion polymer rechargeable in 15-60 min
- Takeoff weight: 1650 lbs
- Speed: 109 mph
- Range: 62 miles

Intel owns a stake in an air-taxi startup called **Volocopter** which develops the 2-seat *Volocopter 2X* light sport multicopter, halfway between a drone and a helicopter. It is equipped with 18-propeller DEP for VTOL and is controlled simply with a joystick (**Figure 7.9**). Demonstrations are slated for 2020. The 450 kg vehicle requires a power of 50 kW and comes with a propeller span of 9 m. It can achieve speeds of up to 62 mph.

FIGURE 7.7 Battery DEP VTOL aircraft designs for urban mobility concept.

2-seat *S2* (**Joby Aviation**) [7.9] 1-seat *Vahana* (**Airbus A³**) [7.10]

FIGURE 7.8 *S2* rotor configurations—vertical takeoff/landing and cruise phases.

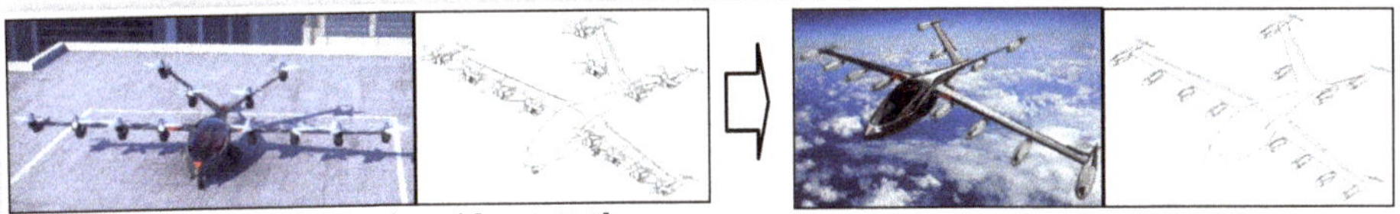

VTOL–tilt rotors deployed [7.11, 7.9]

Cruise–tilt rotors folded [7.9]

FIGURE 7.9 Battery DEP VTOL 2-seater—*Volocopter 2X* (**Volocopter**) [7.12] with simplified commands [7.13].

One-hour range performance is projected with a likely HE solution. Certification as ultra-light aircraft is foreseen with an estimated price around 250,000 EUR (including batteries).

The *Volocopter 2X* features very stable and safe flight characteristics, designed for operating as an autonomous air taxi in megacities. It includes flight control solutions with redundancy and safety features. Smooth flight is enabled by dozens of microprocessors monitoring the environment for turbulences, winds, etc. and sending signals in milliseconds to the rotors. These can react and perform the slightest adjustments instantly due to their battery powered electric motors. Only "on/off" and direction indication from the pilot is required, the remaining being automated. All complex pilot operations are taken care of by onboard vehicle electronics, thereby easing pilot workload. Among other things, this frees the conventional helicopter pilot from having to set vehicle minimum speed, altitude, or angle of attack by moving the cyclic stick and collective pitch controls. As such piloting the vehicle becomes accessible and hassle-free to all.

Specifically addressing autonomous urban mobility, **Airbus** has come up with the *CityAirbus* (**Figure 7.10**). It is a 4-seat multi-rotor VTOL aircraft that can be optionally piloted.

On the *CityAirbus* power comes from an electric drivetrain made up of eight separate motors driving eight propellers producing the required thrust. Drivetrain energy is sourced from four 140 kW battery packs. It has a top speed of about 75 mph, and its batteries store enough electricity for roughly 15 min of operation. That's not much, even for a vehicle designed for short intracity trips, so Airbus will need to develop a clever way to recharge the batteries in record time.

Future urban air transportation concepts discussed in this section blur the line between drones and personal aircraft and call for a regulated airspace yet to be defined. Urban air traffic management will need to be upgraded before vehicles like the *CityAirbus* are deployed en masse. First examples of the *CityAirbus* are hoped to be in the skies as

FIGURE 7.10 Battery DEP VTOL 4-seater—*CityAirbus* (**Airbus**) [7.14].

early as 2023. They will initially be controlled by professional pilots, a solution which will facilitate the certification process and help the general public warm up to the idea of traveling in an electric pod. Later down the line, *CityAirbus* is planned to get entirely autonomous.

7.3 Fuel Cell Electric Propulsion—Commuter Aircraft

Instead of batteries, fuel cells can also be used as primary energy sources and bring aircraft design optimizations. For illustration purposes let us consider the weight and cost-sensitive commuter aircraft segment [7.15]. Figure 7.11 shows a projected baseline for a 20-seater in Y2035 with range capability of 800 nm. It is an advanced twin-engine turboprop aircraft with unit thrust rating of ~3350 lbf for the engines. Cruise mach

FIGURE 7.11 Y2035 reference turboprop commuter [7.15].

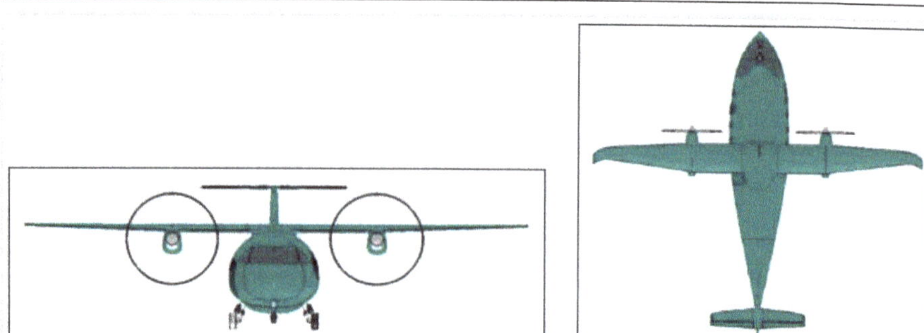

number below 0.6 eliminates compressibility drag and enables the formation of natural laminar flow. The airframe shape enables low drag through laminar flow and high aspect ratio wings.

Many challenges still remain for fuel cell technology and it may take some years to mature. Theoretically, for a thrust rating of 4200 lbf, cruise thrust SFC of electric propulsion may be 70% lower than that of an advanced turboprop engine.

Nevertheless, actual mission fuel burn improvement is expected to be far less due to the significant increase in propulsion system weight and volume. Due to their large volume, the fuel cell powerplants would need to be located in the fuselage, with electric motors on the wing driving the propellers. Liquid hydrogen, due to the spherical, insulated high-pressure tanks, would also be stored in the fuselage. The study in [7.15] makes aggressive estimates for Y2030 fuel cells as well as cryogenic motor power-to-weight ratios and transmission efficiencies.

On a mission basis, compared to the Y2035 turboprop reference, the aircraft with fuel cell propulsion allows for fuel burn and mission energy usage to be slashed by 55% and ~19%, respectively. On top of these savings, essentially zero emissions, with the exception of water vapor, is another major benefit with fuel cell propulsion. On the flip side, fuel cell propulsion based on liquid hydrogen winds up increasing aircraft empty weight by well over 80% resulting in thrust requirements that have to be scaled up by ~20%. This fallout leads to performance deterioration such as increased costs and community noise that only reduced propulsor power requirements, or further improvements in technology, can alleviate. Fuel cell systems also merit further study for longer-range applications.

7.4 Battery Electric Propulsion—Regional Aircraft

The aircraft studied is the Dornier Do 328 aircraft with twin turboprop engines, able to transport 32 passengers over a range of 650 nm (Figure 7.12).

The analysis in [7.16] shows that on the baseline turboprop aircraft, if engines were to be replaced with electric propulsion and aviation fuel swapped with batteries for the same weight, the range would get reduced by a factor of six. Similar range performance

FIGURE 7.12 Dornier Do 328 regional aircraft (conventional vs. electric propulsion) [7.16].

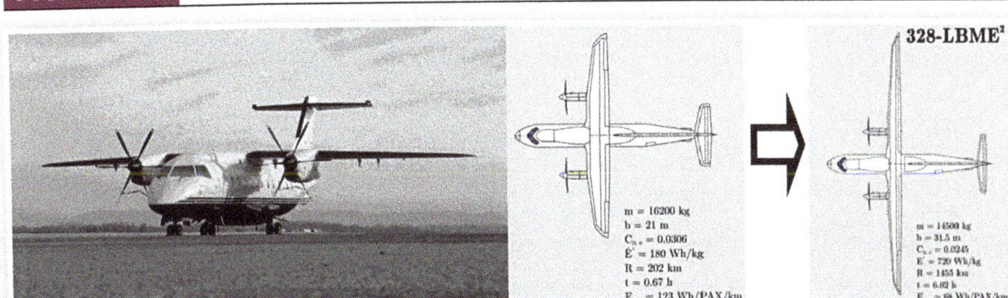

as the baseline with 40% energy reduction is possible when resorting to electrical propulsion, provided the aircraft is redesigned with the following options:

- Electric propulsion with high specific energy 720 Wh/kg batteries
- Propulsion installation, wing/tail laminar flow for zero-lift drag reduction
- Increased wing span reducing induced drag
- Structural lightweight design targeting aircraft empty weight reduction

7.5 Battery Electric Propulsion—Short-Range Aircraft

Following up on the ECO-150R studies of ES Aero, Wright Electric, a U.S. startup, revealed plans for a 150-seat aircraft for short-haul flights by 2027 using modular swappable battery packs and can be either fully electric (335-mile flights) or HE, depending on the advances in energy storage. Interested in that concept, easyJet, a leading European airline, made it part of its strategy to progressively decarbonize and reduce noise from aviation operations. To that end, the airline announced a collaboration with Wright Electric. Figure 7.13 shows a rendition of the electric aircraft intended for short-haul operations (London to Paris, Seoul to Jeju, or New York to Boston).

A recent preliminary design study intended to fulfill the Flightpath 2050 goals was performed by Bauhaus Luftfahrt [7.1]. The proposed 180-200 seater targets an initial range capability of 600 nm for a projected entry into service in 2030. Depending on technological evolutions, mainly regarding batteries, a longer range (1400 nm) may not enter into service before 2040. The aircraft design targets emission-free operations and combines multiple new developments such as C-shaped wings and electric propulsion (Figure 7.14) [7.19].

The twin-engine propulsion configuration relies solely on advanced technology batteries with energy densities of up to 2000 Wh/kg. The powerplant integrates the following features:

- High-voltage (3000 VAC) high-temperature superconducting (HTS) electric motors
- Integrated cryocooler
- Reversible rotation for thrust reversal
- Silent advanced ducted fans
- Translating nozzle plug

FIGURE 7.13 Battery DEP aircraft rendition—Tube-and-split wing short-range aircraft [7.17, 7.18].

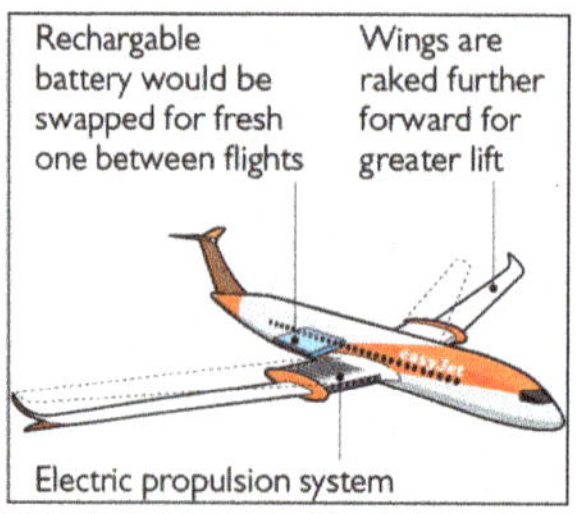

FIGURE 7.14 Battery electric propulsion—C-wing short-range aircraft [7.19].

Fully electrified aircraft systems are supplied in high voltage (540 VDC) via distribution based on solid-state power controllers (SSPC). This aircraft requires batteries to be packed into containers and installed in the cargo bay. As onboard recharging is not deemed time-efficient, batteries would be exchanged during turnarounds.

7.6 Electric Taxiing—Short-Range Aircraft

As early as in the 1970s, several companies started developing external electric tug systems for aircraft pushback at airports, while Airbus provided its conclusions on a study for "motorised wheels for autonomous taxiing for a 76 tonnes subsonic aircraft" [7.20].

Very recently, several aircraft Ground Support Equipment (GSE) manufacturers have offered pilot-controlled semi-robotic electric tractor designs, consisting of an external electric drive attaching itself to the landing gear for ground taxiing purposes. However, on the downside, changes are necessary in airport taxi ways to accommodate the taxi device returning to the gate without interfering with aircraft using the taxiways.

The more electric and electric short-range aircraft can also achieve electric taxiing, without any external assistance, thanks to electric wheel motors installed in the landing gear system. To understand the operational impact and to assess different electric taxiing solutions, a general understanding of the current taxi-in and taxi-out process is needed.

Taxi-in indicates the movement from the runway to the gate after landing and taxi-out from the gate to the runway before takeoff. The classic taxi-in process utilizes all available jet engines near idle mode to gain enough thrust to move the aircraft. Since the 2000s some airlines switched to the so-called single engine taxiing approach with only one engine running during most of the taxiing period. Due to the unsymmetrical propulsion, the aircraft cannot perform sharp turns in the direction of the running engine. To avoid this problem and depending on the aircraft configuration, there is also an approach that 2 out of 4 engines are producing the necessary symmetrical thrust. The overall target of all these new approaches is to save fuel and reduce noise and emissions. One limiting factor is a mandatory cool-down and warm-up phase for each engine. Depending on the engine type, this period is between 2 and 5 min long.

The taxi-out process is an inversed taxi-in plus pushback if the aircraft is parked at a gate. The pushback is performed by a tug via a tow bar or by a tug, which scoops the nose landing gear and lifts it off the ground. This process is accompanied by a walk-out assistant who coordinates the communication between the cockpit and the tug. On many airports, the pushback has to end at a dedicated pad ensuring a safe engine run-up without the danger of engine blast damage.

The eTaxi approach goes even one step further than single engine taxiing: apart from the mentioned warm-up/cool-down phases of the engines, the eTaxi system would move the aircraft without engines running.

7.6.1 Operation

As an electric motor does not have the limitation to provide unidirectional thrust, an aircraft with an eTaxi system can perform the pushback by the aircraft itself without the help of a tug or other devices. Nevertheless, studies and surveys involving airports showed that a walk-out assistant seems to be mandatory. Otherwise, the danger of collisions with other aircraft, vehicles, or persons would be too high. Unlike the current pushback process, the aircraft could use the shortest possible way. Since starting the engines is not needed to move autonomously, the aircraft also doesn't need a dedicated pad after pushback. To optimize the process, an exact estimation of the time from gate to the runway is essential in order to comply with the engine's warm-up phase before being ready for departure at the runway holding point. The taxi-in process works similarly. After the cool-down time, the engines would be shut down and the electric motor moves the aircraft to the gate.

7.6.2 System Configurations and Performance

The term eTaxi describes currently an electric drive on one or several wheels of the landing gear. When considering electric aircraft configurations, there might be the opportunity to use electrically driven fans to perform the same task. Nevertheless, this chapter focuses on landing-gear-wheel motor-based developments with similar system architectures. All these systems consist of a human machine interface (1), electric drive (2), power electronics and cabling (3), and a power source (4). In the following section, the most prevalent approaches are described, and their technical and operational feasibilities are discussed (Figure 7.15).

In general, there are two possibilities for the integration of the electrical drive: the nose landing gear or the main landing gear actuation system. Both systems have advantages but also challenges. The following sections will go into further details as well as explore some more disruptive ideas.

FIGURE 7.15 eTaxi system sub-component location on aircraft (source Lufthansa Technik).

7.6.3 Nose Landing Gear Actuation

Since the end of 2008, first approaches to integrate an electric motor into the nose landing gear were demonstrated (**Figure 7.16**). The most common test platforms are short-range aircraft like Airbus 320, Boeing 737, or Embraer 190.

One of the challenges for nose landing gear actuation is that it carries only a very limited share of the overall weight of the aircraft—around 8% of the maximum takeoff weight (MTOW). This leads to a lag in traction. So, even with very powerful motors, there is a big challenge to accelerate to a certain speed in an acceptable timeframe. Estimations and tests showed that a maximum speed of 7-10 km/h is possible. However, almost all airports require a minimum taxi speed of 20 km/h or even more under all-weather conditions with a maximum slope of 3%. This is due to the limited number of taxiways from the gate to the runway: aircraft have to queue on them and therefore need to be able to operate at set speeds in order not to cause delays. Conventional aircraft with jet engines taxi at speeds of up to 40 km/h. One possibility to increase the speed of a nose wheel-driven aircraft would be to implement traction control. Some of the demonstrated systems showed first approaches to integrate this kind of function.

Another integration aspect is the weight of the electrical motor itself. Studies show that there has to be a major modification of the landing gear to reinforce its structure. This impacts the retraction system, which also has to be scaled-up to ensure its function. This makes a retrofit scenario very unlikely.

The biggest advantage for nose landing gear actuation is that its wheels are not fitted with brakes. Therefore, there is more space for integration and no interference with the heat dissipation of the wheel braking disks.

Concerning the power source, there are two main choices: the gas turbine-based APU and/or a fuel cell in combination with batteries or not. Concerning integration and retrofit scenarios, the APU solution seems to be the most promising. Yet, a lot of aircraft

FIGURE 7.16 Fuel cell-powered nose wheel eTaxi integrated by Lufthansa Technik on DLR A320 test aircraft [7.21].

configurations are not able to provide enough power with their existing hardware. Therefore, an additional battery or even fuel cell may resolve this issue, but at the cost of additional weight.

To use a fuel cell as a power source only for electrical taxiing would not tap its full potential. Therefore, the substitution of the APU by a fuel cell would probably need at least a major aircraft upgrade or even a complete new aircraft design. So-called multi-functional fuel cells allow for the following secondary use cases, thanks to fuel cell byproducts such as water, heat dissipation, and exhaust gas:

- Onboard toilets and galley making use of waste water
- De-icing applications deriving heat from thermal dissipation
- Fuel tank inerting enabled by oxygen-depleted exhaust gas

From several perspectives, despite its potential, fuel cell technology for aircraft use may be a few years behind.

Last but not the least, the cabling between the nose wheel in the front of the aircraft and the APU, located in the tail section, has a relative weight impact, especially because of the high power consumption of the electric motor.

7.6.4 Main Landing Gear Actuation

Another strategy for electric drive integration is into the main landing gear actuation system. This has also been researched and demonstrated on Airbus 320 (**Figure 7.17**), Boeing 737, and Embraer 190 short-range aircraft.

There is a big potential for retrofit on the A320. In fact, its main landing gear was originally designed and certified for steel brakes. Since a couple of years, Airbus is delivering the A320 with weight-saving carbon brakes by default. This modification saved potentially enough weight to now install an electric drive. For the Boeing 737, a carbon brake option is available but a lot of airlines still order steel brakes because of the lower cost. More recent aircraft from Embraer, Bombardier, or COMAC have been designed from the start with the lighter brakes, making major landing gear modifications necessary for the integration of an eTaxi drive.

FIGURE 7.17 Main landing gear eTaxi system demonstrator (Lufthansa Technik).

Due to the higher load on the main landing gear (92% of the aircraft weight), traction should not be an issue. The acceleration and maximum speeds depend mainly on available onboard APU electric power. Estimations and first demonstrators showed that the required 20 km/h taxi speed is feasible even without APU modification, depending on the aircraft type and configuration.

Yet, there are also some technical challenges for this integration. On the one hand, the electric motor is very close to the brakes inducing reciprocal impacts between them. The motor heat potentially increases the brake cool-down time, whereas heat dissipation from brakes may cause the motor to overheat. Therefore, some kind of active cooling would be necessary (**Figure 7.18**). First prototypes used active fan cooling for the motors.

On the other hand, the system has to incorporate some kind of clutch to separate the motor from the landing gear in case of a system failure (Figure 7.18). The main failure case to be considered is a blocked drive also blocking the wheels entailing potential consequences such as an explosion, a runway excursion, or other catastrophic events. Precluding the possibility of blocked wheels is therefore a major requirement for the overall system design. Another reason for a clutch is that the high landing/takeoff speed relative to taxi speed compromises the motor design to withstand the high speed, and generates high voltage.

The power source, power electronics, and cabling are very similar to the nose landing gear integration. As the basic target for a main landing gear drive is faster taxiing, the estimated power consumption of the concept is higher. This is why most concepts favor the utilization of the APU.

According to publications, using such a system can reduce NOx emissions by 50%-75% during taxiing. Furthermore, the average pushback time is reduced by 60%.

Overall the most promising integration would be on the A320 due to the high number of aircraft, the route profiles, and retrofit potential as previously discussed.

FIGURE 7.18 Electric Green Taxiing System (EGTS) prototype—Paris Air Show (2013).

In this chapter, the potential savings of an eTaxi system were examined. In general, there are three big factors for the business case:

- Fuel saving
- Pushback time and costs
- Additional weight

It was assumed that the aircraft can pushback/drive backwards with the help of only a "walk out assistant" at the gate and then is electrically driven to the runway thanks to eTaxi. The engines would be started during taxi taking into account the warm-up time in order to be ready for takeoff directly when arriving at the runway. Upon landing, the engines are idled for the required time to cool and then shut down. The aircraft moves up to the gate with the help of electric motors only. Note that taxi time is longer as compared to single engine taxiing due to the lower maximum speed of eTaxi.

On the other hand, eTaxi components (motor, cabling, and power electronics) add extra weight to the aircraft with a negative impact on fuel consumption during the entire flight. So, eTaxi fuel savings has to compensate this.

All in all, during the taxi phase, with aircraft engines running only during their warm-up and cool-down sequences and the APU powering the eTaxi system, fuel burn reductions of up to 3% can be achieved [7.1]. This is mainly explained by the fact that the APU is more efficient than a turbofan engine. So even when only one engine is used for taxiing, fuel consumption is still higher than that obtained with APU-powered eTaxi.

In addition, there might be savings on pushback costs for airlines. For an accurate calculation, it is important to know that pushback costs are often part of airport handling charges. It is unlikely that all airports will pass the complete savings on to the airlines, that is, there is a certain risk on this benefit.

The perfect mission profile for an eTaxi application would be a very short flight with very long taxi-in and taxi-out times. So the target market is an area densely served by big, rather old and crowded, hub airports. Some coastal regions in America, most parts of Europe, and some parts of Asia fit that profile.

Nevertheless, the complete business case has far more indicators. **Figure 7.19** lists some of the most obvious operational impacts expected and their qualitative financial impacts.

Environmental benefit is directly proportional to the fuel consumption. As shown in **Figure 7.20**, there is a linear relationship between flight time and taxi time below which eTaxi stops to be more environment-friendly than conventional taxiing.

As discussed previously there are mandatory minimum warm-up and cool-down phases for the engines, which moves the graph to the right.

A similar curve can be expected concerning the overall mission business case but with a shift by especially the operational costs versus the savings.

Some concepts advertise further benefits especially concerning new boarding scenarios. These savings result from boarding via two-passenger boarding bridges if the aircraft drives fully electrically to the parking position and stops parallel to the terminal, much like a regular bus in a holding bay. According to several airport managers, however, this procedure is undesirable because it blocks two parking positions in front of the terminal, which may be urgently needed, especially at busy airports. Thus, this solution

FIGURE 7.19 Influence of eTaxi on aircraft operations.

Operational impact	Financial impact
Shorter operating time of the engines	Fuel consumption ↓
	Maintenance costs ↓, maintenance frequency ↓
Longer operating time of the APU	Fuel consumption ↑
	Maintenance costs ↑, maintenance frequency ↑
Less actuation of the brakes	Brake wear ↓, maintenance frequency ↓
No pushback by tug	Push back costs ↓
Use of additional ramp agent when resetting	Handling costs ↑
Longer taxiing time	Crew costs ↑
Maintenance of eTaxi system	Maintenance costs ↑

FIGURE 7.20 Environmental/fuel impacts: Dependency on flight and taxi durations.

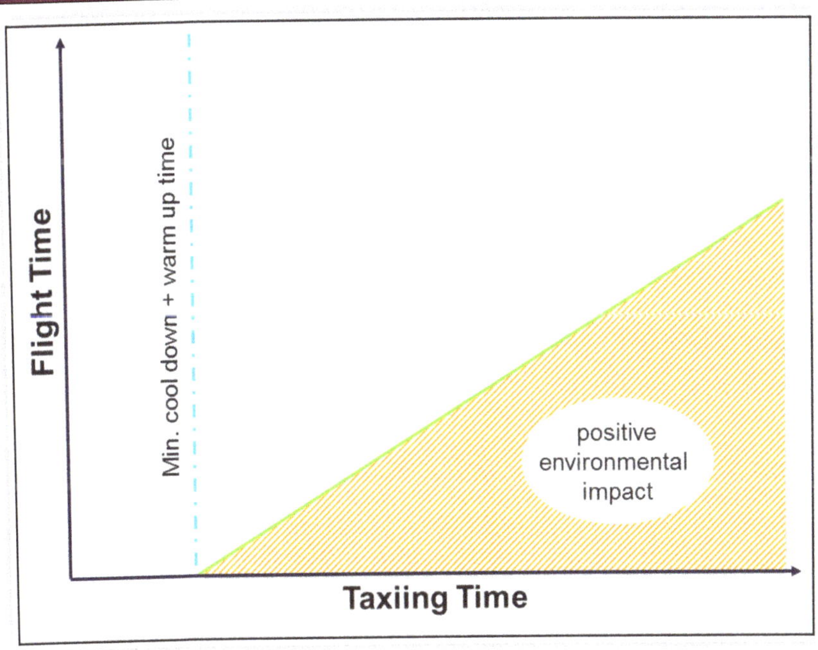

is conceivable only at airports with low flight volumes and sufficient aircraft parking spaces at the terminal, but undesirable at large hubs such as London, Frankfurt, or Paris. On some airports there are gates featuring more than one bridge even for short-range aircraft parked forward. So these savings are more connected to airport infrastructure than to taxiing systems.

7.6.5 Business Models

Different airlines already tried market entry using different business models. Besides the classic buyer furnished equipment (BFE) approach, a lease model was introduced. Most of the airlines currently lease their aircraft and engines. In general, there is not much capacity for consequential capital expenditure in the business. In the lease model, the system supplier would ensure line fit and/or retrofit of the eTaxi system and would cover all the maintenance costs. The supplier would recoup from airlines a certain percentage of savings attributable to eTaxi. The challenge is to accurately track the actual savings as there are a lot of factors in play.

7.7 Conclusions

When aircraft gets larger in size, thrust requirements are pushed higher and electric propulsion is faced with dramatic challenges to do with batteries, fuel cells, motors, and controls, all linchpins to the advent of the electric aircraft.

The case studies above have unveiled how unlocking the potential of electric propulsion, a highly disruptive paradigm shift, is unavoidably tied to giant technology leaps in the fields of energy sources with high specific energy, whether batteries or fuel cells, and power-dense electric motors, and controls. Unfortunately, this may take time to become a reality for collective transportation, given the steep challenges that enabling technologies have to overcome.

Nevertheless, HE propulsion, a stepping stone between an advanced gas turbine engine and electric propulsion extremes, may be able to parlay energy densities lesser than 1000 Wh/kg, achievable within the next twenty years. This compelling solution strikes a balance through seamless and hand-in-hand operation of optimally designed gas turbines on one end and, on the other, electric systems running on batteries or fuel cells. In some cases, energy storage may be altogether forsaken and gas turbines used instead to produce energy required by electrical components of the HE (turboelectric) propulsion system.

References

[7.1] Stumpf, E., Nolte, P., Apffelstaedt, A., Zill, T. et al., *IATA Technology Roadmap*, 4th ed., German Aerospace Center DLR, Georgia Institute of Technology, and IATA, June 2013.

[7.2] http://www.pipistrel-ca.com/alpha_electro.html, accessed March 3, 2018.

[7.3] http://www.algavia.top/cirrus-sr-22/, accessed May 1, 2018.

[7.4] Moore, M.D., "The Forthcoming Distributed Electric Propulsion Flight Era," NASA Langley Research Center, *SAE 2016 Aerospace Systems and Technology Conference (ASTC)*, Power Systems Track Panel—Electric Flight, Hartford, CT, USA, September 27-29, 2016.

[7.5] https://www.nasa.gov/sites/default/files/thumbnails/image/dsc00225_heist_wing1.jpeg, accessed March 3, 2018.

[7.6] https://www.nasa.gov/sites/default/files/thumbnails/image/sceptor_city_nasa_half_res.jpg, accessed March 3, 2018.

[7.7] Moore, M., "Distributed Electric Propulsion (DEP) Aircraft, NASA," https://aero.larc.nasa.gov/files/2012/11/Distributed-Electric-Propulsion-Aircraft.pdf, accessed May 4, 2018.

[7.8] Holden, J. and Goel, N., "Fast-Forwarding to a Future of On-Demand Urban Transportation," white paper prepared by the Uber Elevate team, UBER, October 2016.

[7.9] https://newatlas.com/joby-s2-tilt-rotor-vtol-multirotor-aircraft-concept/40662/#gallery, accessed March 8, 2018.

[7.10] https://vahana.aero/welcome-to-vahana-edfa689f2b75, accessed May 3, 2018.

[7.11] https://newatlas.com/joby-s2-tilt-rotor-vtol-multirotor-aircraft-concept/40662/, accessed March 8, 2018.

[7.12] https://www.connaissancedesenergies.org/volocopter-lhelicoptere-electrique-du-futur-pilotable-par-tous-160413, accessed March 3, 2018.

[7.13] https://www.volocopter.com/en/product/, accessed March 3, 2018.

[7.14] https://www.digitaltrends.com/cars/airbus-cityairbus-quadcopter-news-photos-range-specs-price/, accessed March 3, 2018.

[7.15] D'Angelo, M.M., Gallman, J., Johnson, V., Garcia, E. et al., "N+3 Small Commercial Efficient and Quiet Transportation for Year 2030-2035," prepared by GE Aviation, Cessna Aircraft and Georgia Institute of Technology for NASA Langley Research Center, Contract NNC08CA85C, NASA/CR–2010-216691, May 2010.

[7.16] Hepperle, M., "Electric Flight—Potential and Limitations," prepared by German Aerospace Center (DLR) for NATO STO Workshop "Energy Efficient Technologies and Concepts of Operation", 2012, DOI: 10.14339/STO-MP-AVT-209.

[7.17] https://www.thetimes.co.uk/article/easyjet-promises-electric-planes-ready-for-take-off-by-2027-mpff5x5zc, accessed May 4, 2018.

[7.18] https://mediacentre.easyjet.com/en/stories/11618-easyjet-and-electric-aircraft-pioneer-wright-electric-outline-electric-future-of-aviation#, accessed May 3, 2018.

[7.19] Hornung, M., "Aviation 2050: Potentials and Challenges," Bauhaus Luftfahrt, *Electric & Hybrid Aerospace Technology Symposium 2015*, Bremen, Germany, November 17-18, 2015.

[7.20] Nicolas, Y., "e-Taxi—Taxiing Aircraft with Engines Stopped," Airbus, Flight Airworthiness Support Technology (FAST) Issue 51, *Airbus Technical Magazine*, January 2013.

[7.21] http://www.dlr.de/dlr/en/desktopdefault.aspx/tabid-10204/296_read-931/#/gallery/2079, accessed May 6, 2018.

CHAPTER 8

Aircraft Applications—Part II

Hybrid-Electric Propulsion

Pascal Thalin
Chair and Member - SAE Electric Aircraft Steering Group

This section will cover the topic of hybrid-electric (HE) propulsion (HEP) for which the following case studies will be addressed:

- Parallel HEP—commuter (fuel cell)
- Series HEP—commuter (battery)
- Parallel HEP—short-range aircraft (battery)
- Series HEP—short-range aircraft (battery)
- Distributed HEP (DHEP)—commuter and regional aircraft (battery)
- DHEP—short-range aircraft (no battery, fuel cell, battery)
- DHEP—long-range aircraft (SMES)

8.1 Fuel Cell Parallel HEP: Commuter Aircraft

For HEP let us now consider how fuel cells play out as alternatives to batteries in unlocking new possibilities in propulsion system design, despite some technical and economic issues that have to be overcome prior to their wide utilization in aviation.

For example, in the previous chapter [7.15] analyzes the potential of a fuel cell HEP system on a future reference turboprop commuter aircraft already discussed in Section 7.3 (Figure 7.11).

If a parallel HE concept were to be implemented within a turboprop powerplant, one solution would consist in connecting an electric motor via a gearbox to the same propeller as the gas turbine. The fuel cell would supply electrical energy to drive the motor amounting to additional thrust on top of the thrust produced by the gas turbine. Megawatt-range electric motors are required.

Fuel cells may draw energy from liquid hydrogen acting as the basic energy source. Once again, a liquid-hydrogen solid-oxide fuel cell (SOFC)/gas turbine HE engine offers fuel savings, but with weight and system volume penalties. On the upside, the system has a potential for reducing mission energy usage by approximately 30% and mission fuel weight by 60% compared to an advanced turboprop engine. NOx emissions would be brought to zero for the hydrogen-fueled fuel cell part of the propulsion system.

On the flip side, there would be a fourfold increase in propulsion system weight compared to an advanced turboprop engine. This weight increase results in additional volume that the aircraft would have to accommodate on top of the volume necessary to carry liquid hydrogen fuel. Consequent increase in aircraft size and empty weight (+~90%) may lead to higher aircraft acquisition costs. Moreover, compared to advanced turboprop engines, thrust requirements would have to be scaled up by 20%.

Even in this extremely weight- and cost-sensitive commuter aircraft segment, fuel cell HEP shows some potential for reducing energy consumption, with the added advantage of partially eliminating emissions. But still, at the assumed technology level, it appears less attractive than advanced turboprop engines. Reductions in propulsor power requirements or further improvements in technology may upend this status in the future.

Taking advantage of opportunities in specific power improvement, the penalties at aircraft level discussed above could be completely overwhelmed. Therefore, an SOFC HEP system may offer in the long run a competitive performance edge over advanced turboprop propulsion.

Fuel cell systems also merit further study for larger aircraft applications.

8.2 Battery Series HEP: Commuter Aircraft

For long-range commuter aircraft, airlines and aircraft manufacturers have also showed open interest in partnering with start-ups. With backing from Boeing and JetBlue, startup Zunum Aero is building an initial 12-seat electric jet (**Figure 8.1**) paving the way to scaled-up aircraft designs down the line capable of flying a few dozen passengers over a range of 1000 miles.

The initial 12-seater is slated for entry into service in 2022 with 80% less emissions and producing 75% less noise allowing for environment-friendly short takeoff and landing (STOL) operations of such an aircraft and future larger designs.

But while most companies are banking on huge advances in battery or fuel cell technologies, Zunum has chosen the hybrid route. Today's batteries are still too large and heavy for an all-electric aircraft with the targeted size and range. The *Alpha Electro* aircraft presented in Section 7.1, for instance, has a range of about 80 miles.

FIGURE 8.1 Battery Series HEP—commuter aircraft (Zunum Aero) [8.1].

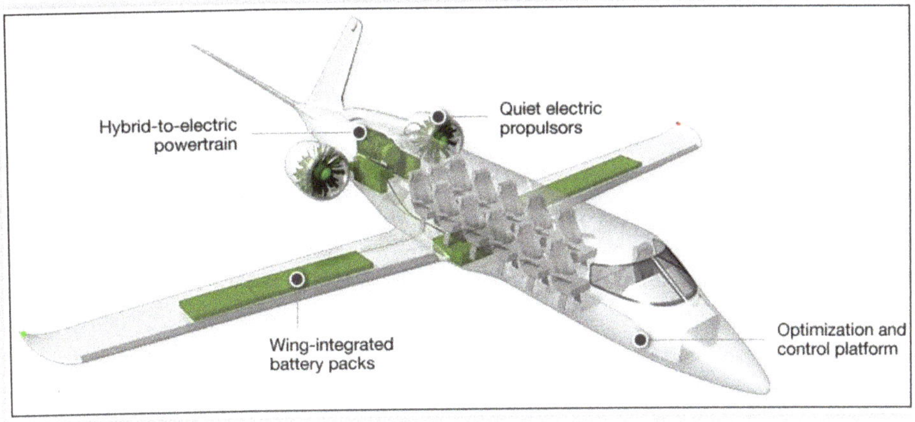

Zunum's airplane design has a "series hybrid" powertrain as outlined in Figure 6.20: propulsion is electric, with only an electric motor powering the drivetrain, but the engine extends the battery's range by generating power for the motor. Most HE cars, by contrast, use parallel hybrid or series-parallel hybrid powertrains. The standardized wing-integrated battery bays are chemistry-agnostic and the fuselage is pre-wired for the eventual transition to fully-electric. The airplane will use a modular 1 MW battery system. Modular battery packs would allow for fast swap or recharge and flight tailoring. They are located in composite box wings structurally optimized to leverage freestream for lightweight cooling.

Design is focused on making the aircraft super efficient. That means including the obvious features such as lightweight parts, a composite airframe, and a sleek shape. But the engineers are also devising ways to use waste heat from the engine to heat the cabin or de-icing systems, and using the motor's windmilling action to recharge the battery during periods of slow flight or on a descent (akin to regenerative braking in cars).

Quiet operation of propulsion system is attributable to electrically driven low-pressure fans (high efficiency), low-noise fine thrust control and variable pitch for strong off-cruise performance. On the other hand, airframe noise such as spoiler noise is attenuated thanks to regenerative braking.

The key performance characteristics of the aircraft developed by Zunum are as follows:

- Range-optimized hybrid powertrain: battery, low maintenance, easy upgrade to future technologies, highly responsive power without altitude lapse
- Quiet variable pitch electric propulsors: short takeoff and climb (40% shorter runways), quiet regenerative braking, all-hours access to small airfields
- Quiet comfortable cabin
- 75% lower community noise
- Intuitive control and optimization: flight optimized energy delivery, integrated control, and one-touch modes. Fine thrust control
- Piloted or drone (pilot optional cockpit, fly-by-wire)
- Optimal regional speeds
- GPS navigation

According to Zunum, one of the keys to delivering this new breed of travel is a radically different air network from the concentrated airliners and hubs of today consisting in leveraging an arsenal of thousands of dormant secondary community airports. This will allow for circumventing the traffic, congestion, and time wasted on the ground, and destinations can be reached 2 to 4 times faster at half the cost… and in a quiet, green aircraft.

The best part about this network: it doesn't require much to get it up and running when using STOL aircraft. These airports would still require facilities for the recharge or swap of battery packs, the massive roll-out underway to support electric ground vehicles is already paving the way. A major upgrade of air traffic control systems that is already in progress supports a distributed electric air network: NextGen in the United States, Single European Skies in Europe, and OneSky in Australia are all due to be live by 2023. These use GPS satellites and onboard ADS-B transmitters to allow for high-density flights, reduction in air-traffic control workload, and all-weather flights to secondary airports without requiring instrument landing systems on the ground.

By 2030, Zunum plans to develop 50-seater airplanes with a 1000-mile range. Announced capabilities include: 8 cents per seat mile of operational cost, 700+ miles max range, 340 mph max speed, and a takeoff distance limited to 2200 ft. The business model is based on short commuter flights between small, regional airports cutting costs and door-to-door travel time. Flying from Boston to Washington, D.C., would cost 140 USD one way, the company predicts, and take around 2.5 h door-to-door, half as long as it takes now.

8.3 Battery Parallel HEP: Short-Range Aircraft

Propulsion performance and sizing are best optimized depending on their location and integration on an appropriately selected aircraft design.

In the short-range segment, the high span "truss-braced tube-and-wing" (TBW) aircraft design has gained traction among research entities and aircraft manufacturers. The concept utilizes a structural support to allow for large span wings without as large an increase in structural weight. By increasing the span, lift can be increased and engine size can be reduced.

The twin-engine configuration considered allows transportation of ~150 passengers over a maximum range of ~3500 nm at cruise Mach number of 0.7, each engine delivering a thrust of ~20,000 lbf. As such, this aircraft design, in combination with two advanced turbofan engines brings a notable 54% block fuel reduction compared to a conventional tube-and-wing aircraft powered by two lower bypass turbofan engines (CFM56). When compared to advanced turbofan engines at fleet level, analysis in [8.2] demonstrates that HEP allows for almost 20% reduction in fuel weight.

The TBW aircraft considered is shown in Figure 8.2 [8.2]. The parallel HEP system down selected for this aircraft comprises two HE turbofan powerplants as detailed in Figure 6.17 and two battery packs in the form of under-wing pods located afar from them.

The resulting design minimizes fuel burn while meeting thrust requirements for the transport of 150 passengers over 900 nm and 3500 nm missions. The propulsion

FIGURE 8.2 Battery parallel HEP—TBW short-range aircraft [8.2].

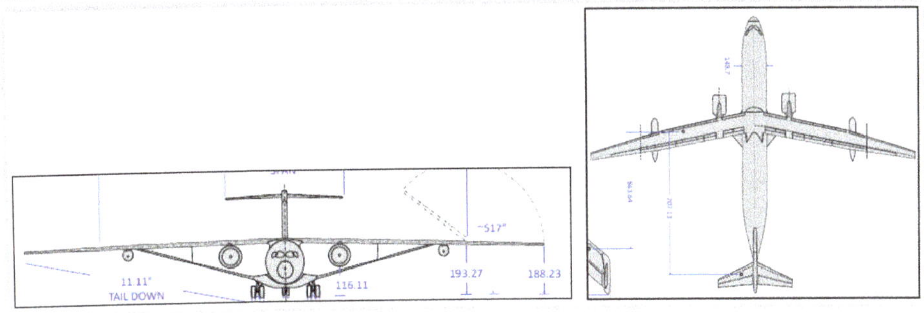

system is also required to satisfy full aircraft thrust requirements while operating strictly as advanced turbofans only, that is, without any aid from the electric motors.

Two distinct HEP configurations are studied:

- "Balanced": both motor and gas turbine provide thrust power in a balanced fashion throughout the flight. Electric motor and batteries are sized for relatively even operation over most of the 900 nm mission. This leads to an aircraft with circa 140,000 lb takeoff gross weight (TOGW), assuming a 1750 hp electric motor and 750 Wh/kg batteries.
- "Core shutdown": the electric motor is sized for cruise on 100% electric power for at least part of the cruise leg and uses more batteries. The TOGW goes up to 190,000 lbs and a larger 7150 hp electric motor is used.

On a 900 nm mission, compared to tube-and-wing aircraft powered by CFM56-like turbofans, the "balanced" HE TBW aircraft reaches its fuel burn reduction goal of 60%, and energy use is reduced by 54%, whereas the "core shutdown" variant is able to reduce fuel burn by 64% and energy use by 46%.

Nevertheless, comparison between the advanced turbofan TBW aircraft and its HE variants shows that while improvement in SFC may not necessarily represent a reduction in specific energy consumption, it does represent a shift from generating the mechanical power in flight to generating it on the ground, where there is significant flexibility and potential for cost and emissions reduction. Even with no energy advantage on the economic mission, the hybrid architecture may allow for a greener source of energy as many environment-friendly ground charging options exist. It is also possible to alter where in the atmosphere emissions are being deposited (cruise altitude or ground power station).

Let us now consider the noise performance of HE configurations. The total in-flight engine noise analysis, without considering airframe noise and for hardwall engine (i.e., no acoustic liners), shows a rather small influence on engine noise due to utilization of electric power. Nevertheless, this may be mitigated by resorting to the following engine fan and jet noise reduction techniques:

- Fan noise reduction: inlet design, advanced/active liners, and active noise control
- Jet noise reduction: chevrons, higher BPR /GTF, and variable area fan nozzle

The picture looks brighter with regards to NOx emissions. It is shown in [8.2] that during takeoff and landing phases, for one, HE engines help slash by more than 40% the

NOx emissions encountered with non-HE engines, thanks to propulsive and thermal efficiency gains in combination with improved thrust lapse characteristics. Moreover, at reductions of ~90% below CAEP/6 levels, set by ICAO's Committee on Aviation Environmental Protection (CAEP), the 80% reduction goal is achieved.

During cruise, engine analysis shows that, when compared to CFM56 turbofan engines, NOx levels of the 1380 and 1750 hp aircraft are close to the 80% reduction goal. On the other hand, the 7150 hp core shutdown aircraft has essentially no NOx emissions over approximately 50% of the cruise segment of the 900 nm mission, since the turbine engine is shut down at midflight and the electric motor takes over.

The HE aircraft therefore maintains a total life cycle emissions advantage (CO_2 and NOx) if the charging electricity is from a clean source.

In the TBW HE aircraft presented in this section, under-wing battery pods used as external components can be quickly exchanged during turnarounds, thus allowing slow charge rates and increasing battery life, thereby avoiding interference with conventional ground servicing operations and increased aircraft turn times.

For the same 150-passenger capacity as above, degrees of freedom possible with design configurations of other types of parallel-hybrid turbofans (Figure 6.19) are illustrated by the following preliminary research outputs [8.3]:

- Geared turbofan: optimized for cruise, ~2.1 MW motor assistance during takeoff and climb phases only. On a 900 nm mission with 1000 Wh/kg batteries, results show 6% reduction in fuel burn and 2.5% reduction in energy usage.
- Three-spool turbofan: parametrically optimized, hybrid climb and cruise phases with ~1-2.6 MW front-mounted motors. Results show 28% reduction in fuel burn for a 900 nm mission, up to a 10% total energy reduction over 500 nm and, optimizing for minimum fuel usage, predict an 18% reduction in total fleet fuel usage.

8.4 Battery Series HEP: Short-Range Aircraft

The series HE principle shown in Figure 6.20 involves operating a thrust-producing fan/motor (or propeller/motor) assembly drawing energy from two remote energy sources: energy storage system and a gas turbine/alternator (GTA) assembly. The latter is comprised of a gas turbine driving a generator rated to produce the power needed by the motor driving the fan (or propeller).

Airbus is intending to test by 2020 such a series HEP system rated at ~2 MW. The project called E-Fan X, is a partnership program between Airbus, Siemens, and Rolls-Royce. The common objective is to validate a new hybrid propulsion system installed on a modified BAE 146, a high-wing aircraft powered by four turbofan engines. After extensive ground testing, it will undergo flight testing from 2020 onwards, with one of the standard turbofan engines replaced by a 2 MW electric motor. The test-bed aircraft configuration is shown in Figure 8.3. Development of the flight demonstrator will combine some of the world's most advanced electrical and propulsion technologies in a modified airframe that will be the largest hybrid flying R&D platform to date. Once the maturity

FIGURE 8.3 Battery series HEP—high-wing short-range test aircraft (E-Fan X) [8.4, 8.5].

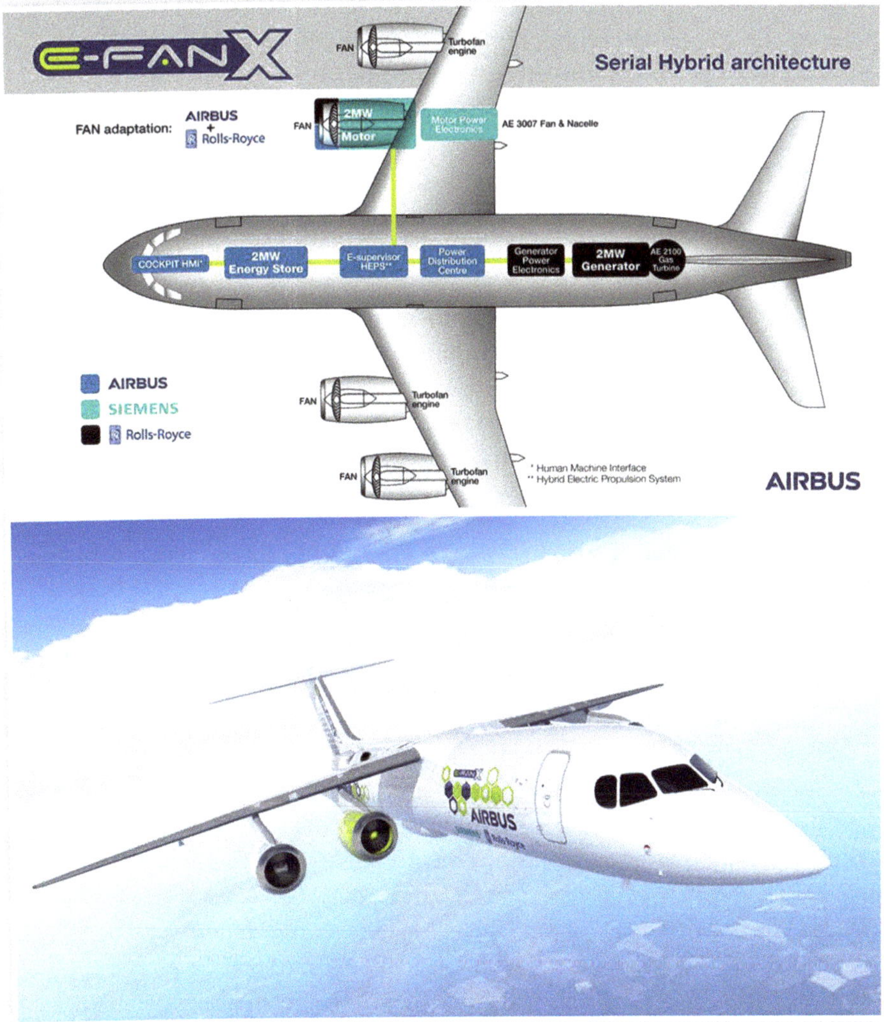

of the system has been proven, provision has been made to replace a second gas turbine engine with another electric motor.

This work and Airbus E-Aircraft Systems House collaboration with Siemens is seen as paving the way for a future hybrid short-range commercial aircraft that is safe, efficient, and cost effective.

The E-Fan X demonstrator will look at and explore all the challenges of developing a high-power HEP system, including thermal effects, electric thrust management, altitude and dynamic effects on electric systems, and other issues such as electromagnetic compatibility. The objective is to push forward and mature the technology, performance, safety, and reliability of HE technology. Other major aspects of the program will be to establish new requirements for future certification of electrically powered aircraft and the training needs of a new generation of designers and engineers to bring such commercial aircraft closer to reality.

FIGURE 8.4 Traditional engine pod (Rolls-Royce Trent XWB turbofan—long-range Airbus A350 XWB) [8.6].

As discussed above, HEP is scalable and invariably applicable to different aircraft sizes and segments. Likewise, the electrical energy source can also be appropriately chosen depending on the target aircraft. Target aircraft designs have stuck close to the familiar tube-and-wing shape. Moreover, starting off from the commuter segment, the HEP considered above, despite partial electrification to provide thrust, is still circumscribed by an aircraft-attached pod we are accustomed to see (Figure 8.4).

In the following section, we will see how powerplant electrification may enable its subdivision into smaller propulsion units allowing for optimization thanks to an integrated aircraft and propulsion system design approach. This ushers in paradigm shifts in propulsion architecture and aircraft aerodynamics, capable of leveraging the benefits of electric and HEP on a larger scale.

8.5 Battery Distributed HEP: Commuter Aircraft

In the parallel HEP cases studied previously, apart from GA, one set of the electric motor/gas turbine combination, producing half of the total aircraft thrust, is installed on either side of the aircraft. Each motor/gas turbine set is integrated and collocated within a traditional powerplant envelope, whereas batteries or fuel cells are remotely located. Moreover, within a given powerplant, the motor and gas turbine drive the same shaft (turbofan LP spool, turboprop propeller shaft, etc.). As opposed to these monolithic parallel HE powerplants, a split solution for parallel operation is possible whereby on the aircraft the electric part of the powerplant is physically separated from its gas turbine part, each part driving its own propeller. Even though the two parts of the powerplant are split, parallel thrust production is still achieved thanks to their combined operation enabled by controls, as with the monolithic powerplant format.

Let us see how on a small scale such a distributed concept can be applied to "parallel" HEP. Fefermann et al. [8.7] illustrates such a "distributed parallel" tri-propeller HEP configuration comprising two downsized under-wing podded gas turbines (turboprops) and one electrical motor, driving its own propeller, mounted in the aft-fuselage section of the aircraft and powered by batteries. Implementation on a 19-passenger 700 nm commuter aircraft is shown in Figure 8.5.

In normal conditions, the electric propulsor is operated during taxi-out, takeoff/climb, cruise, and taxi-in and provides electrical assistance to the mainframe turboprops. During descent, only the turboprops are active and the electric propulsor is relieved of its thrust production function and, in reverse operation, recovers energy for battery recharge. In fact, during descent the propulsor is turned by the airstream (windmilling) and, like wind turbines, is capable of generating electrical energy that can be stored. Recharging batteries during cruise and descent may not be necessary if exchangeable battery packs are used. In abnormal conditions such as rejected takeoff, go-around, and

FIGURE 8.5 Battery DHEP on a commuter aircraft [8.7].

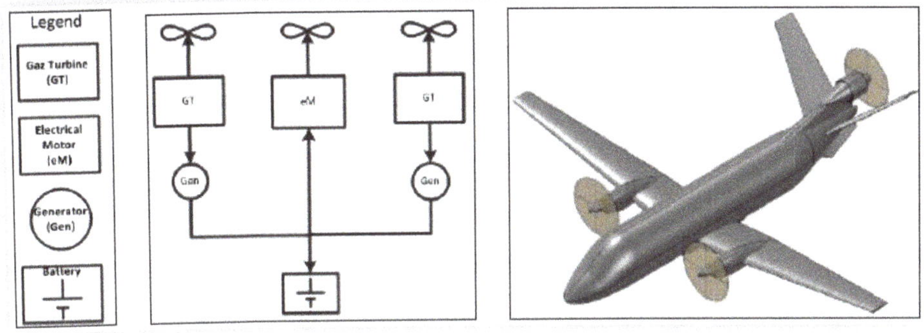

partial cruise with one engine inoperative, the electric propulsor would be systematically switched on as a safety measure.

As can been seen in this approach with three propellers (or propulsors), two are driven by gas turbines in their conventional turboprop format, smaller in size though, and the third one independently driven by a battery-powered electric motor. Other types of energy storage or energy source may also be utilized instead of batteries.

8.6 Battery Distributed HEP: Regional Aircraft

Assuming that high-density motor technologies discussed in Chapter 6 for small general aviation aircraft can be scaled up in power, when addressing larger aircraft markets such as the 50–100 seat regional aircraft segment, the obvious showstopper for better performance and full-range compliance is battery specific energy. Nevertheless, the study in [8.8] shows how on a 50-seat turboprop aircraft (600 nm) design choices could help rendering performance less sensitive to battery specific energy variation.

Firstly, the two turboprop engines of the aircraft are replaced by four electric propulsors powered by batteries alone. Then, the electric propulsion integration is carried out in a synergistic manner allowing for significant aerodynamic improvements. These benefits make aircraft weight, total energy, and costs much less sensitive to battery-specific energy evolutions over time, that is, from 500 Wh/kg deemed achievable by 2025 to ~1000 Wh/kg expected further down the line. Although 500 Wh/kg batteries can help get 82% energy reduction, range performance is brought down to a mere 200 nm, a far cry from what is needed. One way out is to resort to HEP allowing for lesser range reductions while hardly compromising total energy reduction. This will be discussed in more detail further down.

The study in [8.8] underpins the fact that aircraft design with reduced range and a propulsion integration strategy that is synergistic to electric propulsion dramatically improves the economics. Therefore, relatively near-term batteries may be as compelling of an economic proposition as long-term batteries, provided the range reductions are acceptable from an operational standpoint.

So, a battery-powered parallel HE regional aircraft may be feasible by the 2030 horizon. The study in [8.8] considers a regional aircraft the size of an ATR-42

(48 passengers, 600 nm range), optimized for wing loading and thrust-to-weight. The proposed optimal DHEP solution consists in replacing the original turboprops and distributing their thrust production capability over two outboard HE turboprop engines and two inboard electric motors driving propellers (**Figure 8.6**).

Additionally the aircraft analyzed in [8.8] incorporates the following structural changes allowing synergistic integration of the proposed DHEP:

- Induced drag reduction by removal of inboard turboprop and introduction of wingtip propulsion
- Fuselage BLI with increased propulsion efficiency

Let us figure out how the degree of hybridization, or the amount of battery energy out of the total energy requirement, influences key performance metrics. **Figure 8.7** shows how battery, fuel, and total energy vary when the battery portion of the total power requirement is tuned from 0% to 75%, the 100% data point referring to the electric aircraft, wherein HE turboprop engines are replaced by their electric motor variants.

Energy curves in Figure 8.7 correspond to non-distributed HE aircraft propulsion achieved with twin stand-alone HE turboprop engines, without any sort of synergistic integration. Data points encircled in red relate to the more optimized DHEP configuration with synergistic integration. Firstly, the chart shows that, as opposed to the curves obtained with non-synergistic HEP, synergistic integration of DHEP results in low sensitivity to changes in the specific energy of batteries used. Secondly, when the degree of hybridization is set at 60%, total energy and fuel energy consumptions (therefore gas emissions) are knocked down by over 50% for a range of 400 nm with 500 Wh/kg batteries. Bearing in mind that, despite the ~600 nm range capability regional aircraft are usually designed for, 90% of total U.S. regional aircraft trips do not operate beyond the 400 nm range mark; a ready-to-offer value proposition may therefore be on the table with the DHEP solution discussed here. Therefore, the analysis in [8.8] can be summarized as follows for the regional aircraft segment:

- DHEP solution constitutes a compelling sweet spot between range performance and operating economics, provided operators are willing to accept range limitations, nonetheless aligned with real operational requirements.
- Relatively near-term batteries are as compelling of an economic proposition as long-term batteries.

FIGURE 8.6 Battery DHEP on a regional aircraft [8.8].

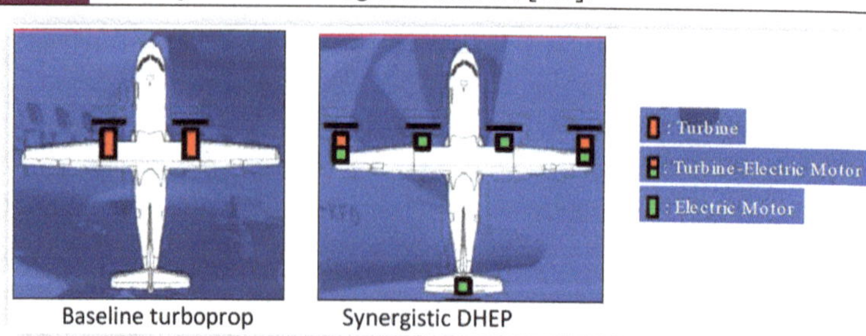

FIGURE 8.7 Propulsion energy vs. degree of hybridization (conventional, HEP and DHEP) [8.8].

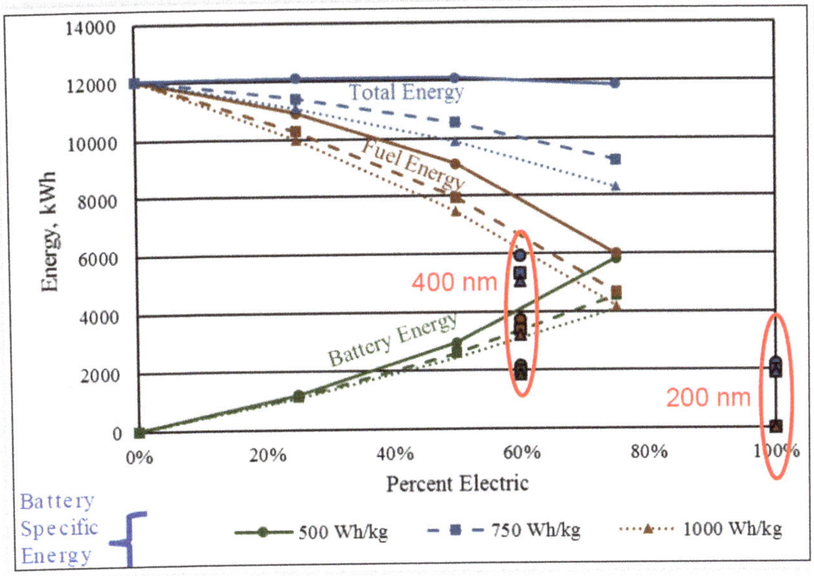

8.7 Distributed HEP: Short-Range Aircraft

8.7.1 No-Battery Partial Turboelectric DHEP

Partial turboelectric DHEP as explained in Figure 6.23 comes here with no batteries. The tube-and-wing NASA concept aircraft considered is shown in **Figure 8.8**. Conventional turbofans are still in place but have to ensure partial turboelectric operation of a tail-mounted 2–3 MW electric boundary layer ingestion (BLI) fan. So, the turbofan engine integrates an electric generator dedicated to the power supply of the BLI fan. The airframe is kept the same except for the BLI integration. In 2016, preliminary studies concluded that expected fuel burn reduction would be in the range of 7%–12%, with no change either in range/speed performance or aircraft infrastructure requirements [8.3]. Further analysis with higher-fidelity assumptions are less optimistic, at ~3%–4% improvement caused only

FIGURE 8.8 No-battery partial turboelectric DHEP—tube-and-wing short-range aircraft [8.3].

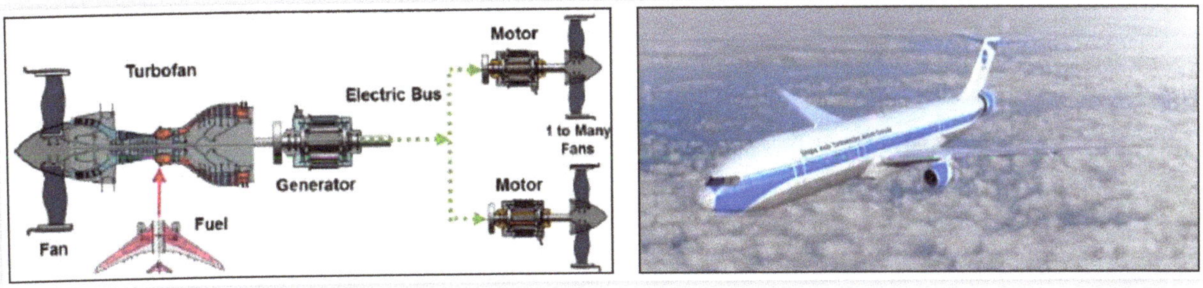

FIGURE 8.9 Fuel cell partial turboelectric DHEP—TBW short-range aircraft [8.3].

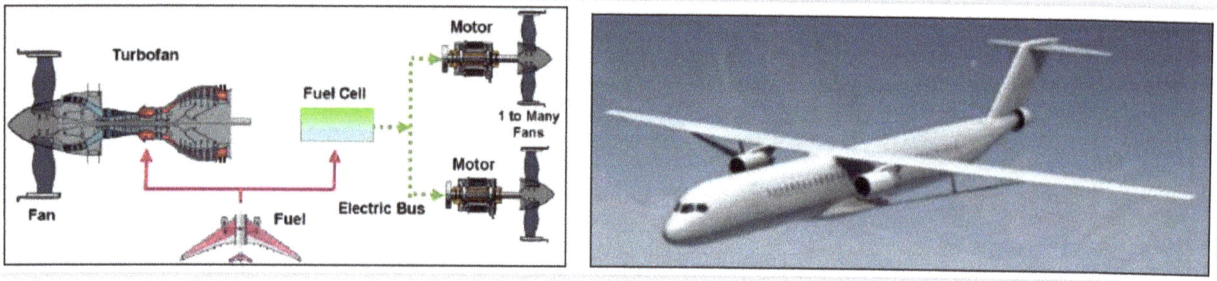

by the electrification of the propulsion system, and much work is ahead of the industry to truly understand and exploit the systems benefits that can be achieved.

8.7.2 Fuel Cell Partial Turboelectric DHEP

The partial turboelectric concept can be applied to the TBW aircraft with BLI. Very advanced turbofan engines are used, and the aerodynamic efficiency is augmented thanks to BLI. A fuel cell is used to operate the tail-mounted BLI fan (**Figure 8.9**) powered by a SOFC topping cycle and driven by a cryogenically cooled superconducting motor. Fuel burn reduction can top out at 56% for a 900 nm mission [8.3].

8.7.3 No-Battery Total Turboelectric DHEP

As opposed to partial turboelectric DHEP, full-scale turboelectric DHEP offers greater propulsion flexibility and design freedom for the overall aircraft design that could result in reduced structural weight and aerodynamic drag, for example, by relaxed engine-out design constraints leading to a smaller vertical tail plane, by being able to better distribute the weight of the propulsion system components and by re-energizing the momentum losses in the "boundary layers" that grow over the wing and fuselage causing a "wake." An additional efficiency gain appears possible if this boundary layer is "ingested" and accelerated by the fans, because it can reduce the aircraft's wake and hence its drag. Therefore, with full-scale DHEP untapped opportunities such as very high bypass ratios and BLI can be exploited to boost overall efficiency.

In the two applications (commuter and regional aircraft) studied above, the distributed propulsion approach consists of adding electric propulsor(s) while downsizing the turboprop engines. If this trend is pushed to an extreme, the ultimate solution would be to remove the turboprop engines altogether, whether hybrid or not, and stick to ideally located multiple electric propulsors instead. One can easily see that this opens the door for what has come to be coined as distributed electric propulsion (DEP), wherein powerplants are broken down into smaller electric propulsor units and distributed over the aircraft. This configuration leverages the efficiency edge that electrical chains have over the thermodynamic cycles in play with gas turbine engines as illustrated in Figure 6.15. Moreover, DEP can provide highly coupled aero-propulsive integration options. Noise and ride quality improvements can be ensured through synergistic integration with aero/control/structures.

Now the issue is how to make the required electrical energy available onboard in order to power the distributed electric propulsors, given the specific energy challenges still remaining with batteries and fuel cells. Remember, when moving to DHEP, we started out increasing the number of electric propulsors while downsizing the conventional gas turbine powerplants, thereby transferring partially the onus of producing thrust from gas turbines to electric motors. Now, some of the gas turbines may be designed, not for driving a propeller or a fan to produce thrust directly but as electrical energy sources thanks to electric generators they would drive. Such gas turbine alternators (GTA), would contribute indirectly to thrust production by supplying electric energy to the motors of the electric propulsors, which now become directly in charge of overall aircraft thrust. This kind of distributed propulsion architecture would still be tagged as DHEP due to the utilization of gas turbines. The basic principle is rooted in the total turboelectric or series HE concept explained in Figure 6.20, the gas turbine/generator thereof now being called GTA and the propulsor/motor setup now coming in the form of multiple smaller units remotely located and distributed.

Therefore, total turboelectric DHEP systems offer the possibility of forgoing batteries altogether and improving overall efficiency by allowing for separate optimizations of the thermal efficiency of GTAs and the propulsive efficiency of fans. In fact, such a system is a powertrain consisting of a gas turbine used mainly to power multiple electrical propulsive fans distributed above, below, or inside a wing.

For instance, on the ECO-150 aircraft studied by NASA and ES Aero, distributed fan/motor assemblies and GTAs are integrated into the wings of a 150-passenger tube-and-split-wing aircraft. The total turboelectric DHEP system architecture used in this application features overall eight fan/motor assemblies and two GTAs, each side of the aircraft wing comprising four assemblies powered by the GTA installed on that side [8.9]. GTAs are assumed to not provide propulsive thrust on their own. The electric power they generate is bussed through the wing to each motor/fan assembly and to the onboard subsystems, if necessary (Figure 8.10).

In the initial ECO-150 studies, while several benefits appeared to show promise of improved mission performance and efficiency, concerns of integration, availability of high-performance electric machines, and the increased propulsion system weight had to be mitigated in order not to jeopardize system feasibility. Depending on the underlying

FIGURE 8.10 No-battery total turboelectric DHEP—tube-and-split-wing short-range aircraft [8.9, 8.10].

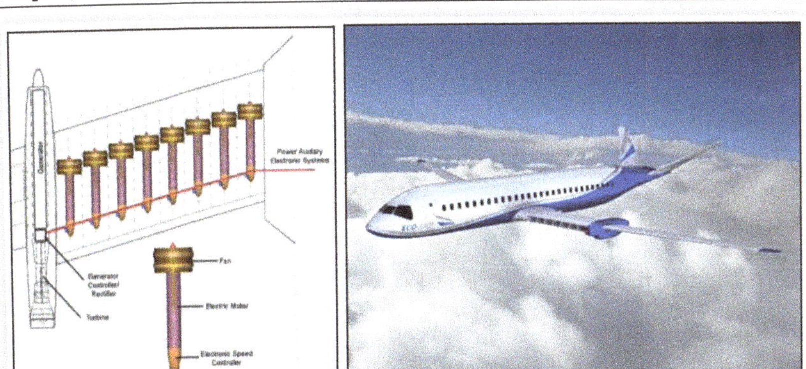

technology assumptions, from superconducting electrical machines cooled with liquid hydrogen to conventional machines at various technology levels, performance ranged between matching and significantly exceeding current aircraft fuel burn. Ultimately, the ECO-150R version (Figure 8.10) was set to utilize midterm electrical machine technology.

8.7.4 Battery Total Turboelectric DHEP

Main and backup gas turbines dedicated to the generation of electricity from fuel, in combination with other electrical sources such as secondary power units (SPUs) and energy storage mix, would allow for reliable and dependable electric energy to be supplied to the motors driving the propulsors (**Figure 8.11**). For the sake of clarity, power conversion equipment and power electronics are not shown in the figure. Such a system can deliver smoothing of the energy peaks of the flight cycle and a boost capability in demanding phases.

On large aircraft, DHEP is set to unlock new aircraft design possibilities while reducing noise, fuel burn, and gas emissions. As explained in 6.3.1.3, DHEP's main function is to enable the major benefits of distributed propulsion, namely, very high bypass ratios and BLI capability. It features numerous small electric fans arranged in clusters along the length of the wing(s). Sleek and aerodynamically efficient airframe designs are possible allowing for reduced weight and drag by decreasing the size of the vertical tail and improving weight distribution. Fans integrated in the airframe instead of a few large wing-mounted turbofans as in traditional tube-and-wing configurations allow for the reduction of total propulsion system noise.

The total turboelectric DHEP for a "blown-wing" short-range aircraft design called "E-Thrust," studied by Airbus and Rolls-Royce, is shown in **Figure 8.12** [8.13]. The array of six electrically powered fans are distributed in clusters along the wing span, with one or more advanced GTAs providing the electrical power for the fans and for recharging the energy storage. The hybrid concept makes it possible to downsize and optimize the GTA(s) for cruise. The additional power required for takeoff will be provided by the energy storage system (batteries). The final number of fans and GTAs will be determined as a result of aircraft level trade studies.

Peak power needed during takeoff and climb phase is delivered by the GTA and the energy storage system. The latter is sized to ensure safe takeoff and landing should the GTA fail during this phase. In the cruise phase, the GTA will provide the cruise power and the power to recharge the energy storage system. In the unlikely event of a GTA failure, power from the energy storage is available to continue the flight to a safe landing. In the initial descent phase, no power is provided to the fans, and the GTA is switched off. The aircraft will be a glider, and the energy storage system will provide the power for the aircraft's onboard systems. During the second phase of the descent, the fans will be windmilling and produce electric power to replenish the charge in the energy storage system. For the landing phase, the GTA is re-started and provides power at a low level for the propulsion system. This is a safety feature to cover a hypothetical loss of power from the energy storage system during this phase.

With DHEP, effective bypass ratios of over 20 are deemed achievable, leading to significant reductions in fuel consumption and emissions. Therefore, compared to limited

FIGURE 8.11 DHEP multi-redundant electrical energy supply to distributed propulsors [8.11].

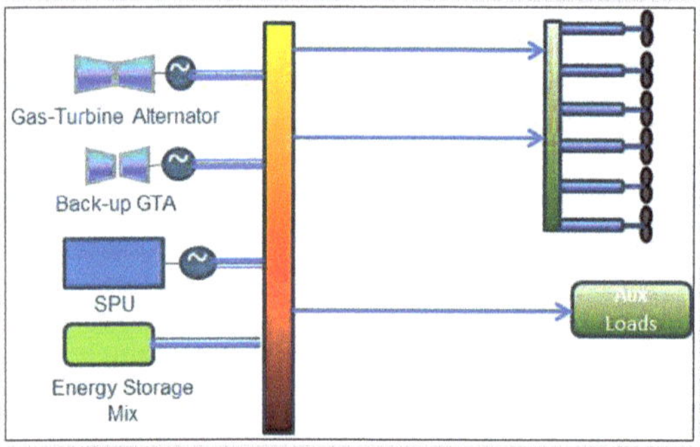

FIGURE 8.12 Battery total turboelectric DHEP with BLI (six fans and one GTA)—"blown-wing" short-range aircraft [8.12, 8.13, 8.14].

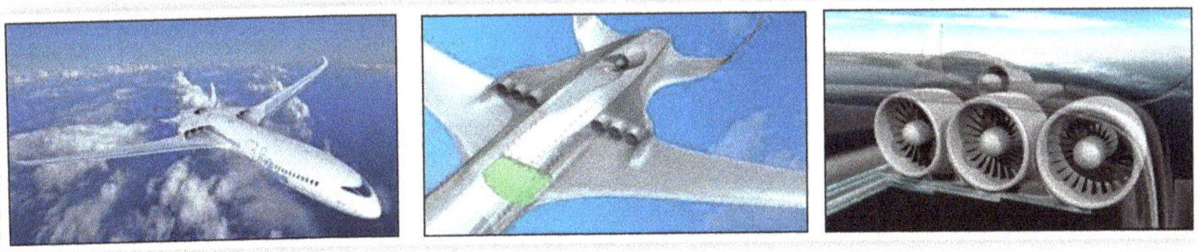

improvements possible in the next years on already stretched existing systems, HE solutions for a distributed fan propulsion system on a short-range aircraft target significantly higher reductions in fuel burn, emissions, and noise. In the years 2030 and beyond, such solutions could achieve 30%–40% reduction in fuel burn.

The electric motors are superconducting and in the megawatt range. They have to provide thrust from electric energy, be able to reduce drag by re-energizing the wake energy (BLI), and also recover electric energy for storage when they are not used to provide thrust. In order to achieve an integrated distributed fan propulsion system design (Figure 8.13) that matches the overall airframe requirements, three key innovative components are required:

- Wake re-energizing fan: captures wake energy by re-accelerating the complex wake
- Structural stator vanes: recovers thrust from the swirling air produced by the fan and accommodates routing of superconducting cables (power and cryogenic coolant) connected to the superconducting machine
- Reversible AC high-voltage superconducting machine: drives the wake-re-energizing fan in motor mode (takeoff/climb, cruise, and landing), and reverts to generator mode (descent windmilling) for recharging batteries

The DHEP system architecture described above for a short-range aircraft calls for breakthroughs in the following areas:

- High specific energy for energy storage (up to 1000 Wh/kg for batteries)
- More efficient power electronics with high power-to-weight ratios
- High-temperature superconducting (HTS) machines, cables, and distribution
- Cryocooling solutions for superconducting components
- Composite and metal technologies

While battery technologies are maturing and specific energy issues are sorted out, DHEP solutions relying exclusively on GTAs without having to resort to energy storage (batteries) may also be up to expectations, provided superconducting components are used to a certain degree. Such an architecture with two GTAs and superconducting machines is shown in Figure 8.14.

FIGURE 8.13 Integrated DHEP fan propulsor/motor unit—short-range aircraft (adapted from [8.13]).

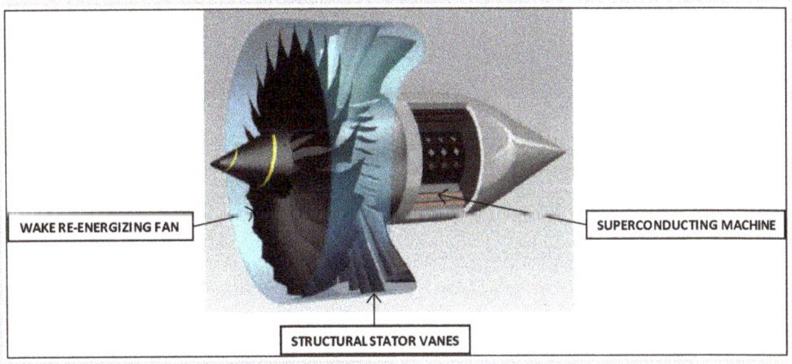

FIGURE 8.14 Battery-less total turboelectric DHEP with BLI (eight fans and two GTAs)—"blown-wing" short-range aircraft [8.11].

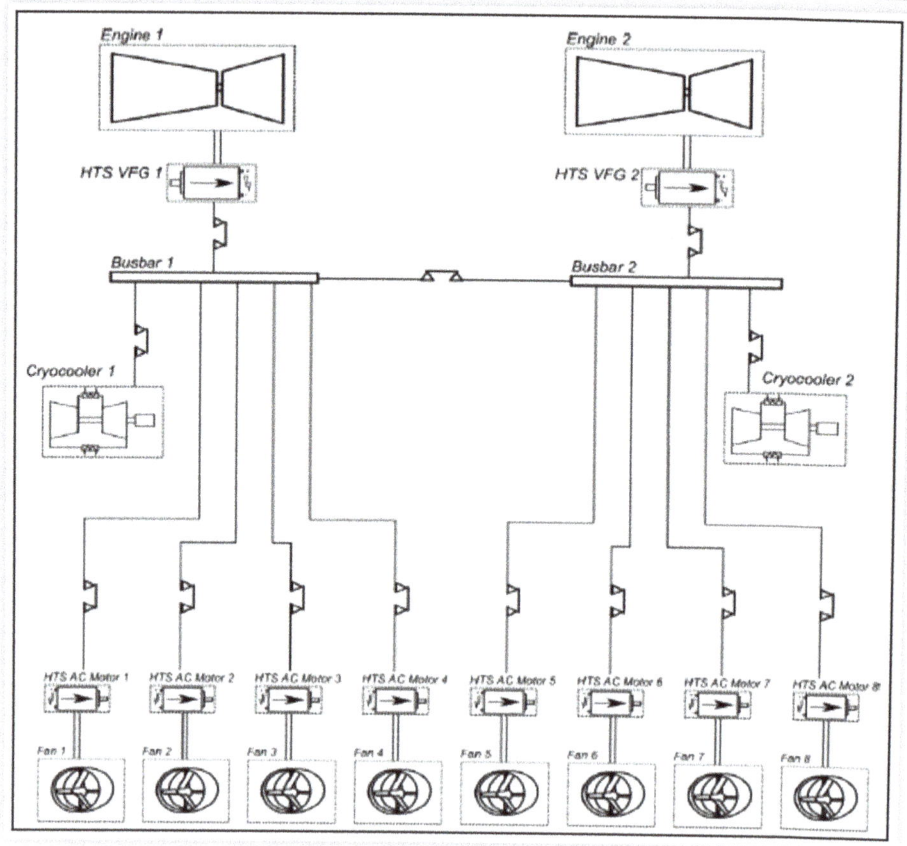

8.8 SMES Total Turboelectric DHEP: Long-Range Aircraft

DHEP enables more aerodynamic aircraft designs such as blended wing body (BWB), also called hybrid-wing body (HWB), for long-range capability (Figure 8.15). This concept has been existing for decades but has had a recent resurgence. As a mix between the traditional tube-and-wing and a flying wing, the HWB seeks to obtain increased fuel efficiencies through the elimination of the tail section and increased structural efficiency. This is probably the most radical departure from the conventional aircraft that is seriously under consideration by aircraft manufacturers, even though several technical challenges remain.

On BWB aircraft with DHEP, having the engines incorporated into the airframe allows for the leverage of BLI on top of very high bypass ratios and compelling noise performance. As discussed in [7.1], the advantages of the HWB have made it very appealing in the face of steep challenges laid out in the NASA N+3 goals.

The difficulties with the BWB concept stem among others from the fact that it does not have a circular cross section like the conventional tube-and-wing. When pressurized at altitude, this can put uneven stress on the fuselage and the ability to address this in a lightweight and economical fashion is an issue under current research. Other challenges are, for example, the vertical acceleration on passengers sitting far off the centerline, gate space at airports, evacuation regulations, and the option to apply a family concept onto the design.

The BWB studied by NASA has a flattened and reflexed airfoil shaped body (Figure 8.16). A low effective wing loading and beneficial trim effect means a complex high-lift system is not required. The outboard wing supports slats and all trailing edge devices are made up of simple hinged flaps that double as elevons. The distributed fan array is mounted atop of the main body, thereby ingesting large portions of the boundary layer [8.16].

FIGURE 8.15 Total turboelectric DHEP with BLI (fan array and two GTAs)—BWB long-range aircraft [8.15].

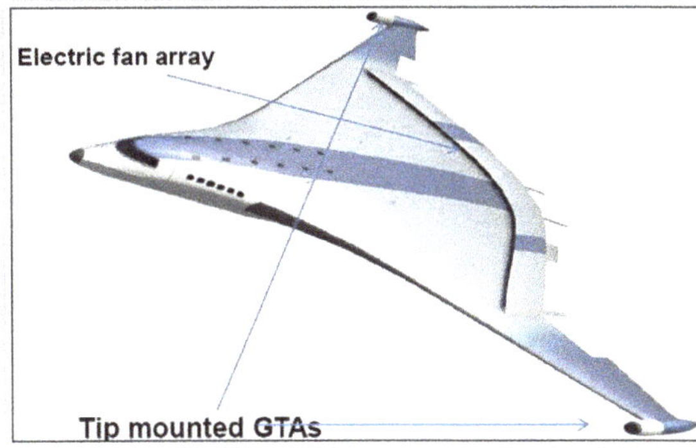

The BWB aircraft depicted in Figure 8.16 was studied by NASA in collaboration with Rolls-Royce in order to quantify the benefits of a DHEP. The study [8.17] demonstrates effective bypass ratios upwards of 30.

FIGURE 8.16 SMES total turboelectric DHEP with BLI (up to 16 fans and 2 GTAs)—BWB long-range aircraft [8.17].

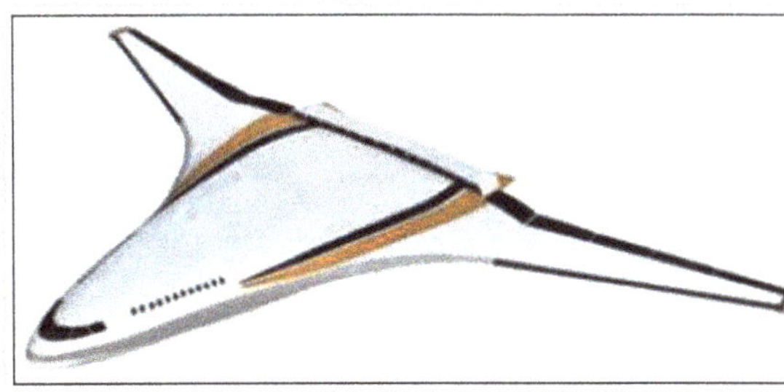

Two tip-mounted GTAs generate all the electric power necessary for propulsion. Depending on aircraft width, fan pressure ratio, boundary layer, and net thrust, up to 16 fan/motor units may be integrated to the airframe. Technology-wise, cryocooled superconducting generators, motors, cables, protective devices, and energy storage constitute the design baseline. Batteries give way to superconducting magnetic energy storage (SMES). Superconducting fan/motor units are rated at 3.3 MW each, and the total 50 MW propulsion power is delivered to these units by the two tip-mounted GTAs.

The BWB thrust rating is ~86,000 lbf (rejected takeoff max), and the thrust-specific fuel consumption (TSFC) is in the order ~0.3 lbm/h/lbf (top of climb max). Compared to the Boeing 777 aircraft, a tube-and-wing long-range aircraft with legacy turbofans, the BWB is found to be heavier but allows for fuel weight to be slashed by a staggering 60%.

On top of promising performance outcomes, the community noise aspect is one area where the BWB aircraft with DHEP particularly excels. Due to the large expanse of its structure, its body provides a much greater barrier to noise from upper surface engines than what could be accomplished with a tube-and-wing configuration. While the engine noise itself would be comparable, it would be shielded from the ground and therefore reduce community interference. In a NASA study [8.18], the BWB was found to have a cumulative −37 dB difference from a 2005 reference tube-and-wing aircraft.

8.9 Conclusions

Chapters 7 and 8 show how conventional aircraft propulsion will slowly give way to disruptive HE and electric configurations. In the former case, current engines may get challenged by hybrid solutions wherein gas turbines (or piston engines) would still remain in the fray but with a lesser (or indirect) contribution to thrust, the required complementary thrust coming from electrical motors powered by energy storage. In the latter case, outright electric alternatives may replace turbine engines (or piston engines). In both cases, the non-propulsive power for aircraft systems will be electric and may be derived to some extent from the propulsion systems.

Following the evolution of state-of-the-art aircraft propulsion towards HE and electric alternatives, commonplace gas turbines may first, in combination with electric motors, morph into HE variants and then also get weaved into the fabric of distributed propulsion. Whether distributed or not, HEP proves to be a compelling transition on the road to full electric propulsion.

The topics addressed portray how aircraft manufacturers and engine manufacturers are constantly unveiling electric and HEP concepts poised to radically transform aviation between now and the 2050 horizon. Projections show that, provided technology evolutions are up to expectations over the required time span, these new aircraft and electric propulsion architectures would slash fuel consumption, emissions, and noise, thereby ensuring greener and more cost-efficient operations. Performance enhancements would ultimately fall in line with the steep efficiency and environmental targets set forth by Flightpath 2050 and NASA N+3. All these approaches come with strings attached, especially energy storage specific energy, much less demanding on small aircraft compared to larger ones.

Taking into account foreseen evolution of technologies and energy storage specific energy, small general aviation aircraft may be early adopters of electric or HE propulsion. Walking in their steps may come commuter aircraft. Regional aircraft may follow suit with DHEP, though with some relatively acceptable range caveats. Further down the line, short- and long-range aircraft segments, enabled by superconducting electrical systems, are poised to adopt DHEP architectures. Future urban mobility concepts under study call for car-sized electrically propelled aircraft in different forms and shapes featuring vertical takeoff and landing (VTOL) and autonomous operation capabilities.

Last but not least, this inexorable move towards electric propulsion will be accompanied by changes in certification rules, currently applicable to reciprocating and turbine engines only. Not to mention totally new regulations required by urban mobility concepts for both urban airspace integration and air traffic management.

Anecdotally, at the end of the very famous 1985 feature film, *Back to the Future*, Doc, Marty, and Jennifer are readying to "time travel" to 2015. Marty expresses his concern that Doc isn't giving himself enough "road" for his futuristic flying car to get up to the 88 mph takeoff speed (required for time travel in that movie). Doc responds with "Roads? Where we're going, we don't need roads." And then the DeLorean flies off. That is certainly a nod to how future urban mobility with VTOL aircraft awaiting urbanites and suburbanites might look like. Thanks to technology advances and air traffic rules and regulations, safe and ultrafast mobility will go in pair with pollution-free and quiet city life.

References

[8.1] http://zunum.aero/technology/, accessed March 3, 2018.

[8.2] Bradley, M.K. and Droney, C.K., "Subsonic Ultra Green Aircraft Research: Phase II—Volume II—Hybrid Electric Design Exploration," prepared by the Boeing Company for NASA Langley Research Center, Contract NNL08AA16B—Task Order NNL11AA00T, NASA/CR-2015-218704/Volume II, April 2015.

[8.3] Jansen, R.H., Bowman, C., Jankovsky, A., Dyson, R. et al., "Overview of NASA Electrified Aircraft Propulsion Research for Large Subsonic Transports," NASA Glenn Research Center, *EnergyTech 2017*, Cleveland, OH, USA, October 31-November 2, 2017.

[8.4] https://community.plm.automation.siemens.com/t5/Simcenter-Blog/The-E-Fan-X-E-aviation-is-about-to-happen-and-sooner-than-you/ba-p/452069?lightbox-message-images-452069=48831i3B0306C0ACA43B7A, accessed September 14, 2018.

[8.5] https://community.plm.automation.siemens.com/t5/Simcenter-Blog/The-E-Fan-X-E-aviation-is-about-to-happen-and-sooner-than-you/ba-p/452069?lightbox-message-images-452069=48826i1AC8A723252C3ABE, accessed September 14, 2018.

[8.6] https://kids.kiddle.co/Airbus_A350_XWB, accessed February 16, 2018.

[8.7] Fefermann, Y., Maury, C., Level, C., Zarati, K. et al., "Hybrid-Electric Motive Power Systems for Commuter Transport Applications," Safran, *30th Congress of the International Council of the Aeronautical Sciences (ICAS)*, Daejeon, South Korea, September 25-30, 2016.

[8.8] Antcliff, K.R., "Turboprop Electric Propulsion with and without Synergistic Integration and Operations," NASA Langley Research Center, *2nd On-Demand Mobility and Emerging Aviation Technology Joint NASA-FAA Workshop*, Arlington, VA, USA, March 8-9, 2016.

[8.9] Schiltgen, B., Gibson, A., Green, M., and Freeman, J., "More Electric Aircraft: 'Tube and Wing' Hybrid Electric Distributed Propulsion with Superconducting and Conventional Electric Machines," Empirical Systems Aerospace, SAE Technical Paper 2013-01-2306, 2013, doi:10.4271/2013-01-2306; *SAE 2013 AeroTech Congress & Exhibition*, September 24, 2013, Montréal, Canada.

[8.10] http://aviationweek.com/technology/when-will-these-civil-aircraft-concepts-fly#slide-2-field_images-1570041, accessed March 5, 2018.

[8.11] Husband, M., Raffaelli, L., Alderman, J., Berg, F. et al., "Distributed Electrical Aerospace Propulsion (DEAP)," Rolls-Royce, Airbus Group Innovations, Cranfield University (UK), Electric & Hybrid Aerospace Technology Symposium 2015, Bremen, Germany, November 17-18, 2015.

[8.12] http://aviationweek.com/technology/when-will-these-civil-aircraft-concepts-fly#slide-0-field_images-1570041, accessed March 5, 2018.

[8.13] "E-Thrust: Electrical Distributed Propulsion System Concept for Lower Fuel Consumption, Fewer Emissions and Less Noise," Airbus, Rolls-Royce, May 2014.

[8.14] http://company.airbus.com/news-media/media~item=2efe334d-1141-403c-8449-1b7180c8e7fa~.html, accessed May 4, 2018.

[8.15] Malkin, P., "Hybrid Electric Distributed Propulsion Aircraft—A Hybrid Like No other?," Newcastle University, *Electric & Hybrid Aerospace Technology Symposium 2015*, Bremen, Germany, November 17-18, 2015.

[8.16] Isikveren, A.T., Seitz, A., Bijewitz J., Hornung, M. et al., "Recent Advances in Airframe—Propulsion Concepts with Distributed Propulsion," Bauhaus-Luftfahrt, Central Institute of Aviation Motors, ONERA, Airbus, *29th Congress of the International Council of the Aeronautical Sciences (ICAS 2014)*, Saint Petersburg, Russia, September 7-12, 2014.

[8.17] Armstrong, M., "Superconducting Turboelectric Distributed Aircraft Propulsion," Rolls-Royce, *Cryogenic Engineering Conference/International Cryogenic Materials Conference*, Tucson, AZ, USA, June 28-July 2, 2015.

[8.18] "Status of Hybrid Wing Body Community Noise Assessments," NASA, AIAA Aerospace Sciences Meeting, Special ERA Session, Orlando, FL, USA, January 6, 2001, http://www.aeronautics.nasa.gov/pdf/asm_presentations_status_hwb_community.pdf.

CHAPTER 9

Maintainability and Operational Overview

Sven Taubert
Lufthansa Technik

This chapter summarizes scenarios of how an upcoming more electric aircraft and especially an upcoming electric aircraft would impact current processes and regulations for aircraft operations. Therefore, the following sections address both ground operations and in-flight operations. As these topics are introduced within a mid- to long-term timeframe and as some key questions defining the most plausible path remain unanswered, the following discussion is an attempt to envision future air transport operations with some caveats though.

9.1 Ground Operations

9.1.1 Maintenance—State of the Art

Maintenance in aviation covers all tasks ensuring the compliance with Airworthiness Directives including Service Bulletins. These tasks are highly regulated and mainly specified by the Original Equipment Manufacturers (OEMs) in order to guarantee a safe use of every component. The regulations are supervised by national regulation agencies like the Federal Aviation Administration (FAA), Civil Aviation Authority (CAA), or European Aviation

Safety Agency (EASA) and internationally coordinated by bodies like the International Civil Aviation Organization(ICAO). To ensure the compliance with these regulations, every maintenance task and every appointed staff has to be licensed. Maintenance is also part of the certification process—during type certificate (TC) or amendments to TC, the maintenance schedule is documented and approved by the certifying authority.

The maintenance market itself is very fragmented. Most of the larger airlines are performing at least some of these tasks themselves. The rest would be subcontracted to either independent Maintenance, Repair, and Overhaul (MRO) companies like Lufthansa Technik, among many others, or the component/aircraft OEMs themselves.

9.1.1.1 **Maintenance Planning:** Currently, maintenance tasks are often planned far in advance. Most checks are repeated in periodic intervals depending either on the flight hours or cycles. One cycle describes one takeoff and one landing. The main categorization is done in A, B, C, and D checks.

A and B checks are performed in line maintenance, and C and D checks performed in base maintenance. Line maintenance includes minor checks and unscheduled incidences, solvable within a few days or which would cause the aircraft to lose its airworthiness. Loss of airworthiness or "Aircraft on Ground" (AOG) prohibits flying with immediate effect.

Planned checks are mainly done during night shifts. During daytime, between flights for aircraft in service, the main task for line maintenance is to service and repair "on-wing" as may be necessary. Base maintenance includes major upgrades like installing a satellite communication system, cabin refurbishment, larger inspections, and almost all "off-wing" repairs.

9.1.1.1.1 A Check. The A check subscribes the smallest planned interval. It has to be performed every 400-600 flight hours or every 200-300 cycles, depending on the aircraft type and the components needing maintenance, and can take up to 50-80 man-hours. Normally, airlines will try to complete an A check during one night shift in a hangar (6h-10h). Some of the tasks can also be postponed or even done earlier to ensure a maximized utilization time of the aircraft by a guaranteed safety level. As there are thousands of components in each aircraft, an ingenious maintenance schedule can easily save millions of dollars.

9.1.1.1.2 B Check. The B check has to be performed every 6-8 months and takes 150-200 man-hours. Again, these estimations depend on the type of aircraft and equipment conditions. All tasks can be done within a downtime of 1-3 days. This is widely considered as the maximum time for line maintenance. Using better planning possibilities, more and more MROs divide B checks into several A checks (Checks A-1 through A-10) to avoid daytime downtime and maximize aircraft utilization time.

9.1.1.1.3 C Check. The C check has to be performed every 20-24 months, depending on either actual flight hours (or cycles) or manufacturer Service Bulletins. This check grounds the aircraft for 1-2 weeks and involves up to 6000 man-hours. This includes inspections of a majority of the aircraft systems and components. The C check cannot be subdivided into several B checks, as a lot of components have to be shipped to off-site suppliers. Due to the long downtime, the C check can be performed at specialized MRO facilities, which are not necessarily near the airline's hubs. Most airlines make use of the check to perform upgrades like cabin modifications, installation of connectivity equipment, or avionic upgrades. For the Boeing 747-400, the C check itself may cost somewhere between $0.7 million and $1.5 million.

9.1.1.1.4 3C Check. The 3C check, sometimes known as "Intermediate Layover" (IL), is necessary on some aircraft to check structural parts and some high-load parts for corrosion. Similar to the C check it can be combined with major cabin upgrades like class changes with new seats or monuments. The combination of these tasks saves again downtime of the aircraft. The 3C check can be incorporated in several C checks or into one D check. The main reason for that incorporation is an improvement of the reliability due to better corrosion protection.

9.1.1.1.5 D Check. The D check, sometimes known as "Heavy Maintenance Visit" (HMV), has to be performed every 6-10 years. It is by far the largest maintenance check and takes 2-3 months. During the 50,000 man-hours more or less necessary, the whole aircraft is disassembled. All parts have to be checked. For some structural visual inspections, even the aircraft paint has to be removed. As this check is very time consuming and workload intensive, it is often performed in countries with low labor and hangar costs. Such a check may easily cost several million dollars. For example, a Boeing 747-400 has a D check every 72 months (6 years) costing around $5 million. Due to the long downtime, D checks are planned far in advance. It is common to move the inspection ahead in time to perform it during the winter flight plan utilizing a smaller number of aircraft. As the costs are significant, an upcoming D check can cause the phase-out of an aircraft because its residual value may be lower than the check costs. Most of the commercial aircraft undergo three D checks overall.

9.1.1.2 **Maintenance Prediction—Condition Monitoring:** The advent of computerized control systems gave rise to the concept of condition monitoring, and computing capacity is a key enabler. As the "condition monitoring" term indicates, maintenance tasks here are not based on fixed flight hours, cycles, or time anymore, but on the actual condition of the components. The condition is mainly determined by sensors. Often, the condition has to be defined by multiple indicators, including:

- Pressure
- Vibration and sound
- Temperature and heat transfer rates
- Speed (axial and rotation)
- Power consumption, current, and voltage
- Stress, pressure, and shock
- Overall position
- Computer outputs

Thanks to increased availability of real-time data today, even non-critical systems can feature condition monitoring. The big advantage of such systems is the reduction in maintenance costs they bring. The MRO is able to focus on inspecting and/or changing components with actual defects or find issues much faster, reducing aircraft downtime.

The challenge of condition monitoring is to find the economic balance between the system costs and expected savings. As condition monitoring systems need advanced instrumentation, the initial costs can be significant. Legacy components often lack data or even a power interface, making retrofit scenarios very unlikely. Wireless data communication may ease the situation. An example for this technology is vibration sensing of rotating parts like a turbine shaft or a pump motor.

Analyzing more complex components requires data fusion. The challenge here is to use the right data combination and interpretation to determine the defective components to be replaced.

One of the hazards of condition-based maintenance (CBM) is the rising complexity of the instrumentation itself. If the sensors are faulty themselves, they may indicate component failures that are inexistent, increasing again the maintenance costs.

All and even more electric aircraft implement electrical systems in lieu of hydraulic and pneumatic systems; all-electric systems also eliminate traditional propulsion systems. Most of the electrified systems require active power control enabled by sensors. So the hardware requirements for condition monitoring may be met. But the difficulty remains in finding what kind of data indicates what kind of failure. This is why data interpretation algorithms often get mature only months or even years down the line thanks to cross-checks of sensor readings with actual maintenance findings.

9.1.1.2.1 Condition-Based Maintenance (CBM) and Predictive Maintenance (PdM). The terms Condition Based Maintenance (CBM) and Predictive Maintenance (PdM) are often used as synonyms. Actually, there is a slight difference, or to be more precise, there is a considerable overlap. PdM uses often the tools and methods of CBM but could also use other data sources to predict life of parts.

CBM should only be performed if the component shows any indication of failure or unacceptable degraded performance. PdM has the goal to predict the failure of a component so that it could be changed or repaired preemptively even if the part is not showing any signs of degradation. Therefore, big data analytics based on statistical principles are used.

As a simple example, let's take an armrest of an economy class aircraft seat. CBM could monitor the stress curve through the part. If this curve changes to a certain shape, this would be interpreted as reduced performance and the part would be changed. Predictive maintenance would use the same information in the same manner, but it could also consider, for instance, the number of passengers having taken that seat. The PdM model predicts that every 30,000 passengers the armrest has to be changed due to scratches. Scratches cannot be detected by any sensor, making it impossible to track the actual condition. An experience-based model, on the other hand, would give a very good indication when a change should be considered. The classic maintenance schedule is based on "simple" experience-based models. The essential difference with modern predictive maintenance models is combining the experience-based data with live data. This extensive data may reveal patterns indicating failure root causes which were unknown or undetected before.

TABLE 9.1 Pros and cons of condition-based monitoring

Advantages	Disadvantages
Increase of utilization time	Increase of CBM-enabled component cost
Improvement of component reliability	Risk of increased maintenance tasks due to sensor failure
Minimizing AOG time	
maintenance cost reduction via optimized maintenance schedules	
CBM/PdM is non-invasive. Equipment can be operated during inspection	May generate considerable retrofit effort
Increase of safety levels	Unforeseen failures in case of false analytics

Another very important aspect of CBM is the "on-wing" capability. Most inspection technologies perform sensing on components during their operation, making a complicated disassembly unnecessary or at least far less mandatory. This is an important factor for cost savings.

Table 9.1 shows the advantages and disadvantages of CBM.

To reap the full benefits of a PdM approach, the dataset of each and every aircraft of an airline has to be connected to a so-called computerized maintenance management system (CMMS). It manages the condition of all components and triggers maintenance tasks and schedules. Otherwise, a fleet with hundreds of aircraft, each with millions of parts, cannot be run efficiently.

To achieve this, several inspection technologies are needed. A part's condition is nondestructively analyzed by sensors measuring different light spectrums, acoustics, vibration, temperature, speed, power, stress, liquid composition, or computer outputs. Often, not one but several indicators are utilized to determine the condition, both analyzing the equipment and its immediate environment. Especially wireless communication technology kicked off the implementation of sensor networks.

In the following, some of the mentioned technologies are described in further detail:

> *Visual inspection*
>
> Visual inspections have been used in aerospace industry for decades. For example, black light and contrast liquids are used to detect cracks, folds, and corrosion. Endoscopes, mirrors, and lenses help maintenance staff to detect failures on hidden and not easily accessible spots like in the engines. Most of these inspections are performed during maintenance checks by MRO staff, but recent developments in camera sensor technology enabled automated live data collection. As visible light camera sensors need at least 300 lumens to deliver good results, infrared monitoring is utilized for more and more inspections. Its versatile utilization possibilities allow detection of mechanical and electrical failures (Figure 9.1). Due to its low price it is considered as one of the most cost-efficient inspection methods.
>
> *Acoustics*
>
> Sonic and ultrasonic real-time analyses are mainly used for moving parts like shafts. Sonic monitoring is used in applications that need less accuracy. It is less expensive and can also be detected by trained staff by intent listening with their bare ears. Ultrasound has enough acoustic resolution to inspect even machines

FIGURE 9.1 Infrared monitoring for failure detection.

Source: Lufthansa Technik

rotating with high frequency like jet engines or turbo pumps. Ultrasonic microphones can "hear" the difference between smooth operation of a rotating machine and one with too much friction or stress by comparing sound profiles. Each type of machine has its own sound profile, similar to an acoustic fingerprint. Changes like friction, stress, or deformation generate additional distinguished sounds visible in the upper ultrasound spectrum. That effect detects an abnormal wear far before visible inspection could. The only comparable method would be vibration analytics. But, unlike the ultrasonic method, there are instances whereby vibration monitoring may fail to distinguish the faulty component from the healthy one (Figure 9.2).

Aircraft OEMs outlined that predictive maintenance based on data analytics and CBM could eliminate all AOG events within the 2025-2035 timeframe. This timeline allows for such advances to be taken into account in future design configurations of the electric aircraft. Meanwhile, implementation of more electric systems, until there is finally a full electric aircraft, will not only benefit the environment but also the maintenance effort. Until now, modern conventional aircraft have decreased fuel consumption and weight at the cost of more system complexity. This trend of more and more complex system architectures could be totally reversed on an all-electric aircraft, electrification allowing for the removal of the pneumatic, hydraulic, and fuel distribution systems along with traditional propulsion systems.

FIGURE 9.2 Vibration monitoring for failure detection (method limitation shown).

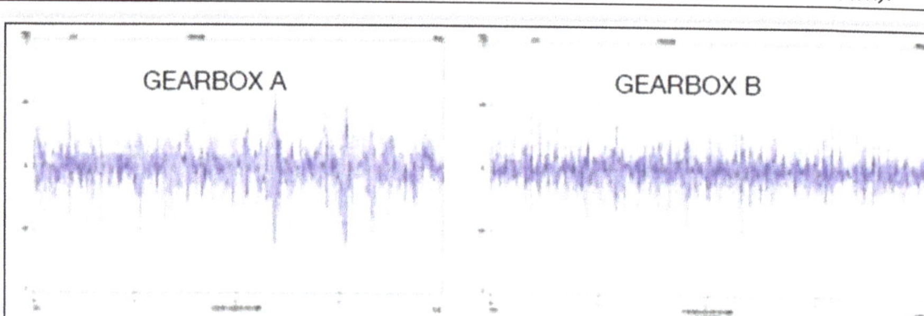

1, 2. Waveforms from the accelerometer are similar, masking the damage inside one of the gearboxes

Source: UE Systems Inc.

9.1.2 Changes for More Electric Aircraft

As a first step, a more electric aircraft may substitute hydraulic and pneumatic systems with electrical ones. Hydraulic systems can distribute large forces over the whole aircraft through cumbersome piping pumps and valves with a huge number of fittings and couplings. However, due to redundancy built into architectures, it is very robust. Unfortunately, it also poses some challenges concerning maintenance. Because of its complexity, leaks are often very difficult to locate. In the meantime the leaking fluid could cause a cascade of other failures. This makes the maintenance of the hydraulic but also pneumatic systems time consuming and expensive.

Hydraulic and pneumatic systems were used because it meant that the actuators supplied by them could be built less complex, lighter, and smaller. As electrical motors made enormous progress in these areas over the past years, this advantage is almost leveled out. Electrical cabling, on the other hand, is far less complex then piping. The functions of today's hydraulic, pneumatic, and power systems could all be integrated into one power distribution network. Instead of maintaining three systems there would be just one. Moreover, cable breakages are much easier to detect than pipe leaks. Wire fault location uses a technology called reflectometry. This consists in sending a discontinued signal into the damaged cable and the break or short would reflect the signal and send it back. The time required for the reflection indicates the distance to the problem.

In general, electric systems are much easier to monitor for health and system status. Even on a more electric aircraft, the reduction in system parts by moving to a primarily electric architecture may be significant. A case in point is the more electric Boeing 787 where the bleedless architecture allows reduction of overall mechanical systems complexity by more than 50% compared to a conventional 767; the elimination of pneumatic systems is a major contributor [9.1]. As a consequence of this reduction in mechanical systems' complexity, airline operations may get less maintenance intensive and more reliable than with conventional systems of bleed architectures, even though the complexity on the electrical side goes up. In fact, the move to electric systems helps in cutting significantly the schedule interruptions compared to a conventional aircraft for the systems affected by the no-bleed/more electric architecture. Other benefits include improved health monitoring and greater fault tolerance.

Moreover, the 787 features greatly expanded and improved systems monitoring capability coupled with an advanced onboard maintenance computing system. This capability combined with e-enabling technologies, which make real-time ground-based monitoring possible, significantly aid in rapid, accurate troubleshooting of the 787. Airplane systems information used in conjunction with fully integrated support products help maintenance and engineering organizations quickly isolate failed components and reduce return-to-service times. No-fault-found (NFF) removals are also drastically reduced compared to conventional aircraft, reducing yet another major cost driver for operators. Therefore, thanks to the extensive system electrification on the Boeing 787, even though hydraulics have not been totally removed, airlines have far less maintenance to perform less often than they are used to. This leads to extended check intervals compared to conventional aircraft as shown in **Figure 9.3**.

FIGURE 9.3 C-check interval - long range aircraft [9.2].

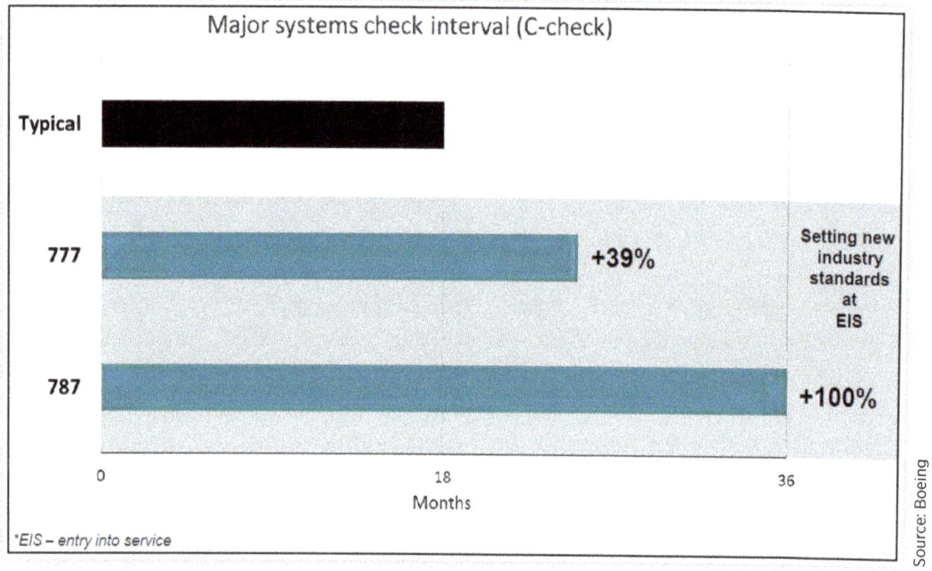

9.1.3 Changes for an Electric Aircraft

With an electric aircraft, the complexity of the engine decreases by orders of magnitude. A current jet engine is one of the most advanced combustion machines ever developed. It consists of up to 30,000 parts. The combustion chamber reaches up to 2000°C and even the exhaust gas can still reach 500°C. Therefore, dozens of blades have to be made of ceramics or single-crystal metal or have advanced cooling outlets for cool air shielding. All this makes it not only the most expensive aircraft system at the moment of purchase but also in terms of maintenance. Current jet engines and their support systems have hundreds of thousands of years of service history, while the electric aircraft is still in its infancy state and will take some learning process to get to the current existing aircraft state.

An electric engine for aerospace applications would in contrast probably have only around 250-300 parts. Also, temperature should not be in the same order of magnitude. So, it is obvious that electric engines will have a much lower share of total cost then today's jet engines.

The electric aircraft also adds a new system: a solid power source. While batteries are already used, the dimension needed for an electric aircraft escalates to an entirely new level. If we imagine batteries, which can store enough energy to use them as primary power source, they would have to be maintained unlike jet fuel today. Battery state of charge and state of health are key measures to ensure safe operations and full battery utilization. That means that it would be mandatory to devise an approach to maintain them. Replacing them after some hundred cycles comparable to today's smartphones or even cars would not be ecologically and financially sustainable. Furthermore, a lot of current battery technology is based on rare earth materials. How that maintenance service could be offered and by whom is currently not clear. Decisive factors are technical expertise, global capacity, price, but also upcoming regulations.

9.1.4 Airport Operations

Next to maintenance, airport operations form the other part of ground operations. In the coming sections, topics like infrastructure, aircraft turnaround, and emergency processes will be addressed.

Airports are currently the foundation of aerospace business. Apart from decentralized helicopter landing areas (helipads), every aerospace mission starts and ends at an airport. Its infrastructure includes runways, taxiways, aprons, terminals, hangars, traffic control centers, and support infrastructure like fuel reservoirs. International airports facilitate border control and customs checks as well.

9.1.4.1 Infrastructure: One fundamental question remains: which future aircraft concepts need what kind of infrastructure? Looking at long- and short-range more electric and electric aircraft, this will not change. Larger aircraft will always need a certain support infrastructure, especially if international travel is involved. The next section on aircraft handling will go into further detail on this.

Urban air mobility concepts like the Volocopter shown in Figure 9.4 will need a much smaller scale of infrastructure. However, the core functions of an airport have to be covered somehow. A combination of helipads and "vertiports" could be a solution.

Assuming that urban air mobility concepts are more competitive concerning pricing than current helicopter services, the number of missions would be much higher. Therefore, a simple and seamless boarding and disembarking process is mandatory. Many studies show that security control is currently the pain point number one at airports. Automated or remote-controlled advanced scanner technology in combination with a kind of background pre-screening similar to systems like ESTA in the United States could minimize the effort. As this process should be done within minutes, a large waiting area like in current terminals would not be needed.

The vehicle itself would need a refueling or recharging facility and maybe a small line maintenance station. Depending on the success rate of predictive analytics, the maintenance could be even coupled with similar services at the airport. While urban air transportation is probably offered in larger cities that are already equipped with at least one full-scale airport, the question whether traffic control of low altitude urban mobility would be managed by local airport(s) is still open. A lot of urban air mobility concepts reckon with the fact that air traffic management principles and its responsibility are yet be addressed. On top of the focus on autonomous operation, navigation, sense, and avoid are also key technologies.

However, that would mean that urban air mobility concepts would be mainly point-to-point connections between localities featuring so-called rooftop "vertihubs" or "vertistops" integrated to roads. First test cases support that prediction.

FIGURE 9.4 Example of urban air mobility vehicle (Volocopter) [9.3].

Future smart ground transportation systems may also be poised to compete directly with urban air mobility concepts. In fact, most of the current ground mobility business models focus on the pain point of gridlocks found in many urban metropolitan areas. Therefore, if autonomous taxi or car sharing concepts were to become reality, with their own traffic management in place, congestion issues would be resolved, and a shadow would be cast on the appeal of air mobility.

From an energy consumption point of view, aerospace applications will always be more demanding and thereby more expensive. Also old applications will have to merge with the new ones seamlessly, which is quite a challenge. In addition, door-to-door transportation depends on ground-based systems to a certain extent. Even if urban air mobility concepts save their customers some time, ground-based solutions offer not only lower prices but also more comfort and onboard infrastructure allowing better utilization of travel time. A similar situation can be currently observed with trains versus aircraft on short-range flights, especially in Europe and Asia.

9.1.4.2 **Aircraft Handling:** Focusing again on larger electric aircraft concepts, these vehicles would need similar aircraft handling services like today with two exceptions—Refueling/Recharging and Pushback/Taxiing.

9.1.4.3 **Refueling/Recharging:** As one of the biggest changes of an electric aircraft would be the replacement of the main energy source, the refueling process would have to be completely redesigned. If the new energy source is liquid or gaseous, current processes might not differ much.

Assuming a kind of battery system, there are normally two solutions—recharging or replacing. Recharging would have the main challenge of time. Today, a turnaround of an A320 is between 30 to 60 min. Refueling is only permitted if no passengers are on board or a fire truck is on standby next to the aircraft. But predicting that recharging would be allowed during the whole turnaround time, it is still technically very challenging to recharge the batteries within that timeframe without losing too much battery life. In addition, battery technology would have to improve a lot from a specific energy standpoint.

If the specific energy challenge could be solved but the recharging timeframe could not, a battery swap would be the most likely solution. Some car manufacturers showed already such concepts of fast and automated battery swapping systems.

Both solutions have one common challenge for the airport—standardization. Today, all aircraft fly with the same fuel. Looking at electric vehicles, there are a multitude of power plug standards. Each car manufacturer defined its own plug with specific standards for voltage and current. For aircraft ground support equipment (GSE), there is a standard connector which is the Euro plug. Nevertheless, for future electric aircraft batteries, the existing Euro connectors may not be sufficient. This may require standardization bodies to develop standards for the battery recharging process. Moreover, depending on the aircraft system architecture, different OEMs are likely to come up with totally different power architectures. So, it will be essential for airports to define a very flexible, but also scalable (up and down), power grid architecture.

Battery pack shapes are mainly driven by the geometry of the powered device/vehicle. Predicting a variety of configurations could also mean that current geometry standards are not useful anymore. For example, the cargo LD-X standard container (Figure 9.5) fits into the cargo hold of almost every long haul aircraft—A380, A350, A330/A340, B747, B777, B787 .... A flying wing configuration may waste a lot of space by adapting to that standard, leading to a new type of container.

FIGURE 9.5 2 LD-3 containers get loaded into a Lufthansa A380 aircraft.

Source: Lufthansa Technik

Accordingly, if all electric aircraft were to have very diverse configurations, airports would have to store large stocks of different kinds of batteries. This would not just mean a storage space challenge but also consequential costs.

This is why standardization bodies like SAE International are already discussing possible solutions with all impacted industry stakeholders to avoid that. The objective is to produce a standard that would minimize variability and provide guidance for standardized design, production, and testing.

Likewise, fuel cell utilization at airports requires storage and/or distribution of hydrogen to airport facilities for aircraft refueling. This requires modifications to infrastructure addressing hazard prevention and safety concerns.

9.1.4.4 **Pushback/Taxiing:** The increasing awareness and attention to environmental concerns drives innovation in aeronautics. One of the issues to be addressed is aircraft maneuvering on ground using jet engines, thus causing a variety of undesired emissions. The general working principle is to use electric vehicles to move aircraft from the gate to the runway and vice versa. For several years, research and demonstrations have been carried out in the short-range segment to validate electric taxiing achieved by the aircraft on its own. This solution consists of integrating electric wheel motors in the landing gear which draw electrical power from the APU (or fuel cell) in combination with batteries, if necessary. With the APU being more fuel-efficient than a turbofan engine, this aircraft-based electric taxiing allows for reductions in fuel burn and emissions. Moreover, with this solution, retrofit on legacy aircraft may be possible. Chapter 7 provides a deep insight into this incremental electrification approach of conventional aircraft.

9.2 In-Flight Operations

This section discusses both cockpit and cabin implications of electric aircraft concepts, bearing in mind that far more changes are expected for the flight deck rather than the cabin. Starting from pilot licenses, an overview of controversial topics such as single pilot operations, autonomous flight, future urban mobility with pilots akin to drone operators, etc. is openly presented. As a general comment, not all these topics are directly linked to the electric aircraft. But this is the future, so when reading through this visioning exercise, remember that the future is predictable only with limited certainty.

9.2.1 Flight Deck Operations

With the introduction of electric aircraft there will be at least some changes in the cockpit.

9.2.1.1 Complex Configurations/Licenses: Similar to aircraft certification challenges, pilot licensing will be a topic for electric aircraft operation. Today, pilots need a certificate issued by the Civil Aviation Authority of the country where they want to operate. Despite several attempts, the licensing process is still different from country to country. The most known agencies regulating pilot certification are the Federal Aviation Administration (FAA) in the United States and the European Aviation Safety Agency (EASA) for the European Union and Switzerland.

As there are different aircraft categories and classes, there are different levels of licenses, further distinguished by ratings:

Private and Commercial Pilot License (PPL and CPL)
For non-commercial and commercial operation of small aircraft, respectively. Depending on the country, different sub-ratings are available, for example, complex aircraft, single-engine/multi-engine aircraft (maximum takeoff weight (MTOW) of 2t), etc.

Airline Transport Pilot License (ATPL)
This license is required in order to fly commercial planes in the range of those usually operated by airlines. ATPL is the highest level license covering PPL and CPL. For each aircraft type flown, an additional type rating is required.

Drone operators
Some countries start regulating drone operations by requiring a Drone Operator License. Especially commercial drone operations are covered by these initiatives. However, a structured international standard is currently not available.

Depending on electric aircraft configuration and its TC, the corresponding pilot license is derived. The biggest gap here is within the urban air mobility concepts. These vehicles are often a combination of ultra-light aircraft, helicopters, and drones. There is, for instance, an urban air mobility prototype which is currently applying for an EASA ultra-light certification. The derived ultra-light pilot license would be just valid for Europe and not for commercial use.

The bigger electric aircraft concepts for commercial short-range applications would be certified under CFR Part 25 so that the normal ATP license with a specific type rating would probably be sufficient.

9.2.2 Single Pilot Operations

A topic that is currently controversially discussed is large commercial aircraft single pilot operations. It would mark a first step towards autonomous flight. Both automotive and aerospace vehicles powered by fossil fuel would be capable to integrate the necessary technology for (greater) autonomy. However, as the topic of autonomous driving is by public perception linked to electrically powered cars, this impression is transferred as well to aerospace. Nevertheless, there are some arguments why an electric aircraft program could foster the implementation of this concept.

As many concepts for even large short-range applications envision multi-impellor or multi-propeller with multi-power sources, thrust control in case of failure is getting much more complex. Moreover, these engines not only produce thrust but also lift or steer the vehicle. It is presumable that, similar to modern unstable fighter jets, these aircraft are not maneuverable without significant computing support. So, as the computer is anyway responsible for a significant portion of flight operation, the question is how effective a human pilot would be interfering in case of a failure. There might be the conclusion that the most effective operation in terms of cost but also safety is an advanced autopilot system supporting one pilot.

The biggest issue in this scenario would be how to compensate the loss of a single pilot in case of medical emergency, for instance. For this particular case, initial studies support a remote pilot solution. As a side effect, these remote pilots could also support aircraft with a twin pilot operation if one of the pilots gets suddenly unfit during flight.

9.2.3 Autonomous Flight

Of course, autonomous flight is discussed with even more controversy. Similar to single pilot operation of commercial flights, autonomous flight is not directly related to electric aircraft but is often mentioned in the same breath.

In contrast to single pilot operation, autonomous operation today is mainly driven by smaller electric aircraft concepts like civil drone applications and urban air mobility vehicles. Here, the business case rests on the assumption that autonomous flight is both technically possible and socially acceptable, and is cost-efficient. Urban air mobility may also need autonomous operation to close the business case just because there won't be enough pilots.

In commercial aviation, autopilot systems have been the norm for years. Automatic landing systems are in place, but the standard procedure is still for the pilot to land the plane manually. Fully autonomous flights were also demonstrated multiple times. In 1988, as one of the most famous examples, Russian spaceship Buran, which looks quite similar to the American Space Shuttle, performed a fully autonomous start, flight in space, and landing. In military drone operations, autonomous or semi-autonomous flights are increasingly becoming the standard procedure. With the Northrop Grumman X-47B Unmanned Combat Air Vehicle (UCAV), even automated air-to-air refueling and landing on an aircraft carrier were demonstrated (**Figure 9.6**).

FIGURE 9.6 UCAV autoland on aircraft carrier [9.4].

Note that these flights have been performed on vehicles with no passengers on board, but the technology is mature. Yet, some industry actors, especially airline related, have suggested that autonomous flight in commercial aviation will take more than one generation to become reality. The Aircraft OEMs Boeing and

Airbus are more optimistic and want to start trials within months not decades. Both sides have valid arguments.

One of the main arguments against autonomous flight concerns safety. The issue is less on the computing power but more on the generation of intelligent decision trees. In unforeseen events, even Artificial Intelligence (AI) isn't capable to come up with "innovative" ideas to solve the issue. Furthermore, AI needs vast amounts of data to train machine learning for the resolution of problems. It is currently not proven that the available data from, for example, simulator emergency situations is sufficient for that approach.

On the other hand, autonomous driving has gained remarkable traction in its development, while a car operates in a far less predictable environment than an aircraft. Therefore, problem-solving algorithms from the automotive industry could also help aviation overcome technical challenges. The near future will show if the OEMs can reach mandatory safety levels and prove that autonomous flight is technically feasible on a commercial aircraft. The outlined and published timelines by various startups and OEMs must be treated with some reservation. In the 1980s, visions of the future also predicted that autonomous flights would become reality no later than the year 2000. Like single pilot operation, the development of a complete new aircraft, especially an electric aircraft redefining current standards, would boost technical development and, therefore, the chances of implementation.

Another aspect is the social acceptance of autonomous flight. Whether or not passengers would board a self-flying plane is just as hotly debated as technical questions. Similar to trains or nowadays cars and trucks, some people believe in autonomy and others do not. History shows that, while it will take some time, people will get used to the new technology. In several cities, like in Paris, for instance, fully automated subways have been in service for years and passengers readily ride them, often without noticing the absence of a driver. Nevertheless, there are still public transportation providers who are convinced that their customers would not accept a driverless train. Self-driving cars triggered an even bigger discussion. That this debate is not always driven by logical arguments becomes obvious when accidents do happen, especially when there are fatalities. Most of these technology demonstrators have theoretically proven that they have much higher reliability compared to human drivers, and the social debate will certainly heat up around whether to adopt the concept or not.

On the other hand, cars with anti-lock braking systems (ABS) or Electronic Stability Control (ESC or ESP) are considered safer, even if this means that a computer overrides the human driver impulse and manages autonomous braking. So, when higher safety levels of an autonomous system compared to a human operator is statistically proven, and the general public is convinced, it could even become a must-have item. Given the emotional rather than logic development of public opinion, it is very hard to predict if or when that will happen.

9.2.4 Pilots as Drone Operators

In unmanned military applications, pilots working as single or even multiple drone operators is something which has already been implemented over the years, but for commercial aviation it is still considered as disruptive. The straightforward reason behind this is that commercial aircraft carry passengers, and as such have to comply with stringent safety requirements.

In the early 2000s, the United States developed weaponized drones from a prototype status to what has come to be known as UCAVs. Combat drone operations are becoming

FIGURE 9.7 Remote and local Ground Control Stations part of the Predator Unmanned Aircraft System (UAS) [9.5].

increasingly commonplace regardless of where around the globe the theater of operation is located. Incidentally, most of the pilots are remotely stationed in Europe (**Figure 9.7**) or in the United States, sometimes several thousand miles away from combat zones. This distance causes a signal delay of up to a couple of seconds between the time a command is issued to the drone and the time when it is received.

Due to that delay, time critical operations like start and landing are performed by drone operators on location. When the drones reach their operational altitude, there is a handover to the remotely located pilots. As most of these operations are highly automated observation missions, drones operate semi-autonomously, that is, fly without human interaction unless a potential finding requires an operator's action. This is why drone operators have the capacity to "fly" more than one drone.

Future combat drones may leverage hybrid-electric propulsion for stealthier VTOL operations like with the LightningStrike from Aurora, a Boeing company, wherein vertical takeoff and landing would be possible without a runway. This would stave off the handover constraint making UCAVs operable remotely right from the onset.

Anyway, the handover principle between remote and local control stations used in current UCAV operations could also be adapted for future commercial aviation or delivery drones. For instance, urban air mobility concepts using electric propulsion like CityAirbus shown in **Figure 9.8** could operate autonomously until a situation arises

FIGURE 9.8 CityAirbus urban mobility concept vehicle rooftop landing (rendering) [9.6].

where the system would need human assistance. This would give the concept the possibility to utilize all four seats for passengers alone and benefit as well from advanced safety measures. Depending on the upcoming scenarios, there could even be a similar split between local operators and centralized facilities like flight control radio stations. Of course, this infrastructure would also have to support single pilot operation or autonomous flight of larger airliners.

9.2.5 Cabin Operations

The other part of in-flight operations is performed on the other side of the cockpit door: cabin operations. Cabin operations consist of both passenger service—a very important brand experience touch point for airlines—and passenger safety, the more important aspect from a regulatory standpoint.

Passenger service processes will slightly differ depending on cabin geometry and flight profile. However, it is one of the top priorities of every airline to offer a consistent service product. The customer should know what baseline to expect, enhanced by the personal touch of every crew member. The current processes are updated regularly based on experience and driven by changing customer expectations. A more electric or electric aircraft is not likely to have a major impact in this area.

Beyond service, flying personnel are required to guarantee the safety of passengers. An electric aircraft with a different power distribution system or even a different main power source could impact today's procedures. Today's procedures require aircraft and ground power supplies to be properly grounded in order to ensure their protection against electrical hazards (static electricity and failures) and that of crew, passengers, and maintenance personnel. These procedures have to be revisited with the advent of more electric or electric aircraft due to the presence of more hazardous high voltage within power systems and energy storage.

One of the questions currently open is how to indicate that the aircraft is electrically grounded. This issue could impact the safety of the crew, passengers, but also first responders in case of an emergency landing. The latter case involves potential hazards unique to onboard electrical systems such as electrocution, fire, and battery electrolyte spillage.

A potentially higher explosion risk of batteries could lead to a shorter maximum evacuation time. Therefore, new ways of disembarking would have to be developed.

Looking at urban air mobility concepts, safety in case of emergency landing could also become a key topic. As most concepts plan for autonomously operated vehicles, there is no trained crew to help passengers leave the vehicle in case of emergency. Digital information technology could potentially compensate portions of these tasks, but the physical support assisting injured or movement-restricted passengers is technically challenging to compensate.

References

[9.1] Sinnett, M., "787 No-Bleed Systems—Saving Fuel and Enhancing Operational Efficiencies," *AERO Magazine* Q4, 2007, published by Boeing.

[9.2] Boeing, "Airline Economics," *2016 Airline Planning Workshop*, Airports Council International (ACI), North America, USA, 2016.

[9.3] https://www.volocopter.com/en/product/, accessed May 5, 2018.

[9.4] https://www.usnews.com/news/articles/2013/06/11/new-military-uav-may-lead-to-commercial-drone-flights, accessed May 4, 2018.

[9.5] https://www.uasvision.com/2011/07/15/view-inside-a-predator-ground-control-station/, accessed May 5, 2018.

[9.6] https://airbus-h.assetsadobe2.com/is/image/content/dam/corporate-topics/publications/press-release/CityAirbus-01.jpg?wid=3626&fit=constrain, accessed May 4, 2018.

CHAPTER 10

Performance and Business Value of Electric Aircraft

Pascal Thalin
Chair and Member - SAE Electric Aircraft Steering Group

As discussed in Chapter 2, fuel consumption is a major cost driver in aircraft operations. In order to quantify its impact, understanding the cost structure of airlines is fundamental.

10.1 Airline Cost Structure

When airlines operate their fleet, operating costs are incurred, such as fuel, maintenance, flight personnel, and aircraft leasing/ownership, among others. Additionally, other charges borne on ground have to be considered, such as servicing and ticketing. Finally, airlines have to pick up the tab for so-called system operating costs. Generally speaking, the following breakdown gives a rough order of magnitude of the cost segments at play:

FLIGHT (DIRECT) OPERATING COSTS (DOC) = 50%

- All costs related to aircraft flying operations
- Includes pilots, flight crew, fuel, maintenance, and aircraft ownership

GROUND OPERATING COSTS = 30%

- Servicing of passengers and aircraft at airport stations
- Includes aircraft landing fees and reservations/sales charges

SYSTEM OPERATING COSTS = 20%

- Marketing, administrative, and general overhead items
- Includes in-flight services and ground equipment ownership

For relative comparisons, **Figure 10.1** shows a rough breakdown of airline direct operating costs (DOC).

As shown, fuel consumption could account for up to one-third of typical airline DOC. Simply put, any reduction made possible in this expense could potentially benefit airline profitability.

Figure 10.2 portrays International Air Transport Association (IATA) data on fuel costs over time of the airline industry. Just to give an idea, it was forecasted that worldwide fuel expenses in 2017 would top out at above a staggering USD 131 billion.

FIGURE 10.1 Airline direct operating costs [10.1].

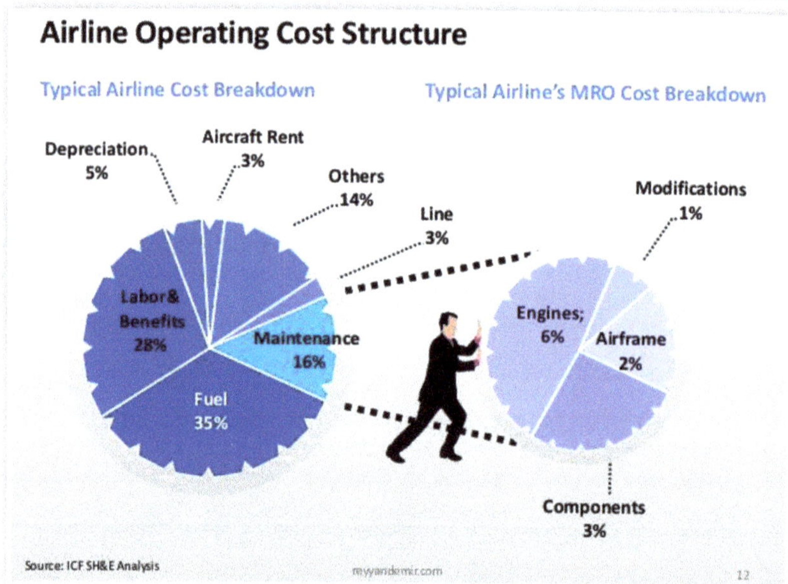

FIGURE 10.2 Worldwide airline industry fuel costs (USD) [10.2].

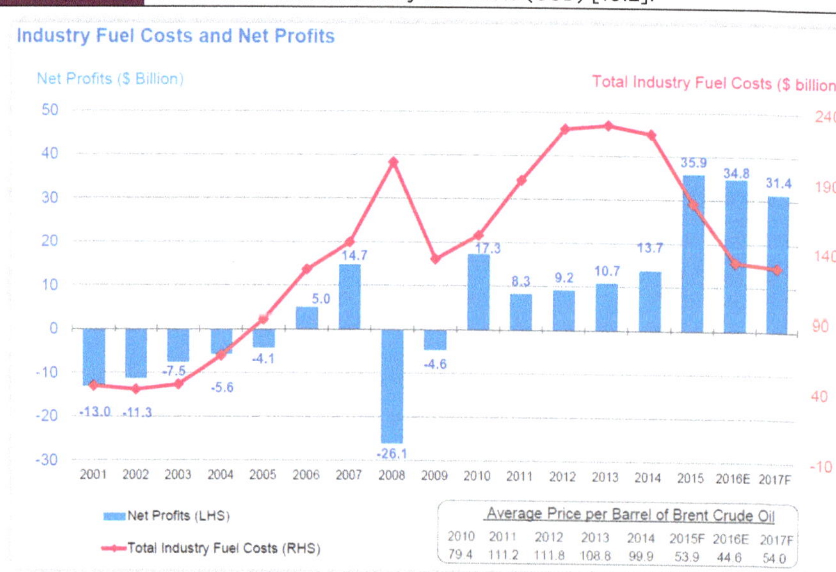

The chart in Figure 10.3 shows the price variation over time of jet fuel used to operate conventional aircraft. Actually, jet fuel price variations mirror the price variations of crude oil price. Unfortunately, airlines have no way to influence these price fluctuations.

Now, let us take a closer look at the share of fuel costs in the total airline operating costs. Figure 10.4 shows how fuel-related costs of the airline industry tend to vary over time. The evolution of the ratio of fuel costs to total operating costs is also highlighted.

As the chart shows, from 1997 to 2001, with relatively low fuel prices of around USD 0.6/gallon, the overall fuel cost on average was a mere 12% of the airline operating cost.

FIGURE 10.3 Jet fuel cost (USD/gallon) [10.3].

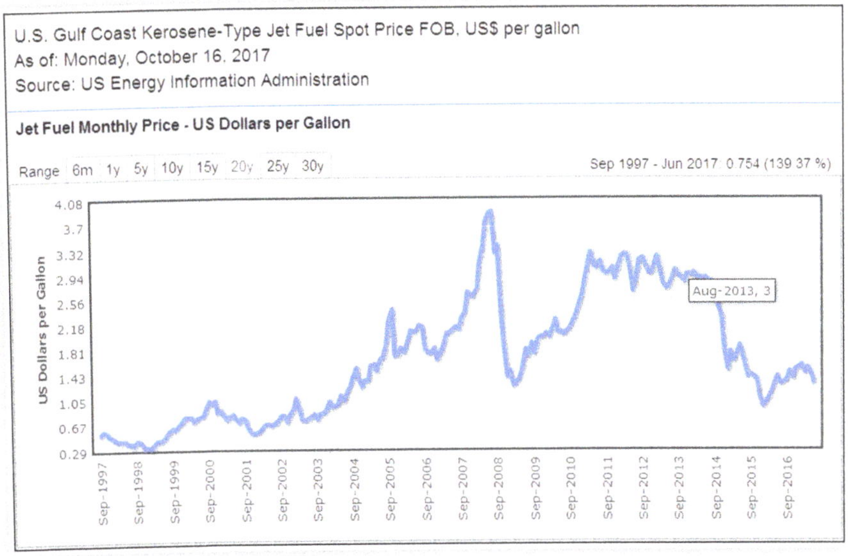

FIGURE 10.4 Worldwide fuel expense (USD) and ratio (%) [10.4, 10.5, 10.6].

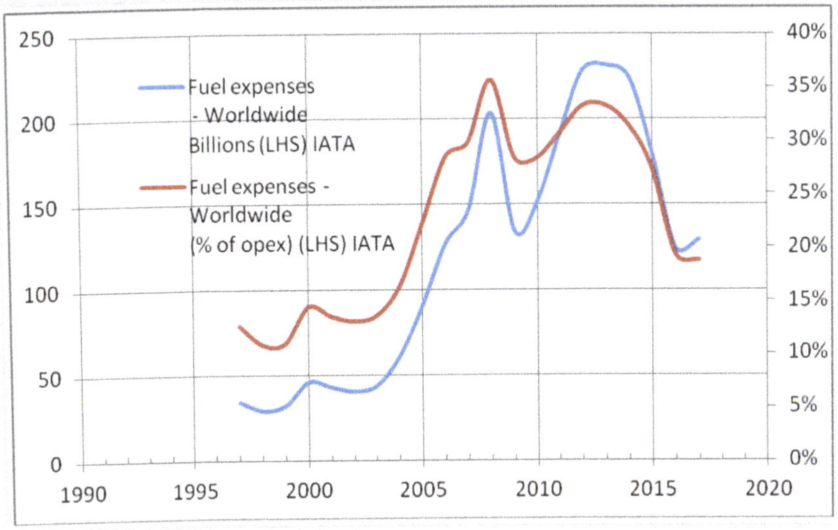

FIGURE 10.5 Direct operating costs across aircraft segments (U.S. carriers—2013) [10.7].

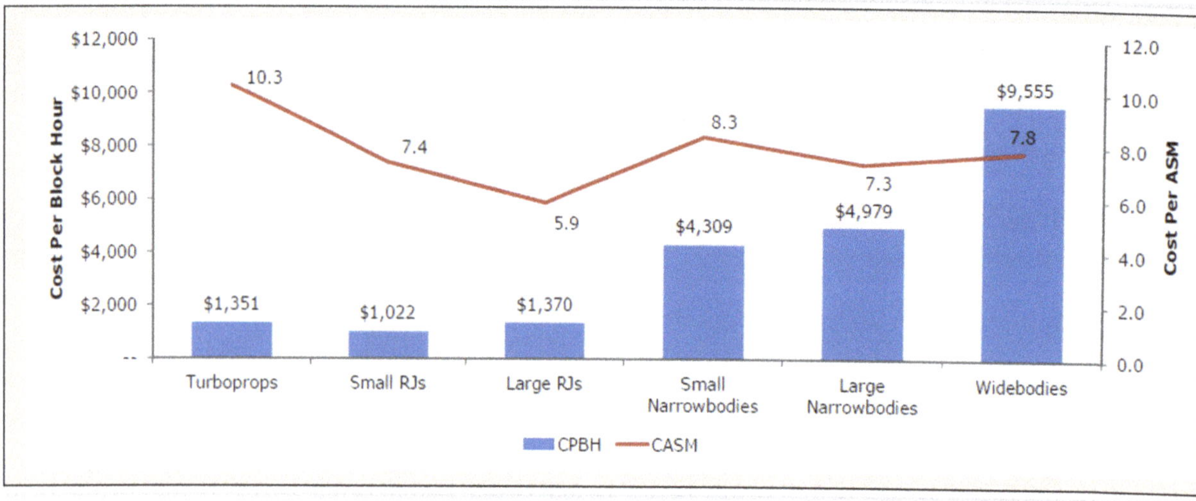

But in 2008, at the height of skyrocketing fuel prices (USD 3.8/gallon), the ratio of fuel cost to airline total operating expense topped out at 36%. Ever since, following two dips in fuel price, the same ratio returned to 19% in 2016.

Now, with regard to DOC, Figure 10.5 also helps us understand the orders of magnitude at the airline level. The chart uses two industry metrics, namely, the block hour and the Cost per Available Seat Mile (CASM). It is worthwhile defining them here. The block hour is the industry standard measure for aircraft utilization. For a given flight, block hours take into account the time from the moment the aircraft door closes at departure until the moment the aircraft door opens at the arrival gate following its landing. The CASM is a common unit of measurement used to compare the efficiency of various airlines. It is obtained by dividing the operating costs of an airline by available seat miles (ASM). Generally, the lower the CASM the more profitable and efficient the airline.

10.2 Aircraft Fuel Costs

Obviously, there is a direct link between the type and size of aircraft being operated and fuel costs incurred. This also holds true for the ratio of fuel costs to DOC which include fuel, maintenance, crew, and aircraft ownership/leasing costs.

The variation of fuel costs across different aircraft segments is illustrated in Figure 10.6, a snapshot for the year 2013 of large certified U.S. passenger air carriers. Costs are expressed in USD per block hour of operation.

Looking at the proportion of fuel in the aircraft DOC, it can be concluded that the larger the aircraft size the more expensive the fuel bill and higher the ratio between fuel costs and operating costs.

This chart shows that, in 2013, for the wide-body aircraft segment, for instance, the fuel ratio was above half of the total costs, just when oil prices stagnated at extremely high levels.

FIGURE 10.6 Fuel cost variation across aircraft segments in 2013 (U.S. carriers) [10.8].

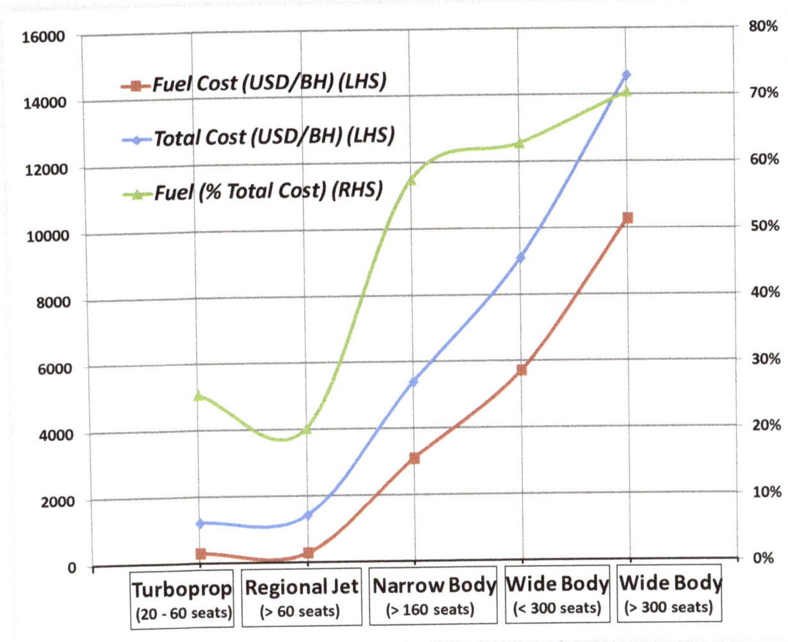

10.3 Airline Fuel Efficiency

Fuel costs of an airline are of course directly linked to jet fuel price in the first place, but there is only so much an airline can do about this. Secondly, fuel costs are tied to the fuel efficiency inherent to the type and age of aircraft making up the fleet, besides the actual efficiency of airline operations. Finally, fuel consumption takes a hit from airspace and airport inefficiencies in the form of tarmac delays and holding delays for instance. But, once again, airlines cannot control these situations.

The impact of jet fuel price could be allayed if the airline opts for shrewd hedging strategies to minimize exposure to price fluctuations. Better operational practices could also help reduce fuel costs further. Last but not least, fuel cost mitigation can be achieved by introducing newer and more fuel-efficient aircraft into the fleet.

Let us see what is behind the fuel efficiency of an aircraft and how it has evolved over time from an airline perspective. For a given type of aircraft and a set of parameters (payload, range, speed), fuel burn is directly linked to the compounded efficiencies of the following:

- Propulsion
- Systems
- Airframe

Figure 10.7 shows for the commercial jet aircraft segment how, notwithstanding fuel price fluctuations, the introduction of new aircraft has helped cut back on fuel consumption. Additional fuel efficiency improvements are expected over the short term

FIGURE 10.7 Fuel burn for new aircraft and fuel prices [10.9].

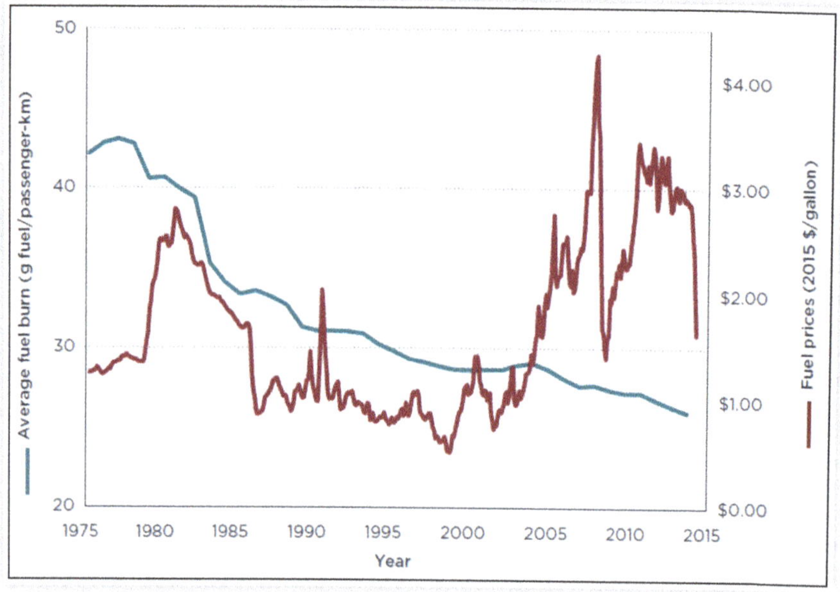

with the entry into service of re-engined versions of conventional aircraft (e.g., Airbus A320neo and Boeing 737 MAX/777X).

One other advantage of aircraft fuel-efficiency gains resides in the improved range performance, that is, the ability of the aircraft to fly farther over longer distances without having to stop to refuel.

Figure 10.8 retraces range improvement of new aircraft over the span of the past 40 years [10.9]. Since 1988, the range has increased by 40%, while operating empty weight

FIGURE 10.8 Structural efficiency and range for new aircraft [10.9].

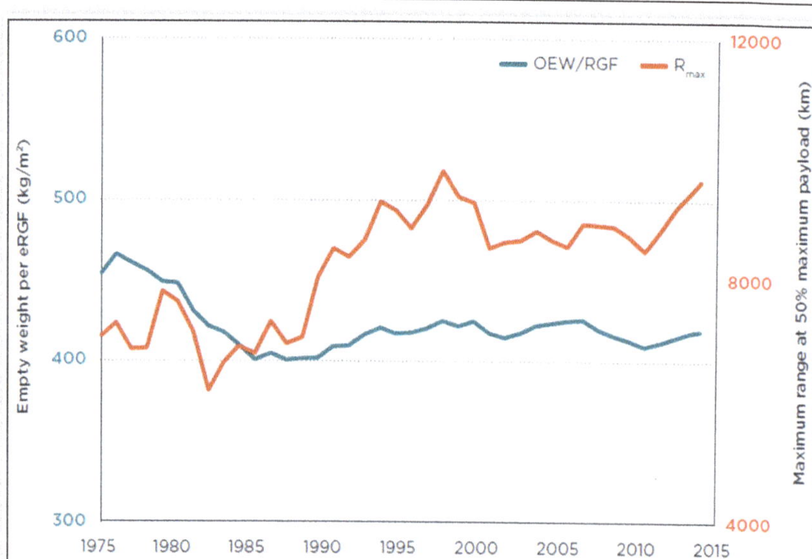

per unit aircraft floor area increased modestly, about 6%. Normally, on a given mission, the longer the aircraft range the lower the fuel efficiency because of more weight. Hence, for the range increase over time shown in Figure 10.8, the fuel efficiency improvement, rather than degradation, is attributable to airframe technology advances (e.g., composite) having been able to offset weight penalties, as noted in ICAO's 2010 fuel efficiency technology review.

All in all, fuel prices alone may not provide a consistent, long-term motivation for fuel efficiency improvements in the aviation sector. The drive to curb carbon emissions through the reduction of fuel consumption has become another key industry driver. Having no handle on fuel price fluctuations, faced with highly demanding environmental targets, global research in the aerospace industry is focusing on an array of solutions.

Simply put, decreasing fuel, maintenance, and other costs with no compromise on performance at the very least is the minimum expectation.

Thankfully, there are solutions ranging from incremental enhancements to all-out aircraft redesigns. An out-of-the-box approach parlaying advancements in architectures and technologies, leveraging innovations drawn from other industries such as automotive, and including cutbacks in operational costs combined with better environment-friendly performance and services could be delivered in the coming years.

The push towards sustainable aviation ushering in cleaner and quieter aircraft technology entails drastic performance requirements that even the most advanced technology on conventional aircraft can hardly cope with. Therefore, rethinking the overall aircraft design and a departure from conventional architectures, technologies, and integration is unavoidable. The move towards the electric aircraft becomes all the more relevant in this context, given the significant added value it could bring in key performance drivers such as fuel efficiency, carbon footprint, and noise while enhancing overall cost effectiveness for the airlines.

In this technology race, industry players are spearheading research to get the picture straight as to how to make the electric aircraft a reality and meet the aggressive challenges of air transport facing the industry.

10.4 Business Aviation

Previous analyses show that major benefits could stem from the electrification of ice protection provided weights are not altered, or (even better) reduced, at the aircraft level. In order to see the benefits of electrification, let us take a look at how some conventional systems perform on a business jet, and what lost performance can be retrieved through electrification and the strings attached to it [10.10].

The conventional pneumatic ice protection system operates from hot bleed air derived from the engines. However, this process impacts engine thrust and the aircraft performance gets negatively affected. Figure 10.9 shows the degradation of engine thrust with the conventional system.

Switching on bleed air decreases thrust by 5%-7% when all engines are operating, and 10%-12% in the case where one engine is inoperative. This thrust degradation not only

FIGURE 10.9 Thrust impact of conventional ice protection [10.10].

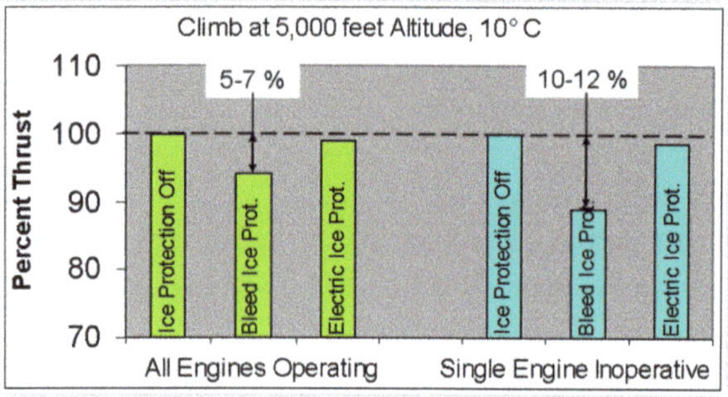

needs additional fuel but, in the latter most critical case, ends up restricting aircraft range performance (**Figure 10.10**).

On a Cessna Citation Jet 2 type of business aircraft with pneumatic ice protection, in order to comply with the minimal climb gradient (2.5% net) of the standard instrument procedure in icing conditions with one engine inoperative, the aircraft takeoff weight has to be reduced by removing almost half of the mission fuel. This drastically degrades range.

Electric ice protection could thwart this drawback and bring real benefit by recovering range performance lost with the conventional system (Figure 10.10). Actually, compared to pneumatic ice protection, the electric version would produce only about one-fifth as much decrease (2%) in thrust.

It is clear that the advantage of electric ice protection lies in lower thrust degradation that allows climbout with more fuel on board. Therefore, in icing conditions, an electric ice protection system can double the usable range of a business jet, compared to a bleed air system.

Let us now consider another pneumatic system, the Environmental Control System (ECS). On business jets, compared to the baseline pneumatic version, although the electric version draws less power from the engines, the positive impact on thrust remains only marginal. This suggests that the greatest energy-saving opportunity of electric ECS, as opposed to the case of ice protection, might not lie in the energy consumed by the system, but rather in the weight reduction of downsizing or eliminating the bleed air components. Reduction in life cycle costs is another area where electrification can add value, thanks to reductions in part count enabling savings on both inventory and maintenance.

On top of these examples, significant performance enhancement could only stem from system weight reductions within electrical architectures, when transitioning from conventional non-electrical systems to their electrified versions.

FIGURE 10.10 Business jet ice protection—range impact [10.10].

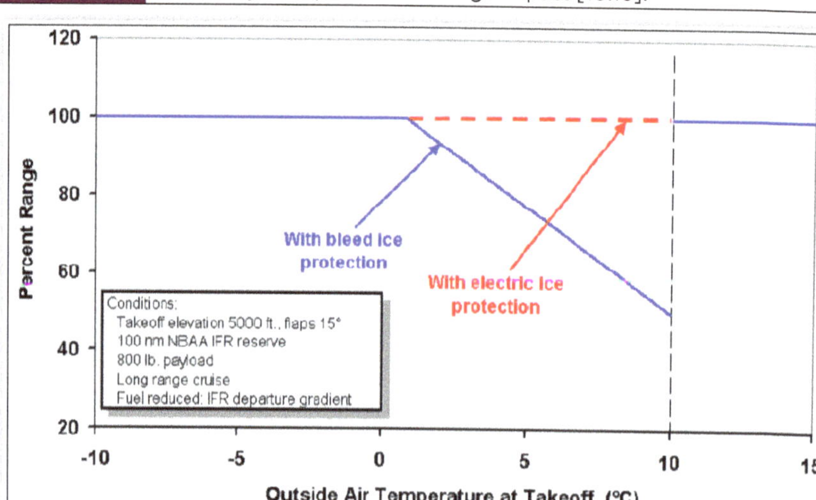

Weight reductions on an aircraft can allow additional fuel to be carried, and therefore extend the range capability. For instance, on the three-passenger Cessna Citation Jet 2, a 100 lb systems weight reduction can take the aircraft farther by around 60 nm.

It is also interesting to check the fuel-burn reduction that may be obtained by parlaying system weight gains into a scaled-down (wing, tail, and engine) redesign to obtain a more fuel-efficient aircraft. In other words, since the aircraft now has 100 lbs less payload, the takeoff thrust requirement is lower, which can translate into smaller tail structure, etc. In this approach, the same 100 lb systems weight reduction, for instance, could help shave 300 lbs off the max takeoff weight on a downsized Citation Jet 2 design, therefore helping reduce the fuel load by around 1.6% on a 1700 nm flight.

But bringing down the weight of systems that undergo electrification is hardly an easy task due to the fact that ongoing research is yet to reach the weight-reduction targets. As explained in Chapter 2, on a more electric version of this type of aircraft, a drastic reduction in the global engine power offtake for systems is possible, even though the electric part is drastically increased. Without improvements in the state-of-the-art electrical components, the overall weight of the airplane can go up, negating any benefit brought upon by electrification. This added weight would then cause a fuel burn penalty or decrease the aircraft's range.

Incidentally, any system-level weight increase gets amplified at the aircraft level. In fact, carrying additional system weight onboard comes down to additional fuel load and volume to be carried up to the destination. This in turn calls for extra fuel, fuel reserves, and space, equating to an overall amplified weight increase. Consequently, the fuel burn also gets amplified in the same manner. This amplification effect is called "spiral" or "snowball effect."

The hurdles behind weight reduction are not trivial. Let us consider a larger, longer-range legacy business jet such as the Falcon 2000 manufactured by Dassault Aviation [10.11]. The analysis in [10.11] uses scaled-up systems and does not take into account weight-optimized engines and systems. Figures 10.11 and 10.12 compare the conventional Falcon 2000 to its more-electric and all-electric versions: partial electrification whereby only hydraulic systems are switched to electrical ones and total electrification of all systems. In Figure 10.11, it can be seen that system weight increase for the all-electric aircraft total is north of 300 kg, which at the aircraft level gets amplified to above a ton due to the snowball effect (Figure 10.12).

These weight penalties hurt fuel efficiency and also the range performance.

In the partial electrification case, even though suppressing hydraulic systems reduces the fuel consumption needed for system power delivery, the reaped benefit is hardly sufficient to counter the overconsumption from weight increase with the electrical replacement.

FIGURE 10.11 System weight [10.11].

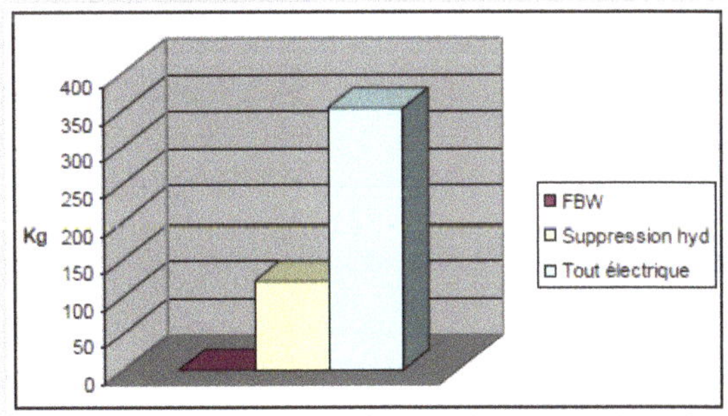

FIGURE 10.12 Max takeoff weight [10.11].

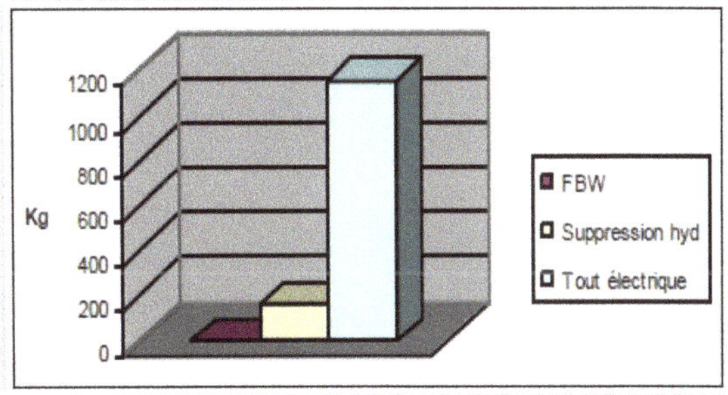

The all-electric version, for a given "engine cycle," does not bring efficiency improvements and, therefore, leads to an impressive fuel burn increase compared to the conventional baseline.

Even in the case of partial aircraft electrification (hydraulic systems going electrical), capable of better fuel efficiency and cost of ownership with only a minor range penalty, fuel savings get drowned by the fuel penalty of additional system weight.

In summary, based on the above analysis of the Falcon 2000, fuel penalties stemming from weight increases during electrification may get exacerbated by the snowball effect, and end up compromising the expected performance and cost benefits of electrification.

This view shows that the stakes are high in the research of power-dense and fuel-efficient solutions when transitioning from legacy aircraft to their electrified versions, whether incrementally or through a complete redesign. The power-to-weight ratio has to be drastically enhanced in order to make the electric business aircraft competitive.

When sticking to conventionally designed fixed-wing aircraft and turbofan engines, the way forward proposed by several business jet manufacturers is to develop newer and more efficient engines and systems and implement well-thought-out integration strategies.

For instance, engine integration of upsized electric power offtake could be dealt with either by embedding power generators into the engine core or, for that matter, by encapsulating electric power generators into the gearboxes that drive them. Another approach could involve sharing oil cooling functions between the engine and the generator (s) installed on them.

One other strategy could consist of seeking alternatives to the conventional sourcing of electric power from the engines. To this end, resorting to fuel-efficient and power-dense sources available on board may alleviate the burden on the engines while bringing economies of scale in fuel consumption and helping resolve the weight penalty issue. When retooled to that purpose, complementary sources such as the Auxiliary Power Unit (APU) could be up to the job and allow the aircraft to benefit from better fuel efficiencies and power-to-weight ratios.

With regard to the electrical network, 28 VDC or 115 VAC electrical networks are found on many legacy business jets for the power supply of systems. If intensive system electrification is carried out, legacy networks ought to be transformed into high-voltage networks. In fact, with high-voltage DC, the levels of current carried by cables and used by various equipment are lower; thereby weight reduction is possible through the downsizing of cabling onboard. Moreover, the fact that high voltage is centrally created and distributed to the end systems downstream relieves the burden on these systems to locally create their own HVDC bus. Lastly, moving up in speed, the power generation for a HVDC network is more weight-optimized compared to legacy networks.

As recent research on business aircraft suggests [10.12], the legacy electrical network can be switched to a lighter 270 DC network thanks to high-speed starter-generators and power-dense power conversion equipment. From thereon, the electrification of pneumatic systems helps garner estimated weight savings of more than 100 kg. As explained previously, this could be translated into either range enhancement or fuel cost savings.

As developed in Chapter 2, let us bear in mind that on an electric aircraft the power electronics needed for the power conversion and motor control constitute a large portion of the overall weight. In the process of electrification, there is a tendency to dedicate one motor controller to each system. Unfortunately, this worsens the dead weight being carried by the aircraft because of the dead time during which controllers do not have to operate.

Logically, multi-purpose motor controllers could be the panacea for reducing dead time. During flight, such a controller can be switched from one system to another when these systems are operated in a sequential manner. But their coverage usually is limited to two dedicated applications.

Taking this approach one step further, dead time and dead weight can be drastically reduced by using standardized power electronic modules addressing multiple systems, as explained in Chapter 2. This modular paradigm, on top of the versatility it brings, helps eliminate unnecessary margins that have to be built into dedicated individual or multi-purpose motor controllers.

10.5 Short-Range Aircraft

Similar to the analysis conducted on the Falcon 2000 business jet [10.11] and developed in the previous paragraph, let us now consider another study [10.13] performed on a short-range aircraft.

The benchmark aircraft for this comparative study is a conventional 165-passenger, short-range (3500 nm) twin jet, with a baseline A320 design but including technologies derived from programs not too far in the past like A318, A380, and A350. In this aircraft, pneumatic systems are a mainstay, whereas hydraulic systems are limitedly used.

The study compares the benchmark to a more-electric version based on the removal of pneumatic systems, and corresponding engine offtakes while sticking to reduced usage of hydraulic power. The turbofan engine technology and performance are kept unchanged between the two versions. The benefits of "more-electric" systems have also been assessed at the aircraft level from the perspectives of maintenance and environmental impact based on fuel consumption.

The weight breakdown of systems analyzed on the benchmark aircraft is reflected in Figure 10.13. This can be compared to the weight picture shown in Figure 10.14 for

FIGURE 10.13 Systems weight (benchmark) [10.13].

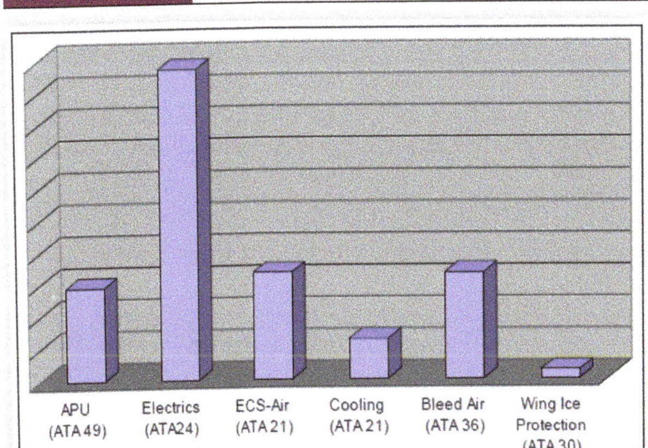

FIGURE 10.14 Systems weight (more-electric) [10.13].

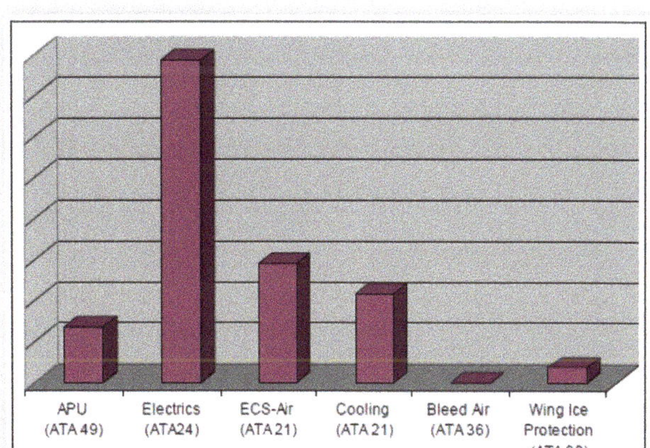

FIGURE 10.15 Aircraft weight (more-electric vs. conventional) [10.13].

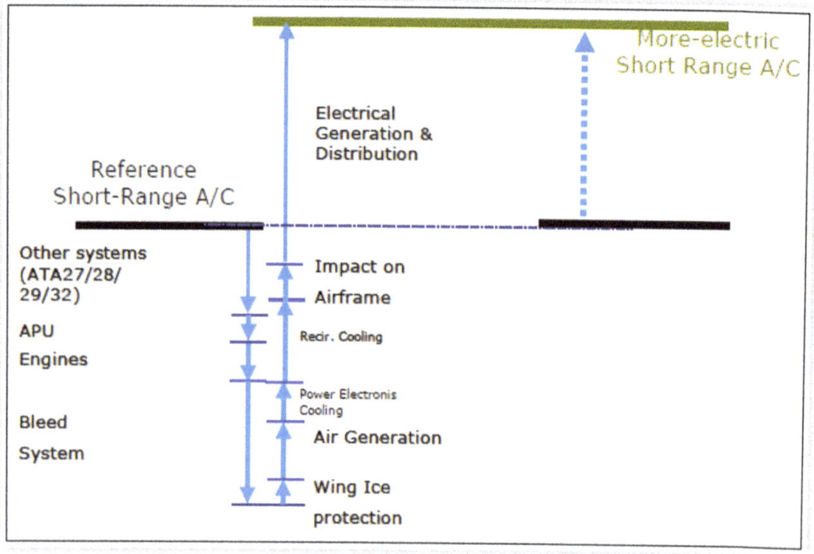

the more-electric version. In short, apart from the bleed air part of the pneumatic systems, a general weight increase is noticed on the more-electric aircraft.

The weight assessment of systems shows that the more-electric short-range aircraft is heavier than the reference short-range aircraft. As shown in Figure 10.15, when shifting to the more-electric aircraft by way of system electrification, the weight reduction benefits, reaped thanks to several systems, are "outweighed" by the weight penalties encountered in the remaining systems. As illustrated in Figure 10.15, major penalties trouncing weight benefits are attributable to the following consequences of electrification, their contributors, and key impacting factors:

- Scale-up of electrical power needs on board: electrical power generation/ distribution technology and power density
- Replacement of conventional systems with electrical versions: ice protection and ECS technology and power density as well as additional cooling technologies and their power density
- Integrating electrified systems: weight- and volume-related additional impacts on airframe linked to the above contributors

Let us focus on how electrification may impact aircraft performance from a drag perspective. On the reference aircraft, fresh air supply needed for the operation of the ECS is ensured by the engines thanks to the engine bleed system. On the contrary, the same engines when fitted to the more-electric aircraft cannot deliver the air required by the ECS because the bleed system is removed in the electrification process. Therefore, alternative air supply has to be sourced from the outside of the aircraft, requiring the implementation of the following design features:

- Cabin air re-circulation: supply of fresh cabin air from outside the aircraft using scoops
- Electrical ECS and cooling system: ram air from outside the aircraft

As a result, compared to the reference aircraft, these additional implementations on the more-electric aircraft wind up creating a drag penalty. Nevertheless, research of possible mitigation means could help alleviate this drawback.

Fuel burn comparisons in [10.13] are based on a 500 nm mission. Firstly, the more-electric aircraft has reduced engine-specific fuel consumption compared to the reference aircraft. This stems from engine efficiency gains brought by the removal of bleed components.

Based on this and after further computation, notwithstanding the weight disadvantage of the more electric aircraft, the study in [10.13] arrives to the conclusion that there is quite no difference in fuel burn between the short-range reference aircraft and its more-electric version.

In addition to the fuel burn assessment, the study evaluates the Direct Maintenance Cost (DMC) for a system perimeter limited to the APU, electric, bleed, ECS, and cooling systems.

The result shows a slight decrease in DMC in favor of the short-range more-electric aircraft.

Ever since the study was performed, leveraging latest breakthroughs, the aerospace industry has developed and tested a full suite of technologies that could be readily integrated on a more-electric aircraft, provided the required maturity levels are reached. Even though system weight and integration remain an outstanding challenge, conceptual design studies conclude that the more-electric aircraft may readily deliver airplane benefits in terms of maintenance, operational flexibility, and technology growth potential without, at the very least, any fuel burn penalty. Better, system simplification and weight savings could lead to potential fuel efficiency enhancement. Furthermore, as more-electric technologies are friendlier to the environment than conventional solutions, they appear as key enablers for sustainable growth of the aerospace industry.

But on the flip side, due to its smaller size, the weight and integration stakes facing the short-range more-electric aircraft are far more challenging compared to a long-range more-electric aircraft.

In summary, compared to its more-electric version, the conventional metallic narrow body has a head start from a weight standpoint, whereas both versions are on a par when it comes to fuel efficiency. Therefore, weight reductions on the more-electric version could make it more competitive by reducing fuel costs. Hence the deep focus in ongoing research on power-dense solutions. Yet, the situation is not that negative. Global research being carried out targets not only the alleviation of system weight penalties found in initial studies, but also addresses weight cutbacks with respect to conventional systems.

To make the case for the more-electric aircraft, one may think of resorting to a lighter composite airframe. But, keep in mind that weighty corrections would have to be implemented in order to render such an airframe compatible with the consequences of system electrification, such as incorporating missing features like noise barrier and lightning protection that come free with a metallic airframe. Once again, the size of the aircraft influences the level of optimization that composites could offer.

While the industry supplier base is scrambling to get power-dense more-electric system options on the table, aircraft manufacturers took the approach of investigating the potential of quick-win incremental changes on existing aircraft. In order to deliver step reductions in fuel burn, they continued to work with system suppliers, and also turned to engine manufacturers and put the onus on them as well.

Actually, while the industry waited for technology promises to spring up on the systems side, engine manufacturers invested heavily in research to improve drastically the fuel efficiency of turbofan engines. Ultimately, engine manufacturers ended up offering fuel-efficient turbofan engine alternatives very close in form, fit, and function to legacy ones. On a secondary level, rework of some systems and structural areas allowed extra efficiency gains.

The re-engining approach also became compelling in the face of sky-high fuel prices. Aircraft manufacturers heeded market calls and resorted to incremental re-engining developments on existing platforms, allowing lower risks and costs and shorter time to market compared to a full-blown aircraft development. The re-engining solution for legacy short-range aircraft platforms, despite a higher price tag, was readily endorsed by airlines under pressure from hurting fuel prices. Moreover, in such an approach, airlines continue to benefit from proven maturity in service for all the other parts of the re-engined aircraft except, of course, the engines themselves.

In such newly developed turbofan engines, still comprising the bleed system, tremendous improvement in fuel efficiency has been possible thanks to leapfrog technologies and materials in the engine redesign, based on following two competing architectures with comparable performance:

- Traditional turbofan architecture on the LEAP engine from CFM International
- Disruptive turbofan called "Geared Turbofan (GTF)" on the PurePower® engine from Pratt & Whitney

Both developments have materialized into ready-for-service engines that can be purchased by airlines as replacements for engines hitherto offered by aircraft manufacturers. They offer drastic fuel-burn reductions at aircraft level. Figure 10.16 shows

FIGURE 10.16 Aircraft fuel burn (re-engined vs. conventional) [10.14].

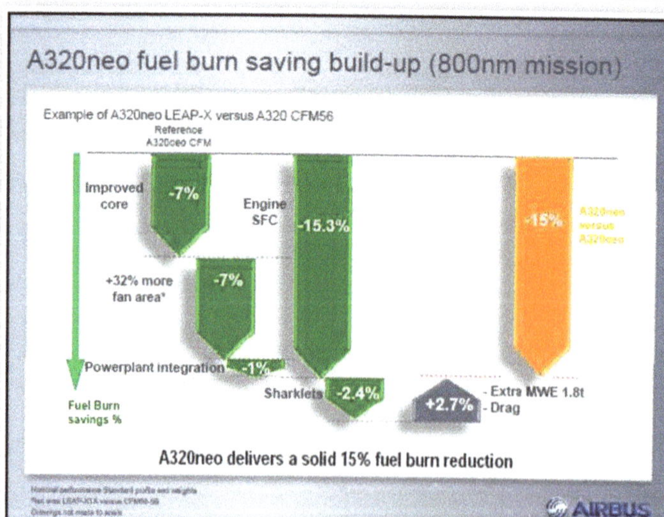

how, compared to the original aircraft, on the re-engined version the various fuel efficiency gains stack up.

All in all, simply re-engining the short-range aircraft brings a steep 15% reduction in aircraft fuel burn.

Helped by the technology overhaul that conventional turbofan engines have gone through in the re-engining strategy, fuel efficiency wise the situation tips in favor of the re-engined version of the conventional short-range aircraft. Logically, this puts into question the performance assumptions, mainly fuel efficiency, behind the reference aircraft used for benchmarking its more-electric version.

Nothing to do with a more electric aircraft, the advent of re-engining options has all but challenged the development of narrow-body more-electric aircraft by raising the bar higher with regard to the reference aircraft against which the more-electric version has to compete.

Factoring in the newly available engine performance into the conventional aircraft architecture serving as the benchmark widens the fuel burn gap that the more-electric aircraft would have to offset first, and prior to offering additional benefits in its own right.

Therefore, making the case for the more-electric short-range aircraft gets trickier from a timeline and technology perspective. This is the main reason why decisions on the more-electric versions of the short-range aircraft segment are ever more tied to power-to-weight and efficiency improvements that aircraft systems research can deliver.

In the meantime, out-of-the-box approaches have gone full swing in the research of alternatives to both more-electric and re-engined versions of the conventional short-range aircraft. When observing the trends behind, both power-dense electrified systems and electric propulsion, whether hybrid-electric or totally electric, appear to be mainstays in the next paradigm shift in electric aircraft design.

As discussed previously, hybrid-electric propulsion will require onboard rechargeable energy storage devices such as battery packs. Global leapfrog research advances are speeding up large-scale shrinking of batteries while making them as energy dense as called for by the electric aircraft design. In the meantime, sweet spots are under investigation using downsized conventional turbine engines, matched and operated hand in hand with a certain degree of electric propulsion. This gives birth to the concept of hybrid-electric propulsion, wherein a parallel may be drawn with HEV powertrains already commonplace in the automotive industry.

Isikveren [10.15] provides performance comparison between a future medium-range 180-passenger reference aircraft with advanced electrical systems and its hybrid-electric version using 1500 Wh/kg batteries. This means a tenfold increase in specific energy, compared to the most recent battery technology available on the Tesla electric car. Results obtained for a range of 1100 nm show that hybrid-electric propulsion, utilizing electric energy during 50% of cruise, could theoretically wind up delivering a 20% reduction in fuel burn compared to the reference aircraft fitted with turbine engines.

Hornung [10.16] analyzes hybrid-electric propulsion using 2000 Wh/kg batteries on a reference 189-passenger short-range aircraft with a self-trimming nonplanar C-wing (Figure 10.17).

FIGURE 10.17 C-wing hybrid-electric aircraft [10.16].

Moreover, aircraft systems are also fully electric and powered solely by batteries. Batteries, lodged in cargo containers, would not require recharge during turnaround because used battery containers would simply get swapped with pre-charged ones during turnaround. The electric propulsion system comprises large high-temperature superconducting electric motors with an integrated cryocooler and high-voltage power electronics dedicated to motor control. On the systems side, high-voltage DC is the adopted standard for the electrical network supplying loads via power electronics or power supplies.

For a range of 900 nm, **Figure 10.18** shows the performance of such an aircraft depending on the degree (or ratio) of propulsion electrification. A ratio of "1" stands for fully electric propulsion, a ratio of "0" applies to the most recent bleedless turbofans; any value in between pertains to hybrid-electric operations using both fuel and battery energy.

Because no jet fuel is used by the fully electric aircraft, its operation entails zero emissions. When the electrification ratio is set to 50%, efficiency gains are around 30%, even though aircraft weight is increased by around 25%. This weight increase is mainly due to the following factors:

- Battery system, despite a challenging specific energy assumption of 2000 Wh/kg.
- Wing redesign to keep wing loading constant. In fact, wing structural weight increases when compensating for lower wing bending moment relief from lower quantity of fuel onboard.
- Lower amount of fuel mass burn-off during mission.

Yet, still sticking to conventional "wing-and-tube" aircraft shapes, another study in [10.17] shows that a tri-fan morphology, comprising two under-wing podded gas-turbines (GT) and one aft-fuselage mounted serially configured motor (M) driven by

FIGURE 10.18 Hybrid-electric aircraft performance vs. degree of electrification [10.16].

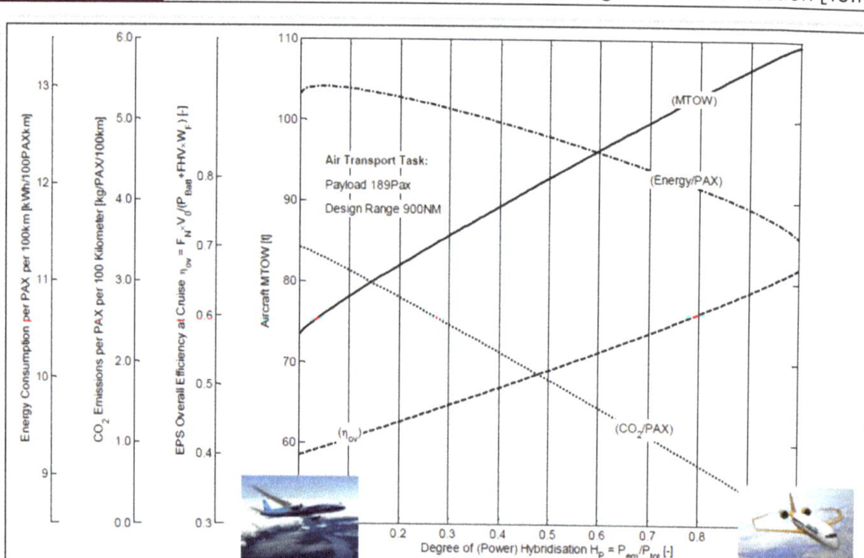

batteries, could be an appropriate choice for a 180-passenger hybrid-electric aircraft (**Figure 10.19**).

The study concludes that, even though resorting to a high-temperature superconducting motor could deliver 2%-4% reduction in aircraft weights, a normal conducting motor is deemed to be a pragmatic choice. In the latter case, the tri-fan morphology motor would have to deliver 8.5 MW of shaft power for a 180-passenger aircraft. The reference aircraft used for performance comparisons is a year 2000 A320-200 aircraft with evolutionary technologies: advanced ultra-high bypass (~20) geared turbofan, all-electric systems, high wing aspect ratio (~12%), reduced zero-lift drag, and advanced structural materials. It achieves a 39% reduction in block fuel, compared to the A320-200, demonstrating the high level of optimization taken into account for the reference.

FIGURE 10.19 "Wing-and-tube" hybrid-electric aircraft [10.17].

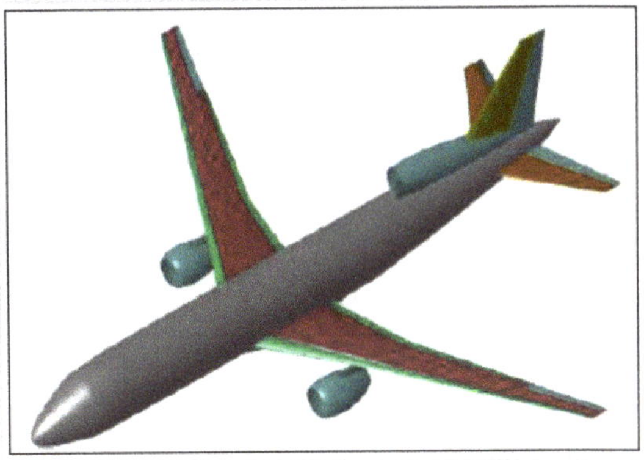

Over a range of 1100 nm, **Figure 10.20** shows that in order for the hybrid-electric aircraft to achieve 15% block fuel reduction compared to the reference aircraft, batteries with 940 Wh/kg specific energy would be necessary. Likewise, a 20% block fuel reduction would be possible over ranges of 900, 1100, and 1300 nm, with a battery specific energy of 920, 1100, and 1290 Wh/kg, respectively.

Let us consider the operating costs, excluding certain costs linked to cost of ownership (depreciation, interests, and insurance) and including additional noise and emissions-related charges. From the perspective of operating costs, the hybrid-electric aircraft is ~10% more expensive than the reference aircraft. Concerning energy costs, that is, fuel and electricity, they go up by ~6%. These figures were estimated using a price of USD 3.30/g

FIGURE 10.20 Hybrid-electric aircraft fuel savings vs. range and battery specific energy [10.17].

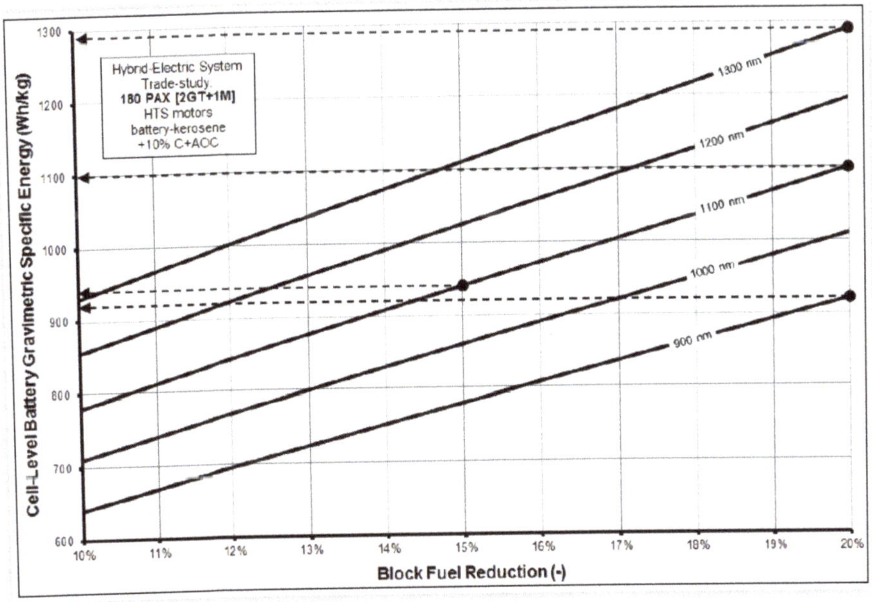

FIGURE 10.21 Trip operating cost breakdown—reference vs. hybrid-electric aircraft [10.17].

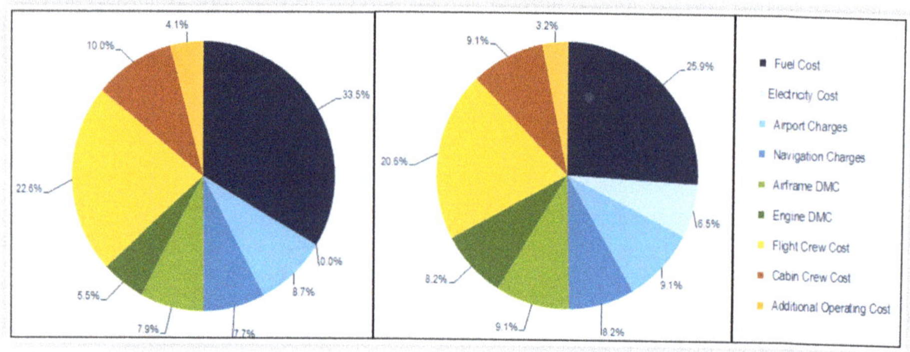

for kerosene and USD 0.1109/kWh for electricity required for battery charging. Fuel price fluctuations considered in the sensitivity analysis show that with USD 2/g, the operating costs go up by 4-5%, compared to the reference architecture, whereas the same costs are cut back by 5%-7% with USD 6/g. Given these cost variations, achieving a cost-neutral situation, thanks to further optimizations, is deemed realistic. For a 1100 nm mission, using 940 Wh/kg batteries for hybrid cruise, the operating cost breakdown for both the reference and the hybrid-electric aircraft is shown in **Figure 10.21** [10.17].

Now, building from a baseline Boeing 737 type of aircraft fitted with CFM56 engines (not the LEAP-1B ones), the study in [10.18] has devised a reference aircraft design called SUGAR High operating solely from fuel. From a structural standpoint, the reference design incorporates various improvements based on high-span truss-braced tube and wing morphology allowing high L/D ratio. The engines of the reference aircraft incorporate engine technology enhancements, spanning from the baseline CFM56 to the latest available engine options for the short-range aircraft segment.

FIGURE 10.22 High-span truss-braced tube and wing hybrid-electric aircraft [10.19].

The study goes further by checking the potential of hybrid-electric propulsion combined with 750 Wh/kg batteries on the SUGAR High reference aircraft leading to its hybrid-electric version called SUGAR Volt (**Figure 10.22**).

Results obtained in [10.18] for a 900 nm mission demonstrate in the first place that the future reference aircraft would per se deliver block fuel reduction of up to 54% compared to the baseline 737 (**Figure 10.23**). Secondly, the hybrid-electric variant brings a fuel-burn reduction of 14% compared to the reference which in reality amounts to a 60% reduction when compared to the conventional short range with baseline turbofan engines (not including the most recent advances available on the 737 MAX or the A320neo). These conclusions are illustrated in Figure 10.23 showing comparisons with respect to the baseline 737 of the future all-fuel reference (SUGAR High) on the one hand and, on the other, its hybrid-electric variants. The latter concerns two different electric motor sizings (1380 and 1750 hp), both being utilized in balanced hybrid operation throughout the duration of the mission.

For hybrid designs, it is important to look at reductions in fuel, but also reductions in total energy use in order to take into account energy expended in charging the batteries if they are charged on ground prior to flight. In this way, and by examining the costs of each type of energy source, the airliners can make informed choices. For emissions benefits, one must look at the fuel burned while flying and also go deeper and look at any carbon expended during the manufacturing of batteries and production of the electricity used to charge them.

FIGURE 10.23 Hybrid-electric aircraft performance vs. reference and baseline 737 aircraft [10.18].

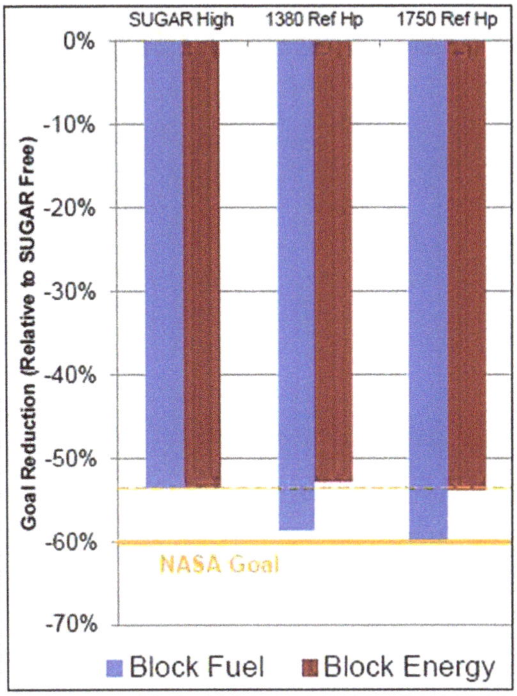

10.6 Long-Range Aircraft

As discussed, re-engined versions of short range are already in service. With regard to legacy long-range aircraft, re-engining with minor structural and system modifications is also in the making.

But, compared to what happened with the electrification efforts on short-range aircraft, the story has been quite different on the long-range aircraft. In fact, prior to the race towards power-dense systems solutions and the re-engining strategy, cleansheet more-electric versions have been successfully brought to market through the Boeing 787 and Airbus A350 long-range aircraft developments. These programs also benefited from more efficient engine developments and, in the particular case of the Boeing 787, were afforded a bleedless version to match the large scale of system electrification.

On the Boeing 787, the bleedless system architecture brings fuel burn reduction in the order of 3%. The bleedless engines allow for circa 15% reduction in specific fuel consumption. Structural changes bring their share of efficiency gains thanks to decreased weight and lower maintenance. In fact, massive usage of lighter composite structures helps leverage these benefits at the aircraft level, and also compensate weight penalties caused by system electrification.

In essence, economies of scale in the operation of the more-electric Boeing 787 could only be delivered through the balanced combination of system electrification, bleedless engines, and composite airframe structures.

When targeting better operational efficiency with the more-electric design, there are differences in the relative constraints of the short-range and long-range aircraft, the latter being largely helped by its bigger size, longer range, and larger engines:

- Larger size facilitates the integration and thermal management of large-scale electrification.
- Longer flight cycle durations allow better fuel optimization.
- Larger engines are more fuel efficient than smaller ones.

Once again, without resorting to composite materials on the structural side, these benefits may not have materialized.

In summary, with a long-range more-electric composite aircraft with bleedless systems architecture and turbofan engines, such as the Boeing 787, airlines may expect fuel burn reduction in the order of 15%-20%, depending on route distances. Lower costs for base checks, thanks to the composite structure and its monitoring, among other factors, allow for at least 15% reduction in maintenance costs. All this contributes to the Boeing 787's ~20% operating cost advantage over similar legacy aircraft.

With this new type of more-electric long-range aircraft, airlines expect to cash in on the opportunity to reduce the fuel and maintenance portions of the operating costs. But, for a given trip, the total trip cost has to take into account aircraft financing charges as well. However, airlines often still find that aircraft financing charges could make their total unit costs higher than the older aircraft they are replacing. Thanks to carbon fiber technology, composite aircraft life is expected to double, and so its financing terms and lease rate may be lower than for conventional aircraft. Luckily, newly purchased aircraft enjoy maintenance cost write-offs during the initial years of operation thanks to manufacturer warranties helping bring down operational costs. All in all, when using more-electric aircraft to replace older conventional aircraft, financing cost burden could be allayed by large-enough reductions in cash operating costs [10.20].

The cost performance estimations presented in **Figure 10.24** corroborate the competitive edge of the more-electric Boeing 787 over other legacy Boeing aircraft in the long-range segment. The comparison is based on a 4000 nm mission, like on a trans-atlantic Paris-New York trip, for a fuel cost of USD 2/gallon.

From an operational standpoint, this analysis of Boeing aircraft shows that the baseline more-electric B787-8 stacks up against the conventional B767-200 with a steep 24% cut in the unit cost per available seat-mile (CASM) for a barely higher trip cost (+3.3%). Similarly, when comparing the larger B787-9 to the legacy B767-400ER, the CASM plummets by 17% along with a 4% dip in the trip cost. Incidentally, with the B787, airlines not only benefit from the operating savings but could potentially offer new routes where the B767 doesn't go and where the B777 is deemed too big. Cases in point for such ultra-long-haul

FIGURE 10.24 Trip cost and CASM (more-electric vs. conventional) [10.21].

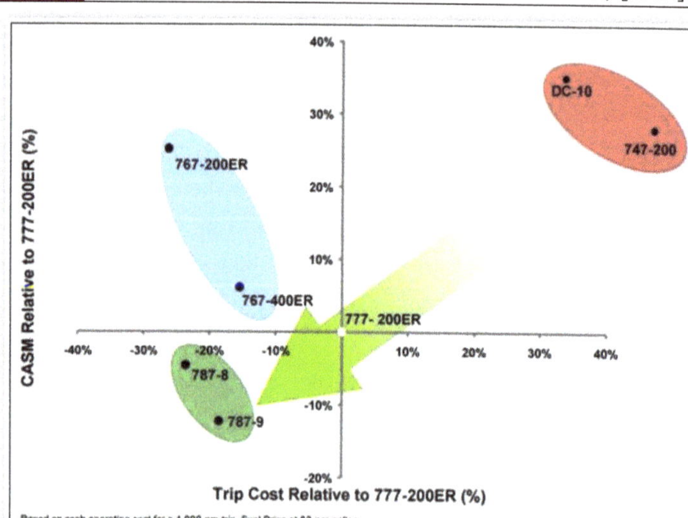

operations are longer transatlantic (e.g., London-San Francisco or Europe–South America), trans-Pacific or Europe–Asia Pacific flights with an average range of 5500 nm.

Therefore, it is obvious that the long-range more-electric aircraft sets a new standard in operating economics by providing airlines with an aircraft that has a significantly improved CASM compared to their current fleet. In addition, thanks to its ultra-long-range capability, it provides the bonus of an opportunity to transform their long-haul networks.

The more-electric Boeing 787-8 has been in service since 2011. Airline operational data for the year ending in 2013 [10.7] shows that the DOC per seat per mile for this aircraft is ¢7.2. This is almost 8% below the average ¢7.8/unit cost registered for the overall long-range aircraft segment.

Revamped system architectures, more efficient engines relieved of system constraints (bleedless), and composite airframes have been the most recent key enablers in the design of more-electric aircraft all the way up to entry into service in the long-range market. Depending on the aircraft segment, when shifting conventional systems to the electric domain, performance could either be leveraged as shown above or get undermined by weight issues stemming from electrification in the absence of design solutions alleviating them.

Economies of scale possible on a large aircraft may get watered down when the aircraft size goes down. The more-electric short-range or business aircraft requires ever more stringent optimizations and ups the ante when it comes to integration. While power-dense system research is in full swing to meet these expectations, engine manufacturers have been able to drastically improve turbine engine performance. This allows existing aircraft to be re-engined at reduced development costs within a shorter time-to-market, thereby undercutting fuel cost impacts on operational economics.

Nevertheless, the re-engining wild card may have reached its limits. Any further attempt to sizably enhance fuel efficiency of turbine engines might prove unsuccessful, for all their potential and margins may already have been squeezed out.

Therefore, aircraft and engine manufacturers may have run out of viable incremental optimization opportunities on existing aircraft platforms. Even though operational cost and carbon footprint reductions do make the case for more efficient conventional aircraft and engines, they still come with noise and greenhouse gas emissions. Industry consensus for business growth, competitiveness, and environmental friendliness is grounded in the prospects of radically different aircraft designs resorting to alternatives to jet fuel. This longer-term approach simultaneously targets steep cutbacks in operational costs, driven not only by fuel efficiency but also by the cost of ownership, and environmental impacts of aircraft design and operations.

Taking stock of the performance ceiling reached with turbine engines using jet fuel, research is doubling down on the electric propulsion paradigm, breaking new ground in energy efficiency and emissions by getting rid of the reliance on fuel altogether. This opens new avenues for innovative configurations where energy is transmitted advantageously around the aircraft to achieve aerodynamic advantages, and reduce cost of operations and opens up avenues for using other sources of ground power.

This entails the replacement of fuel by an alternative electric energy source or storage carried on board. When the replacement is carried out only partially, we end up having a hybrid-electric aircraft powered by hybrid-electric engines running on both fuel and electricity. A direct parallel can be made with the hybrid-electric vehicle (HEV) revolutionizing the automotive industry. In this case, the original fuel load gets split into two parts, a downsized fuel part and another part made up of batteries. The hybrid-electric engine can be schematized as a downsized turbine engine working in tandem with a highly efficient electric motor. Similar to the more-electric aircraft, the electric and hybrid-electric aircraft systems come in their electrified versions.

It is important to note that though electric motors are more efficient than turbofans, the aircraft system also has to take into account how the energy sources are stored on board: jet fuel versus battery cells, typically. Similar as well to the power-to-weight ratio in systems electrification, gravimetric energy density (Wh/kg) and volumetric energy density (Wh/L), commonly named "mass specific energy" and "volume specific energy," respectively, are key characteristics of energy storage on electric or hybrid-electric aircraft. Unfortunately, these parameters for even today's most advanced battery storage systems fall well short of those obtained with kerosene: by a factor of 18 for the volume ratio and a factor of 60 for the weight ratio. This core problem tends to restrict the extent to which batteries may be utilized for powering systems and engines.

Space and volume issues previously raised on smaller-sized aircraft may come in the way of battery implementation. Nevertheless, the lower volumetric ratio may be less critical as long as the aircraft is not limited in space available for integration. Otherwise, the aircraft would require larger wings, fuselage, or additional external "energy pods" which would lead to losses in overall aircraft efficiency due to larger wetted surface. Ideas such as tightly integrating batteries and other storage systems with structural elements, like in the Tesla electric vehicle, are also coming through in the aerospace sector.

In summary, electric propulsion could help ratchet up overall aircraft performance, provided the specific energy of batteries replacing fuel is up to the challenge. Therefore, at the aircraft level, efficiency of turbine engines using jet fuel, though with highly competitive specific energy, may be outstripped by novel hybrid-electric or electric propulsion architectures powered by a new generation of batteries with dramatically increased specific energy.

10.7 Regional Aircraft

Regional aircraft are powered by either turbofan or more fuel-efficient turbo-propeller engines. Let us see how battery supplied electric propulsion stacks up against the latter type of fuel-dependent turbine engine.

The Dornier 328, a regional aircraft fitted with turboprop engines, can transport 32 passengers over a range of 1200 km. On such an aircraft, if fuel is simply replaced by baseline battery technology (180 Wh/kg), and engines are switched to electric ones, range capability would plummet to 202 km [10.22] for the same payload. But, if the mass-specific energy of batteries were to be pushed higher to, for instance, 720 kW/h, range performance would get extended to 800 km, but still fall short of the conventional aircraft capability of 1200 km.

Figure 10.25 portrays how, depending on the specific energy of batteries, payload variations impact range performance for both the conventional aircraft (Dornier 328 TP)

FIGURE 10.25 Regional aircraft performance (electric vs. conventional) [10.22].

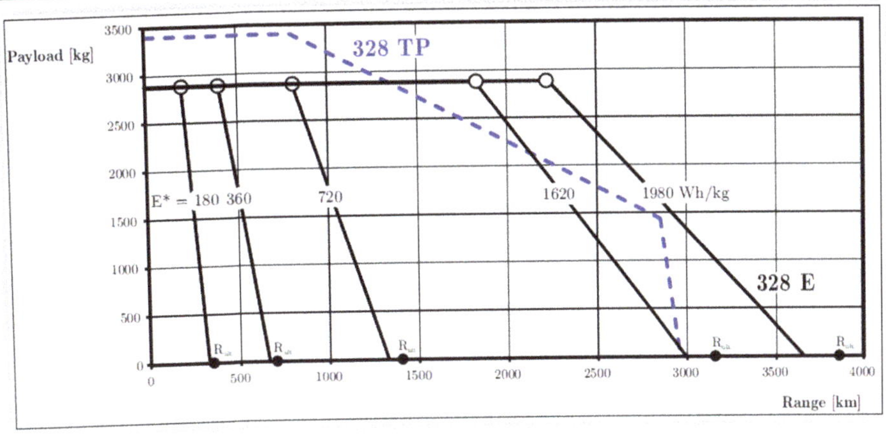

and its electrical version (Dornier 328 E) with drop-in battery and electric propulsion replacement of fuel and turboprop engines.

Due to the comparatively higher specific energy of kerosene, from Figure 10.25, it can be concluded that when it comes down to range performance, trading payload for fuel is much more beneficial than trading payload for batteries, the payload-range gradient depending on the specific energy of batteries.

Thereupon, without taking advantage from worthwhile structural modifications, the battery-powered aircraft might seem less flexible. Nevertheless, payload-range capability equivalent to that of the baseline aircraft is reached when batteries with specific energy exceeding about 1500 Wh/kg are used.

This puts the bar very high for battery performance when resorting to electric propulsion on large aircraft. In order to power large aircraft, a dramatic improvement in battery technology would be required. Hence, the success of scaled-up electric aircraft hinges on lofty levels of specific energy for battery technology that ongoing research is yet to deliver. To attract commercial interest for larger (regional) aircraft, the specific energy of today's battery technology, in the range of 150-200 Wh/kg, would have to be increased by a factor of around ten.

Nevertheless, practically speaking, the specific energy of current battery technology is such that airplanes powered by electric propulsion are today limited in size to small planes carrying up to two passengers over rather short ranges and limited endurance. Neglecting costs, the current technology is suitable for small ultra-light aircraft, but not yet for commercial aviation.

10.8 General Aviation

As with previously discussed aircraft segments, the drive for electrification in general aviation is once again motivated by technical and cost performance enhancements. Reducing carbon emissions through transformative aircraft and airspace operations is among the industry goals, together with the alleviation of specific general aviation's concerns like the far higher accident rate compared to other aircraft segments and road transportation.

Electric propulsion offers a threefold improvement in efficiency compared to turbine engines, whereas the leap in energy efficiency can top out at 4% compared to piston engines commonly used in general aviation. Incidentally, high efficiency of electric propulsion is achieved across more than half of the operating speed range. Additionally, with electric motors being six times more power-dense compared to piston engines, electric propulsion could deliver power-to-weight ratings drastically improved by more than 500%. All the above add up to lower energy consumed and incurred costs. Electric propulsion in general aviation, thanks to the lower passenger count, size, and range, may help cut costs by up to 10%.

Electric propulsion, be it hybrid-electric or totally electric, offers lower community noise. Since recourse to air breathing is either alleviated or totally suppressed, greenhouse gas emissions are either cut back or, better, reduced to zero. Moreover, operational pitfalls of general aviation such as power lapses with altitude or hot weather conditions, directly related to the reliance on air-breathing, may also be circumvented by electric propulsion.

Fewer moving parts with electric propulsion wind up offering more reliable designs. Also, the inherent integration benefits of electric propulsion allow compact aircraft sizing applicable to all aircraft segments, including the small scales in play in general aviation. On top of this, when Distributed Electric Propulsion (DEP) is implemented, the additional integration benefits enable closely coupled synergies across aerodynamics, propulsion, control, acoustics, and structures.

Conventional general aviation aircraft are only aerodynamically efficient at low speeds in cruise due to the wing oversizing necessary to meet constraints related to stall conditions and airfield lengths. This unfortunately compromises the lift-to-drag ratio. When converting to an electric aircraft, wing downsizing in conjunction with distributed-electric propulsion offers better wing loading, more resilient aerodynamics, lower drag, and higher lift; therefore allowing higher speeds during cruise (**Figure 10.26**) [10.23].

In summary, to the extent that required battery specific energy and cost reductions are made viable, with the help of bespoke certification and standardization procedures yet to come, electric aircraft in general aviation may become a reality in the near future. Ongoing research targets both Conventional TakeOff and Landing (CTOL) and Vertical Takeoff and Landing (VTOL) types of aircraft, carrying 4-9 passengers and one or more passengers, respectively, with tentative entry into service slated for 2025.

FIGURE 10.26 Conventional GA aircraft redesigned into an electric aircraft with DEP [10.23].

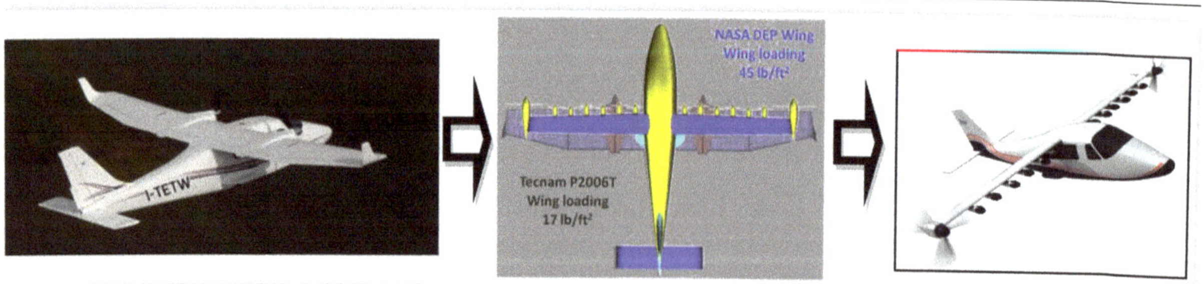

FIGURE 10.27 Upward scalability of CTOL electric aircraft design [10.23].

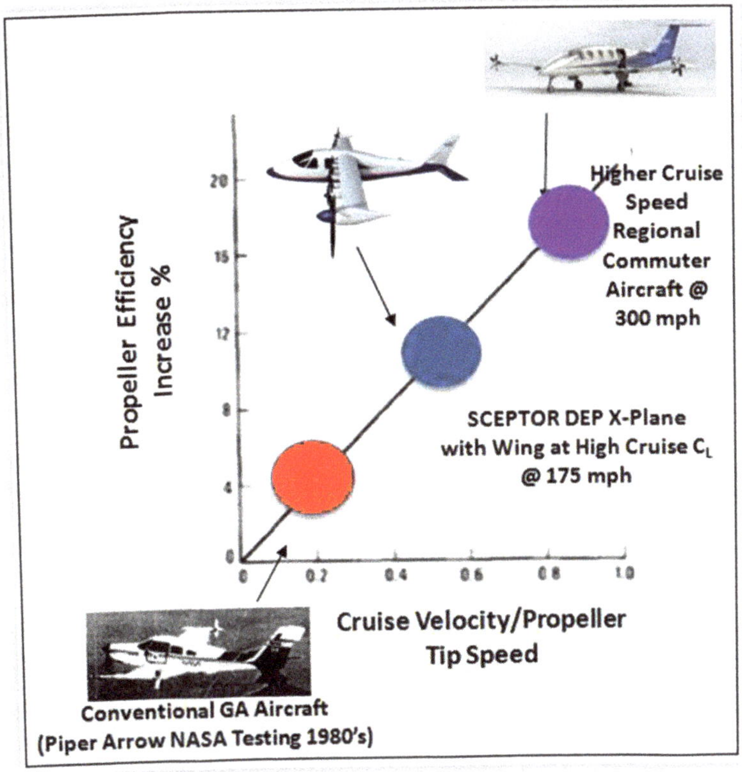

Although aircraft range performance is independent of speed, up to now many general aviation flight demonstrators have concentrated on low speeds only. Nevertheless, research focus is now being devoted to the proof of concept of higher speed electric aircraft, paving the way for upward scalability in electric aircraft design. Therefore, the general aviation electric aircraft is poised to become the stepping stone for scaled-up aircraft designs, thereby allowing the advent of larger electric-aircraft platforms in the future (**Figure 10.27**) [10.23].

A tentative entry into service timeline of CTOL electric aircraft could be drawn up as follows:

2020: General aviation
2025: Commuters (~9 passengers)
2030: Regional airliners
2035: Large aviation

Distributed electric propulsion when applied to vertical flight (VTOL) could dramatically improve key performance characteristics, thereby allowing attractive value propositions (**Figure 10.28**) for mobility solutions including future autonomous air travel.

Just like with CTOL, the VTOL aircraft incurs low marginal costs of operation by cutting energy costs by a factor of more than ten, and alleviating maintenance costs by more than half. The increase in speed by more than half allows quicker travel times and high productivity at fleet level, thanks to high-utilization rates (>1500 h/year).

FIGURE 10.28 Distributed electric propulsion VTOL vs. conventional aircraft [10.23].

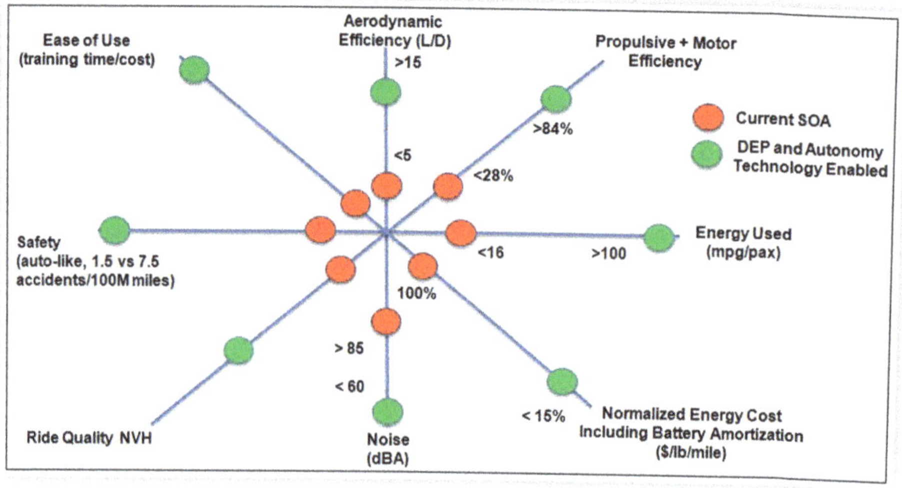

FIGURE 10.29 Total vehicle operating cost per mile vs. cruise speed [10.23].

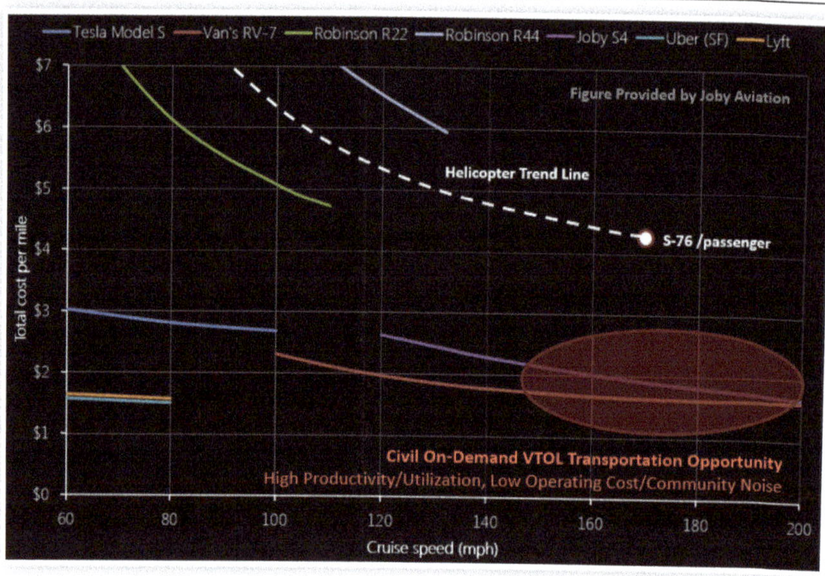

The expected total vehicle cost per mile for distributed electric propulsion applied to vertical flight is compared across various on-demand transportation choices in Figure 10.29.

Enabled by its quieter operation, the electric VTOL aircraft may therefore open new markets for more competitive on-demand urban air travel, compared to current ground transportation services offered by networks such as Uber or Lyft. Nevertheless, such an air travel solution cannot become a reality without bespoke air traffic control and infrastructure yet to come. Provided adequate air traffic management is in place ensuring safe, seamless, fluid, and quick turnarounds, VTOL urban air travel may offer quicker and hassle-free journeys by circumventing road congestions, totally upending travel experience with zero emissions over the city.

10.9 Cost of Ownership

On top of trip cost and CASM comparisons, airlines use the metric of cost of ownership to evaluate the economics of newly marketed aircraft. This parameter allows the benchmarking of all costs incurred by this or that aircraft an airline intends to purchase. This information can be fed into the value analyses it usually performs in parallel. The particularity of cost of ownership, also going by the name of "Life Cycle Cost (LCC)," as opposed to trip cost or CASM, resides in the fact that it is estimated not just at a given point of time but over the lifetime of an aircraft. It is based on cost projections starting from entry into service until the disposal of the aircraft.

Obviously, the electric aircraft, which puts together more expensive newly developed technologies and materials, comes with a higher price tag when it enters the market. This is similar, absent the government incentives attached, to what is observed in the automotive sector.

As discussed previously, maintenance cost savings can potentially contribute to an overall reduction in cost of ownership. The more the aircraft gets electrified the more its maintenance costs could be cut back, especially when health monitoring is implemented on board. Nevertheless, prior to the fruition of the benefits of the latter down the line, a certain amount of time lag is induced by factors such as an initial learning period, recoup time of health monitoring overheads, etc. As presented in [10.11], Figure 10.30 compares the health monitoring and aircraft downtime costs of a conventional, more-electric, and an all-electric business jet, whereas Figure 10.31 displays the relative break even times of the cost of ownership index.

FIGURE 10.30 Health Monitoring (HM) and downtime cost comparison (business jet) [10.11].

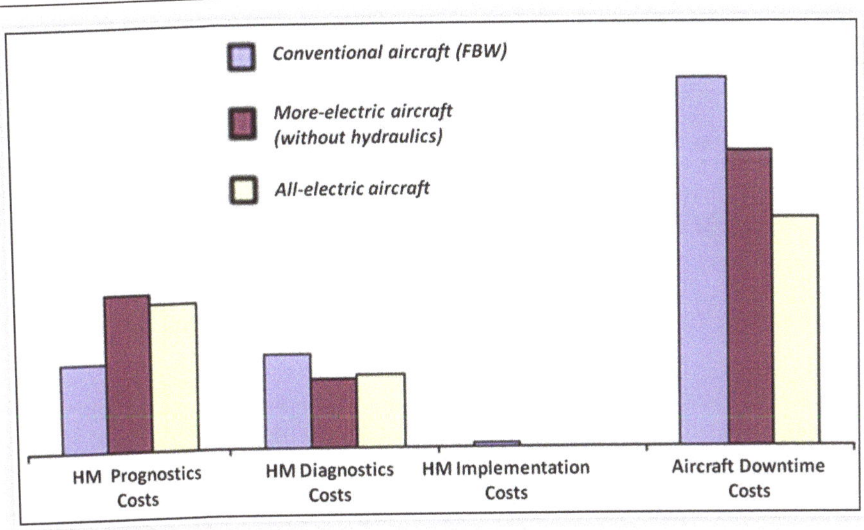

FIGURE 10.31 Cost of Ownership index comparison (business jet) [10.11].

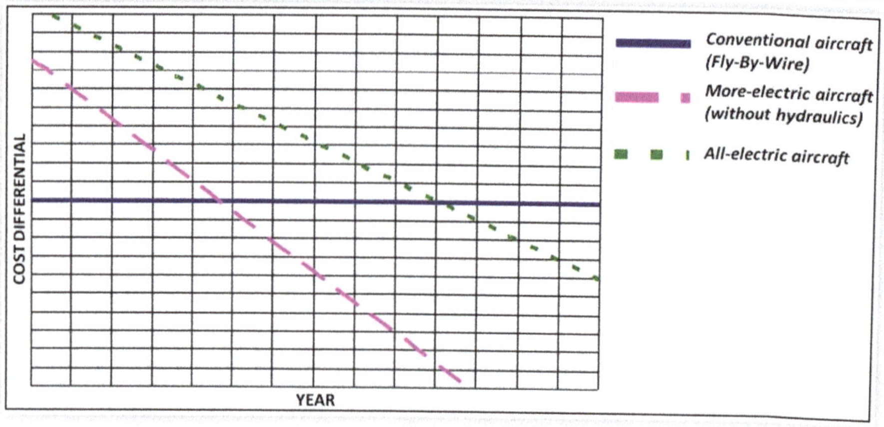

10.10 Environmental Footprint

From an environmental perspective, a lot of progress has been made thanks to the more-electric aircraft in service, such as the Boeing 787.

The Boeing 787 bleedless systems, more fuel-efficient and easier to maintain, are also good news for passengers. Since bleed air, used in the air-conditioning of conventional aircraft, is a byproduct of the engines, it might still contain traces of combustion products which at very, very low concentrations don't pose a health risk. On the contrary, thanks to its electrified air-conditioning system without reliance on engine bleed air, the Boeing 787 uses cleaner outside air that has never been in contact with the engines.

Noise is also a major concern, whether perceived by passengers seated inside the aircraft or by communities along the flight routes.

Passenger aircraft noise is hardly attributable to engines only, for it is also generated by the aerodynamics of some structural and system components. During approach conditions, with landing gears deployed and flaps extended, airframe noise predominates; whereas during other flight phases, the engines are to be blamed for the majority of the noise.

On conventional aircraft, bleed air is also used for de-icing: hot air is simply blown over the wings via exhaust holes but create an unpleasant "whistling" noise. In the bleedless system, an "electric blanket" is built directly into the wings, which heats the wings to keep them free from ice. This is a completely silent process, providing a nicer experience for passengers, but also for communities living near flight routes.

A lot more of these "more electric" optimizations are possible by replacing other pneumatic and hydraulic systems with electromechanical ones. Pneumatics and hydraulics are responsible for a lot of the weird noises that passengers hear during a flight; the switch to electrical solutions would result in quieter operations.

As concluded by the study in [10.18], considering a 154-passenger short-range aircraft, the total aircraft noise footprint hardly changes between the reference aircraft and its hybrid-electric variant. This is no surprise considering that similar noise performance is obtained for both reference turbine engines and their hybrid-electric variants.

FIGURE 10.32 General Aviation noise performance (turboprop vs. DEP) [10.25].

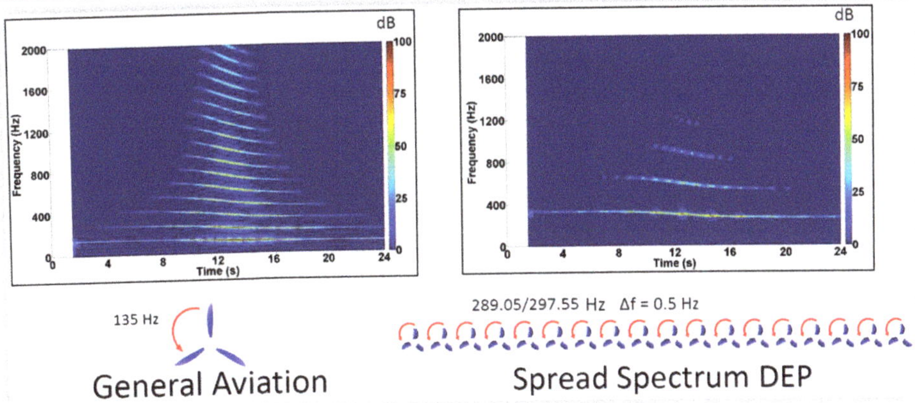

Nevertheless, there is still room for further noise reductions by tackling the root causes of dominant subcomponents in noise generation. As it turns out, this approach comes down to rethinking the design of landing gear and flap systems on the airframe side, along with jet and fan optimizations on the engine side.

All-electric aircraft allow outright elimination of combustion engine noises offering drastic noise reductions. Figure 10.32 shows how, in the case of general aviation, distributed electric propulsion helps alleviate noise issues by enabling sound reduction methods such as "reduced propulsor tip speed" and "spread spectrum," among others [10.25].

Let us now consider the situation concerning the nitrous oxide emissions (NOx). Bradley and Droney [10.18] assessed NOx emissions for hybrid-electric engine variants designed to power the "high span truss-braced" short-range aircraft reference described in Section 10.5. Results obtained show that during takeoff and landing phases, hybrid-electric engines could achieve, thanks to propulsive and thermal efficiency gains in combination with improved thrust lapse characteristics, diminished NOx emissions in comparison with the CFM-56, the baseline turbofan powering the Boeing 737. So much so that NOx levels achieved by hybrid-electric engines, estimated in the range of 7.5% to 11% of CAEP[1]/6 levels, literally outperform the "not-to-exceed" goal of 20% of the same levels.

As far as cruise conditions are considered, for the hybrid-electric aircraft with engines operating in the "balanced" hybrid mode, wherein both turbine engine and electric motor sections are operated in a balanced manner throughout the mission, NOx emissions are close to the 80% reduction goal. When resorting to a hybrid-electric engine capable of the "core shutdown" mode, meaning that the turbine engine section gets cut off at a certain point during the mission leaving the electric motor to operate on its own, the aircraft has essentially no NOx emissions over approximately 50% of the cruise segment of a 900 nm mission.

As of 2010, direct greenhouse gas emissions from aviation account for more than 2% of global emissions. Nevertheless, this hardly represents a significant share of the 14% of global emissions attributable to the entire transportation sector.

[1] Committee on Aviation Environmental Protection of ICAO.

FIGURE 10.33 Initial aviation carbon emissions reduction roadmap [10.27].

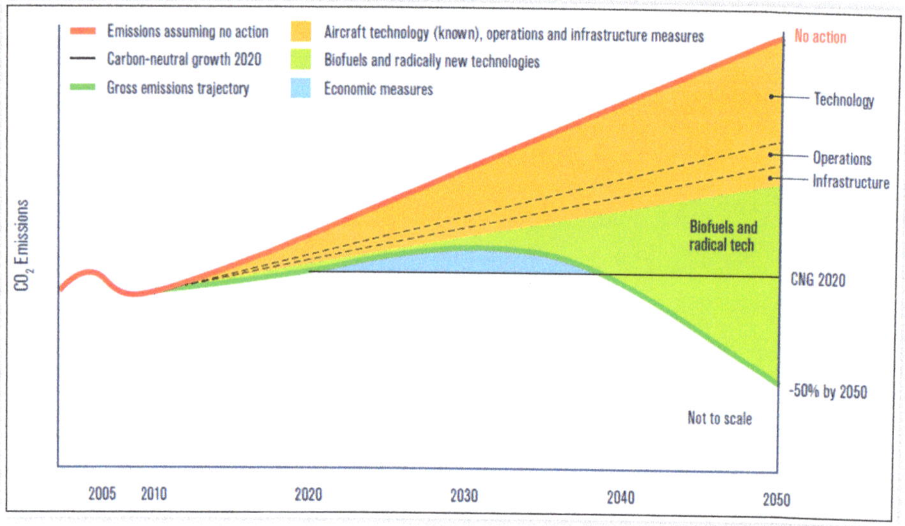

Climate change is a global issue that needs to be tackled from many fronts. In the aerospace industry aircraft and engine manufacturers, their supply chain, airlines, airports, air traffic management services, research institutes, and civil aviation authorities have been working towards a common objective of reducing the overall impact of aviation on the environment.

In order to reach ambitious environmental goals in a limited timeframe, government policy [10.26] and funding have been concentrated on research in this area. As such, Europe is at the forefront of atmospheric research and has taken the lead in the formulation of a prioritized environmental action plan, and the establishment of global environmental standards.

In 2011, a European group of experts set out a vision of European aviation with the publication of Flightpath 2050. In response to this, the Advisory Council for Aviation Research and Innovation in Europe (ACARE) produced a Strategic Research and Innovation Agenda (SRIA) in 2012 that defined the path to reach these ambitious goals. This vision would draw upon airframe, engine, system, ATM/infrastructure, and airline operation optimizations. To spur this into action, the ACARE council set the following steps for the industry to follow (**Figure 10.33**) [10.27]:

- An average improvement of 1.5% per year in terms of fuel efficiency to reach a carbon-neutral situation by 2020 in the first place as an intermediate goal.
- From 2020 onwards, ensure a carbon-neutral growth, assisted by economic measures, to ultimately achieve in 2050 the following reductions compared to 2005 levels:
  - 50% reduction in CO_2 (Figure 10.33)
  - 80% reduction in NO_x emissions
  - 50% reduction in noise

Recently the European Commission has established a 2050 target goal in its highly ambitious Flightpath 2050 Vision for Aviation. Pushing the environmental agenda even more aggressively further down the line, Flightpath 2050 sets forth steeper industry goals for 2050 in comparison to capabilities of typical new aircraft available in 2000:

- 75% reduction in CO_2 emissions per passenger kilometer
- 90% reduction in NOx emissions
- 65% reduction in perceived noise emissions of flying aircraft

This sets the bar very high on additional new-generation technology to come on top of advances in existing technology including significant shifts in the design approach of both aircraft and engine systems.

Also included are goals calling for aircraft movements to be emission-free when taxiing.

As discussed in Chapter 9, taxiing is an area where manufacturers have already been looking to increase fuel efficiency and reduce emissions and noise through electrification. Currently, aircraft use their main engines to move around on the tarmac. As you can imagine, this consumes a lot of fuel. By adding electric motors to the wheels of the plane and powering them from the onboard Auxiliary Power Unit (APU), significant fuel burn reduction can be achieved on ground operations. This allows an aircraft to be able to taxi without using its main engines, offering an attractive fuel saving of around ~3% on ground operations for a short-/medium-haul aircraft.

Sometimes, green solutions do come with economic benefits, but this may be tied to the type of energy used and the related price levels. One such instance is the electric green taxiing system, an incremental electrification solution. Unluckily, relatively low oil prices are not helping the entry into service of such solutions. Indeed, the development of the electric taxiing system described above is currently on hold, despite wide industry support, the reason being that oil prices are so low that it doesn't make economic sense, at least for now.

On one hand, airlines may undeniably get more profitable in case of oil price drops. But on the other hand, low oil prices weaken the economic prospects of incremental developments based on conventional aircraft platforms, irrespective of whether they use electric or conventional technology, as long as their focus is on fuel-burn reduction. In line with that rationale, in the face of low oil prices, re-engined aircraft platforms may suffer weakened sales growth due to their higher price tag.

On conventional aircraft, as fuel is a major cost driver, bringing fuel consumption down helps indirectly to reduce gas emission levels. But we can't rely on the industry's natural desire to decrease fuel consumption in order to reduce emissions, because during periods of low fuel prices the impetus may slacken, or worse, simply grind to a halt.

For the future of aviation, the electric aircraft, whether fully electric or in a hybrid-electric version, is the centerpiece in finding the sweet spot between reining in emissions, boosting energy efficiency, and offering a compelling business case.

References

[10.1] https://www.slideshare.net/reyyandemir/aviation-industry-and-mro-sector-trends, accessed October 18, 2017.

[10.2] https://www.iata.org/pressroom/facts_figures/fact_sheets/Documents/fact-sheet-fuel.pdf, accessed October 18, 2017.

[10.3] https://www.indexmundi.com/commodities/?commodity=jet-fuel&months=240, accessed October 2017.

[10.4] https://www.iata.org/publications/economics/Documents/Financial_Forecast_Presentation_Dec07.pdf, accessed October 28, 2017.

[10.5] https://www.iata.org/whatwedo/Documents/economics/Central-forecast-end-year-2016-tables.pdf, accessed October 28, 2017.

[10.6] http://web.mit.edu/airlinedata/www/Expenses&Related.html, accessed October 28, 2017.

[10.7] Aviation Daily, Market Briefing, June 30, 2014, Page 6, http://aviationweek.com/site-files/aviationweek.com/files/uploads/2014/06/avd_06_30_2014_cht1.pdf, accessed October 21, 2017.

[10.8] FAA, "Aircraft Operating Costs," https://www.faa.gov/regulations_policies/policy_guidance/benefit_cost/media/econ-value-section-4-op-costs.pdf, accessed October 22, 2017.

[10.9] Kharina, A. and Rutherford, D., "Fuel Efficiency Trends for New Commercial Jet Aircraft: 1960 to 2014," The International Council on Clean Transportation (ICCT), White Paper, August 2015.

[10.10] Ensign, T.R. and Gallman, J.W., "Energy Optimized Equipment Systems for General Aviation Jets," Cessna Aircraft Company, *44th AIAA Aerospace Sciences Meeting and Exhibit, AIAA 2006-228*, Reno, NV, USA, January 9-12, 2006.

[10.11] Stoufflet, B., "Towards an All Electrical Falcon," Dassault Aviation, *The More Electrical Aircraft—Achievements and Perspective for the Future—ICAS Workshop*, Cape Town, South Africa, September 2, 2013, http://www.icas.org/media/pdf/Workshops/2013/Towards All Electrical Aircraft Stoufflet.pdf, accessed July 12, 2017.

[10.12] Le Peuvédic, J.-M., "High-Performance HVDC Starter/Generators for the More Electric Aeroplane," Dassault Aviation, *Electric & Hybrid Aerospace Technology Symposium 2015*, Bremen, Germany, November 17-18, 2015.

[10.13] Jomier, T., Technical Report of "More Open Electrical Technologies" (MOET) Project (European Commission 6th Framework Programme), Airbus Operations S.A.S., December 14, 2009.

[10.14] https://leehamnews.com/2012/07/05/no-plateau-on-737ng-boeing/, accessed October 19, 2017.

[10.15] Isikveren, A.T., "Hybrid-Electric Aircraft: The Necessary Waypoints in Fulfilling Flightpath 2050," Bauhaus Luftfahrt, *IQPC 2nd International More Electric Aircraft Conference*, Hamburg, Germany, December 3-5, 2014.

[10.16] Hornung, M., "Aviation 2050: Potentials and Challenges," Bauhaus Luftfahrt, *Electric & Hybrid Aerospace Technology Symposium 2015*, Bremen, Germany, November 17-18, 2015.

[10.17] Isikveren, A.T. and Schmidt, M., "Conceptual Studies of Future Hybrid-Electric Regional Aircraft," Bauhaus Luftfahrt, Munich Aerospace, *22nd International Symposium on Air Breathing Engines*, Phoenix, AZ, USA, October 25-30, 2015, ISABE-2015-20285.

[10.18] Bradley, M.K. and Droney, C.K., "Subsonic Ultra Green Aircraft Research: Phase II—Volume II—Hybrid Electric Design Exploration," prepared by the Boeing Company for NASA Langley Research Center, Contract NNL08AA16B - Task Order NNL11AA00T, NASA/CR-2015-218704/Volume II, April 2015.

[10.19] http://www.boeing.com/aboutus/environment/environment_report_14/2.3_future_flight.html, accessed December 14, 2017.

[10.20] "Can the 787 & A350 Transform the Economics of Long-Haul Services?," *Aircraft Commerce*, Issue No. 39, February/March 2005.

[10.21] https://www.sec.gov/Archives/edgar/data/319687/000095012310086414/h76206e425.htm, accessed October 20, 2017.

[10.22] Hepperle, M., "Electric Flight—Potential and Limitations," prepared by German Aerospace Center (DLR) for NATO STO Workshop "Energy Efficient Technologies and Concepts of Operation", 2012, DOI: 10.14339/STO-MP-AVT-209.

[10.23] Moore, M.D., "The Forthcoming Distributed Electric Propulsion Flight Era," NASA Langley Research Center, *Power Systems Track Panel—Electric Flight, SAE 2016 Aerospace Systems and Technology Conference (ASTC)*, Hartford, CT, USA, September 27-29, 2016.

[10.24] Moore, M.D., "Convergence of Market, Technology, and Regulation," NASA Langley Research Center, NASA On Demand Mobility (ODM) Roadmap presentation, *Transformative Vertical Flight Concepts—3rd Joint (SAE/AHS/AIAA/NASA) Workshop on Enabling New Flight Concepts Through Novel Propulsion and Energy Architectures*, Hartford, CT, USA, September 29-30, 2016.

[10.25] Rizzi, S.A., "Tools for Assessing Community Noise of DEP Vehicles," NASA Langley Research Center, Highly Integrated Distributed Electric Propulsion Tools and Testing Panel Discussion, *AHS-AIAA Transformative Vertical Flight Concepts Joint Workshop on Enabling New Flight Concepts through Novel Propulsion and Energy Architectures*, Arlington, VA, USA, August 26-27, 2014.

[10.26] http://blogs.edf.org/climatetalks/2015/05/01/airlines-biofuel-ambitions-must-not-increase-emissions/, accessed November 8, 2017.

[10.27] Stumpf, E., Nolte, P., Apffelstaedt, A., Zill, T. et al., "IATA Technology Roadmap," prepared by German Aerospace Center DLR and Georgia Institute of Technology and IATA, 4th Edition, June 2013.

conclusion

As the reader may have found, *Fundamentals of Electric Aircraft* provides basic knowledge of how the aviation industry is getting impacted by the electrification trend. This paradigm streamlines conventional aircraft using several types of energy (fuel, hydraulic, pneumatic, mechanical) into ones that parlay the enhanced efficiency of a single type of energy, namely, electric energy. Scaled-up energy storage on an electric aircraft, sometimes in combination with other sources, is poised to supply power to electrified loads, ranging from a variety of systems to propulsion itself.

The automotive sector has made significant strides in the field of electrification. Nevertheless, the efficiency of carbon fuel is difficult to beat. Market penetration has been slow due to higher production costs linked to smaller volumes, limitations in range performance, and the cost of energy storage. But that is changing every day and the tipping point for massive ramp-up is just a couple of years ahead, and technologies developed will certainly prevail in other sectors.

In fact, leveraging the latest advances in battery technology, especially with regard to specific energy, a bigger number of electric general aviation aircraft are poised to take to the skies at the turn of the decade. From there, concerning larger aircraft for short- and long-haul flights, it will probably be at least a decade or two before mature and power-dense electrical architectures get to acceptable levels. Even then, these aircraft might probably wind up being hybrid-electric only.

Global investments and research are in full swing, ushering in breakthroughs that will help reshape the landscape for greener travel with the electric aircraft. Replacing carbon fuel in aviation with an alternative energy source is not that easy. We may wonder if the weight of centuries is not what bestows kerosene with its unmatchable specific energy, but it would take reasonable time for any alternative cost-effective energy-storage technology to completely unseat its predominance.

What can be seen is that electrification, whether via fuel cells or batteries, is slowly outpacing carbon fuels and is already an enabler for autonomous travel in the automotive sector. The same kind of situation is likely to be encountered in aviation for urban mobility concepts, even if autonomous air travel is not for tomorrow.

We all know how paramount safety is for aircraft design and operations. That means, by no way, safety rules may end up getting transgressed. Icarus, in Greek mythology, would probably have survived his flight to freedom if only he had heeded Daedalus' cautionary warnings for "safe flight". But, he went too high in altitude and dangerously close to the sun, and faced the tragedy of his own loss. If only his wings had featured the right level of autonomy and authority, they would have prevented him from doing what he did and saved his life. Reassuring though, lessons learned from ancient times along with more recent ones remain so hardwired in the minds of aviation experts that safety is never a second thought. When designing future air transport solutions, that wisdom of hindsight will help ensure compliance to regulations with no compromise. Otherwise, a third rail will undoubtedly be hit with certification authorities, penalizing entry into market.

As the electric aircraft advances, more and more technologies will deliver but will also set their own limitations depending on when and how they are implemented and operated. Regarding timescales, the reliance on kerosene may gradually decrease thanks to electric and hybrid-electric aircraft developments down the line, but its ultimate phasing out may happen only in the long run.

Future urban and general aviation, regional and short- or long-haul aircraft are poised to be greener and quieter while maximizing passenger mobility and flight experience in a sustainable manner. Electric aircraft are the mainstay of this inexorable push. Cities over which they may fly, and communities close to airports or "vertiports" they would operate from, would ultimately enjoy relief from air traffic pollution and noise.

Pascal Thalin

about the authors

Pascal Thalin is Chair of the Electric Aircraft Steering Group of SAE International. His major focus is on delivering innovative and optimized solutions for electric aircraft and coordinating related standardization activities. In various leadership positions within Airbus, Safran, and Thales for almost two decades, he has gained longstanding experience in research and development and program management of power, propulsion, and actuation systems for a wide range of aircraft. Prior to that, he worked at Electricité de France and General Electric on power systems for industrial and medical applications. He holds a master's degree in engineering from INP-ENSEEIHT, Toulouse, France, and a Ph.D. in electrical engineering from the National Polytechnic Institute of Toulouse, France. He is a regular invited speaker and moderator of panel sessions at conferences on electric Aircraft. He also chairs such conferences and takes part in their organization committees.

Dr. Ravi Rajamani is an independent consultant with many years of experience in aerospace propulsion and energy, specifically in data analytics and model-based methods for controls, diagnostics, and prognostics. He has written three books, including *Electric Flight Technology: The Unfolding of a New Future*. Prior to his current job, Ravi worked at Meggitt, United Technologies, and General Electric. He is active within various SAE technical committees and with the PHM Society, serving on its board of directors. Ravi is a Visiting Professor at Cranfield University and Editor-in-Chief of SAE's *Journal of Aerospace*. He has been elected a fellow of SAE and of IMechE.

Jean-Charles Maré is a Professor at Institut National des Sciences Appliquées (INSA), Toulouse, France. He holds a degree in mechanical engineering from INSA and a doctorate from Université de Lyon, France. He teaches mechanic and fluid power systems with special focus on architecture design and system-level modeling and simulation. In charge of research activity at Institut Clément Ader in Toulouse, France, he has created a team working on embedded actuation systems and components. His expertise includes preliminary design, virtual prototyping, and testing of safety-critical actuators for aerospace applications. Since the early 2000s, he has been a regular contributor to numerous research projects dealing with more electric actuation for aerospace. He is a member of the SAE A-6 technical committee, the author of *Aerospace Actuators*, a three-volume book series, and chair of the International Conference on Recent Advances in Aerospace Actuation Systems and Components (R3ASC) that he initiated in 2001.

Sven Taubert leads Corporate Foresight and Market Intelligence at Lufthansa Technik (LHT), a leading provider of MRO services. In addition to worldwide trend and competitor analysis, he is responsible for electric aircraft initiatives within Lufthansa Group. He is an active member of the SAE Electric Aircraft Steering Group. Prior to his current role, he was responsible for LHT's strategic cabin research program and managed the company's contribution at the Center of Applied Aeronautical Research in Hamburg, Germany. Sven joined LHT four years ago, after several roles in R&T at Airbus. His experience encompasses 3D printing, digitalization, Internet of Things, artificial intelligence, connectivity, interactive surfaces, and human-centered design. Sven studied at the University of Stuttgart, Germany, and Tokyo University, Japan, and holds a degree in aerospace engineering.

index

www.ingramcontent.com/pod-product-compliance
Lightning Source LLC
LaVergne TN
LVHW071342210926
848486LV00003B/206

9780768093223